6th edition

# UNDERSTANDING CONSTRUCTION DRAWINGS

Mark W. Huth

**6th edition**

# UNDERSTANDING CONSTRUCTION DRAWINGS

Mark W. Huth

DELMAR
CENGAGE Learning

Australia • Brazil • Japan • Korea • Mexico • Singapore • Spain • United Kingdom • United States

**Understanding Construction Drawings
Sixth Edition**
Mark W. Huth

Vice President, Editorial: Dave Garza

Director of Learning Solutions: Sandy Clark

Senior Acquisitions Editor: James DeVoe

Managing Editor: Larry Main

Senior Product Manager: Sharon Chambliss

Media Editor: Deborah Bordeaux

Editorial Assistant: Aviva Ariel

Vice President, Marketing: Jennifer Ann Baker

Marketing Director: Deborah Yarnell

Senior Market Development Manager: Erin Brennan

Senior Brand Manager: Kristin McNary

Senior Production Director: Wendy A. Troeger

Production Manager: Mark Bernard

Content Project Manager: David Barnes

Senior Art Director: Bethany Casey

Technology Project Manager: Christopher Catalina

Production Technology Analyst: Joe Pliss

Cover Image: ©iStockphoto.com/Valerijs Kostreckis

© 2014 Delmar, Cengage Learning

ALL RIGHTS RESERVED. No part of this work covered by the copyright herein may be reproduced, transmitted, stored or used in any form or by any means graphic, electronic, or mechanical, including but not limited to photocopying, recording, scanning, digitizing, taping, Web distribution, information networks, or information storage and retrieval systems, except as permitted under Section 107 or 108 of the 1976 United States Copyright Act, without the prior written permission of the publisher.

> For product information and technology assistance, contact us at
> **Cengage Learning Customer & Sales Support, 1-800-354-9706**
> For permission to use material from this text or product,
> submit all requests online at **www.cengage.com/permissions**
> Further permissions questions can be emailed to
> **permissionrequest@cengage.com**

Library of Congress Control Number: 2012942012

ISBN-13: 978-1-285-06102-3

ISBN-10: 1-285-06102-0

**Delmar**
Executive Woods
5 Maxwell Drive
Clifton Park, NY 12065
USA

Cengage Learning is a leading provider of customized learning solutions with office locations around the globe, including Singapore, the United Kingdom, Australia, Mexico, Brazil, and Japan. Locate your local office at **www.cengage.com/global**

Cengage Learning products are represented in Canada by Nelson Education, Ltd.

To learn more about Delmar, visit **www.cengage.com/delmar**

Purchase any of our products at your local bookstore or at our preferred online store **www.cengagebrain.com**

**Notice to the Reader**
Publisher does not warrant or guarantee any of the products described herein or perform any independent analysis in connection with any of the product information contained herein. Publisher does not assume, and expressly disclaims, any obligation to obtain and include information other than that provided to it by the manufacturer. The reader is expressly warned to consider and adopt all safety precautions that might be indicated by the activities described herein and to avoid all potential hazards. By following the instructions contained herein, the reader willingly assumes all risks in connection with such instructions. The publisher makes no representations or warranties of any kind, including but not limited to, the warranties of fitness for particular purpose or merchantability, nor are any such representations implied with respect to the material set forth herein, and the publisher takes no responsibility with respect to such material. The publisher shall not be liable for any special, consequential, or exemplary damages resulting, in whole or part, from the readers' use of, or reliance upon, this material.

Printed in the United States of America
6 7 8 9 10 11   20 19 18 17 16

# Contents

**Preface** .................................................................... vii

## Part I  DRAWINGS: THE LANGUAGE OF CONSTRUCTION

| Unit 1 | The Design–Construction Sequence and the Design Professions | 2 |
| Unit 2 | Views | 8 |
| Unit 3 | Scales | 15 |
| Unit 4 | Alphabet of Lines | 19 |
| Unit 5 | Use of Symbols | 25 |
| Unit 6 | Plan Views | 35 |
| Unit 7 | Elevations | 41 |
| Unit 8 | Sections and Detail | 47 |
|  | Part I Test | |

## Part II  READING DRAWINGS FOR TRADE INFORMATION

| Unit 9 | Clearing and Rough Grading the Site | 60 |
| Unit 10 | Locating the Building | 65 |
| Unit 11 | Site Utilities | 71 |
| Unit 12 | Footings | 76 |
| Unit 13 | Foundation Walls | 84 |
| Unit 14 | Drainage, Insulation, and Concrete Slabs | 90 |
| Unit 15 | Framing Systems | 95 |
| Unit 16 | Columns, Piers, and Girders | 108 |
| Unit 17 | Floor Framing | 115 |
| Unit 18 | Laying Out Walls and Partitions | 127 |
| Unit 19 | Framing Openings in Walls | 132 |
| Unit 20 | Roof Construction Terms | 138 |
| Unit 21 | Roof Trusses | 142 |
| Unit 22 | Common Rafters | 152 |
| Unit 23 | Hip and Valley Framing | 159 |
| Unit 24 | Cornices | 168 |
| Unit 25 | Windows and Doors | 177 |
| Unit 26 | Exterior Wall Coverings | 185 |
| Unit 27 | Decks | 194 |
| Unit 28 | Finishing Site Work | 199 |

| **Unit 29** | Fireplaces | 207 |
| **Unit 30** | Stairs | 215 |
| **Unit 31** | Insulation and Room Finishing | 220 |
| **Unit 32** | Cabinets | 232 |
| **Unit 33** | Lake House Specifications | 236 |
| | Part II Test | |

## Part III  MULTIFAMILY CONSTRUCTION

| **Unit 34** | Orienting the Drawings | 250 |
| **Unit 35** | Town House Construction | 258 |
| **Unit 36** | Plumbing, Heating, and Air Conditioning | 263 |
| **Unit 37** | Electrical | 274 |
| | Part III Test | |

## Part IV  HEAVY COMMERCIAL CONSTRUCTION

| **Unit 38** | Heavy Commercial Construction | 285 |
| **Unit 39** | Coordination of Drawings | 296 |
| **Unit 40** | Structural Drawings | 304 |
| **Unit 41** | HVAC & Plumbing Drawings | 312 |
| **Unit 42** | Electrical Drawings | 324 |
| | Part IV Test | |

## APPENDICES

| Appendix A | School Addition Master Keynotes | 347 |
| Appendix B | Math Reviews | 357 |
| Appendix C | Material Symbols in Sections | 373 |
| Appendix D | Plumbing Symbols | 374 |
| Appendix E | Electrical Symbols | 375 |
| Appendix F | Abbreviations | 376 |

## GLOSSARY            378

## INDEX            382

## DRAWING PACKET

Two-Unit Apartment Building Drawings

Lake House Drawings

Town House Drawings

School Addition Drawings

# Preface

## Intended Audience

*Understanding Construction Drawings* is designed for students in construction programs in two- and four-year colleges and technical institutes, as well as apprentice training. Designed for a course in print reading focused on both residential and commercial construction, the book helps you learn to read the drawings that are used to communicate information about buildings. It includes drawings for buildings that were designed for construction in several parts of North America. The diversity of building classifications and geographic locations ensures that you are ready to work on construction jobs anywhere in the industry. Everyone who works in building construction should be able to read and understand the drawings of the major trades.

## How to Use This Book

The book is divided into four major parts and several units within each part. Each part relates to the prints in the separate drawing packet:

- **Part I, Drawings—The Language of Construction,** introduces you to the basics of print reading by covering views, scales, lines, and symbols, as well as the various plan views, elevations, and sections and details.
- **Part II, Reading Drawings for Trade Information,** provides information on how to interpret drawings for project specifics–everything from footings and foundation walls to room finishing and cabinets.
- **Part III, Multifamily Construction,** details more advanced residential print reading and applies the skills learned in Parts I and II to other types of construction, as well as mechanical and electrical trades.
- **Part IV, Heavy Commercial Construction,** presents the need-to-know information on interpreting prints for large commercial construction including structural drawings, mechanical drawings, and electrical drawings.

## Features of the Units

The individual units are made up of four elements: Objectives, the main body of the unit, Using What You Learned, and Assignment.

- The ***Objectives*** appear at the beginning of the unit so that you will know what to look for as you study the unit.
- The ***main body*** is the presentation of content with many illustrations and references to the prints for the building being studied in that part.
- ***Using What You Learned*** gives you an opportunity to do a practice exercise that is similar to the exercises found in the assignment questions for that unit. The real-world need to be able to do the exercise is explained first. The exercise is presented and followed by a detailed explanation of how to find the specified information. ***Each unit contains 10 to 20 Assignment*** questions that require you both to understand the content of the unit and to apply that understanding to reading the drawings. There are moreover than 600 questions in all.

The book is divided into four parts, corresponding with the four buildings. At the end of each part there is a test of 50 to 80 questions. Additionally units include one or more **Green Notes**, which provide insights and suggestions for green home construction.

## The Drawing Packet, Glossary, and Appendix

At the back of the text you will find several helpful aids for studying construction drawings.

- The **drawing packet** that is conveniently packaged with the book contains 22 sheets with separate drawings that relate to each of the parts within the book. The drawing packet contains prints for four buildings: a simple two-family duplex that is very easy to understand, a more complex single-family home, one unit of a townhouse that uses different materials and methods than the first two buildings, and an addition to a school.
- The **Glossary** defines all the new technical terms introduced throughout the textbook. Each of these terms is defined where it is first used, but if you need to refresh your memory, turn to the Glossary.
- The **Math Reviews** in the Appendix are an innovative feature that has helped many construction students through a difficult area. These are concise reviews of the basic math you are likely to encounter throughout the building construction field. As math is required in this textbook, reference is made to the appropriate Math Review. All the math skills needed to complete the end-of-unit assignments in this book are covered in the Math Reviews.
- The **Appendix** also includes a complete list of construction abbreviations used on the prints, along with their meaning. There is also a section that explains the most commonly used symbols for materials and small equipment.

## New to this Edition

*Understanding Construction Drawings,* Sixth Edition, represents a major revision of the book, including:

- Up-to-date information on the latest materials and construction methods in the industry.
- Significant changes to the materials used in the construction of the two-unit apartment and lake house.
- A new unit on specifications has been added at the end of Part II.
- Green Notes have been added throughout the book.
- Using What You Learned sections have been added to each unit to give the reader guided experience for solving the problems presented in the corresponding unit's Assignment.

## Supplements to the Text

Along with the *Understanding Construction Drawings, Sixth Edition* book, we are proud to offer supplemental offerings that will help support classroom instruction and engage students in learning.

The **Instructor Resources** available on our Companion website contains *free* helpful tools for the instructor teaching a course on reading and interpreting construction drawings. Each component follows the chapters in the book and is intended to help instructors prepare classroom presentations and student evaluations. To access these helpful tools, please visit www.cengagebrain.com. At the home page, search for this Companion website by typing in the ISBN of the book in the search box at the top of the page. On the page illustrating this book, click on the "Access" button next to "Free Study Tools" and this will direct you to the following resources:

- An **Instructor's Guide** provides answers to all the Assignment questions and test questions in the textbook, and explains how the answers were found or calculated. In addition, it contains more than 500 additional questions that can be used for test, supplemental assignments, and review. The answer to each of these questions is given, along with an explanation of the answer.

- **PowerPoint Presentations** include an outline of each chapter along with photos and graphics to help illustrate important points and enhance classroom instruction. These presentations are editable, allowing instructors to include additional notes and photos/graphics from the Image Gallery included on this CD.
- **Computerized Testbanks** are available in ExamView format. Questions are organized by chapter, and may be edited, added, or deleted to suit individual class needs. Questions can be used in current format, or instructors may create their own exams based on these existing questions with ease.
- **Image Gallery** containing graphics and photos from all the chapters in the book provide an additional option for classroom presentations. Instructors may choose to add to the existing PowerPoint on this CD, or may wish to create their own presentations based on the book.

## Acknowledgments

I am grateful to all who contributed to this textbook. The instructors and their students who have used the previous five editions have given me valuable feedback that has played an instrumental role in shaping this edition. Several companies provided expertise and contributed illustrations—including many of the figures that illustrate this book.

I would especially like to thank the architects and engineers who supplied the drawings for the drawing packet, namely:

Robert Kurzon
*A.I.A., for the Duplex and Lake House.*

Carl Griffith
*A.I.A., at Cataldo, Waters, and Griffith Architects, P.C., and HA2F Consultants in Engineering for the School Addition.*

Terrel Broiles did extensive work to update the Two-Unit Apartment drawings and the Lake House drawings, including correcting code compliance issues and updating materials.

Lastly, I would also like to thank the instructors who reviewed the manuscript for the previous editions and for this new edition. They have provided guidance in making it the best print reading textbook it could be.

Scott Bretthauer
*Teacher, E.M.P.S., College of Lark Count, Grayslake, IL.*

Joe Dusek
*Professor, Construction Management Department, Triton College, River Grove, IL.*

Judy Guentzler-Collins
*Building Technology Department, Cochise College, Sierra Vista, AZ.*

Seongchan Kim
*Department of Engineering Technology, Western Illinois University, Macomb, IL*

## About the Author

The author of this textbook, Mark W. Huth, brings many years of experience in the industry to his writing—first as a carpenter and then as a contractor, building construction teacher, and construction publisher—his career has allowed him to consult with hundreds of construction educators in high schools, colleges, and universities. He has also authored several other successful construction titles, including *Basic Blueprint Reading for Construction, Residential Construction Academy: Basic Principles for Construction,* and *Practical Problems in Mathematics for Carpenters.*

# A Word about Math

Construction requires the use of mathematics. Whether you are a carpenter planning stairs, a plumber calculating pipe lengths and fitting allowances, or an estimator preparing for a contract bid, you need math to do your job. The math required in this textbook is basic, so you probably have learned enough math to do all of the work required. Most of the math required on a construction job can be done quite easily with a construction calculator, such as the one shown here. Today's construction calculators are preprogrammed to do everything from converting decimals to fractions, calculating the lengths of rafters, figuring cubic yards of concrete, and other standard industry computations.

If you are studying construction, you probably own a construction calculator now or will soon. However, as you progress in your learning and spend more time working on construction sites, you will soon find that you do not always have your trusty calculator handy. If you have learned to do the basic math required with a pencil and paper (or scrap of wood), you will not be hampered by not having your calculator. Also, it is easy to make big mistakes with a calculator—any kind of calculator. With one wrong press of a key, you can add when you meant to multiply or add an extra zero. If you learn the math, you can check your work and sure that it is close to what the calculator got, so there is less likely to be an catastrophic error

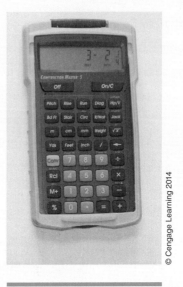

A typical construction calculator.

For these reasons, you are urged to complete the assignments at the ends of the textbook units by doing the math *without a calculator*, at least until you feel confident in your ability. If you have trouble doing the math, check the Math Reviews in Appendix B at the back of the book. They give easy, step-by-step directions for doing all of the types of math needed in the book.

# DRAWINGS: THE LANGUAGE OF CONSTRUCTION

Part One helps you develop a foundation upon which to build skills and knowledge in reading the drawings used in the construction industry. The topics of the various units in this section are the basic concepts upon which all construction drawings are read and interpreted. The details of construction are explored in Parts Two, Three, and Four.

Many of the assignment questions in this part refer to the drawings of the Duplex included in the drawing packet that accompanies this textbook. The Duplex was designed as income property for a small investor. It was built on a corner lot in a small city in upstate New York. The Duplex is an easy-to-understand building. Its one-story, rectangular design requires only a minimum of views; you can quickly become familiar with the Duplex drawings.

# UNIT 1

# The Design-Construction Sequence and the Design Professions

## Objectives

After completing this unit, you will be able to perform the following tasks:

○ Name the professions included in the design and planning of a house or light commercial building.

○ List the major functions of each of these professions in the design and planning process.

○ Identify the profession or agency that should be contacted for specific information about a building under construction.

The construction industry employs about 15 percent of the working people in the United States and Canada. More than 60 percent of these workers are involved in new construction. The rest are involved in repairing, remodeling, and maintenance. As the needs of our society change, the demand for different kinds of construction increases. Homeowners and businesses demand more energy-efficient buildings. The shift toward automation in business and industry means that new offices are needed. Our national centers of commerce and industry are shifting. These are only a few of the reasons that new housing starts are considered important indicators of our economic health.

There are four main classifications of construction: residential, commercial, industrial, and civil. *Residential construction* includes single-family homes, small apartment buildings, and condominiums (see **Figure 1–1(A)**). *Commercial construction* includes office and apartment buildings, hotels, stores, shopping centers, and other large buildings (see **Figure 1–1(B)**). *Industrial construction* includes structures other than buildings, such as refineries and paper mills, that are built for industry (see **Figure 1–1(C)**). *Civil construction* (see **Figure 1–1(D)**),

**Figure 1–1(A).** This hotel under construction is an example of commercial construction.

**Figure 1–1(B).** Commercial construction.

**Figure 1–1(D).** Civil construction. At 726 feet, Hoover Dam is the highest dam in the United States.

**Figure 1–1(C).** Industrial construction. Delta Energy Center, water treatment tanks and buildings in the foreground.

is more closely linked with the land and refers to highways, bridges, airports, dams, and the like.

## The Design Process

The design process starts with the owner. The owner has definite ideas about what is needed but may not be expert at describing that need or desire in terms the builder can understand. The owner contacts an architect to help plan the building.

The architect serves as the owner's agent throughout the design and construction process. Architects combine their knowledge of construction—of both the mechanics and the business—with artistic or aesthetic knowledge and ability. They design buildings for appearance and use.

 GREEN NOTE

*Green construction can be defined in many different ways, and sometimes the definitions sound complicated. In its simplest terms, green construction is the process of designing and constructing a building that minimizes its impact on the environment both during construction, over its useful life, and, ultimately, the recyclability of its materials—or their safe and proper disposal—when that life comes to an end.*

*A green home is built from environmentally sustainable materials using practices that reduce material use and waste. A durable, long-lasting home has lower maintenance requirements and less overall impact on the environment than a home that needs to be replaced sooner or requires frequent repairs. A green home is also designed to conserve resources such as heating and cooling energy and water.*

The architect helps the owner determine how much space is needed, how many rooms are needed for now and in the future, what type of building best suits the owner's lifestyle or business needs, and what the costs will be. As the owner's needs take shape, the architect makes rough sketches to describe the planned building. At first these may be balloon diagrams (see **Figure 1–2**), to show traffic flow and the number of rooms. Eventually, the design of the building begins to take shape (see **Figure 1–3**).

Before all the details of the design can be finalized, other construction professionals become involved.

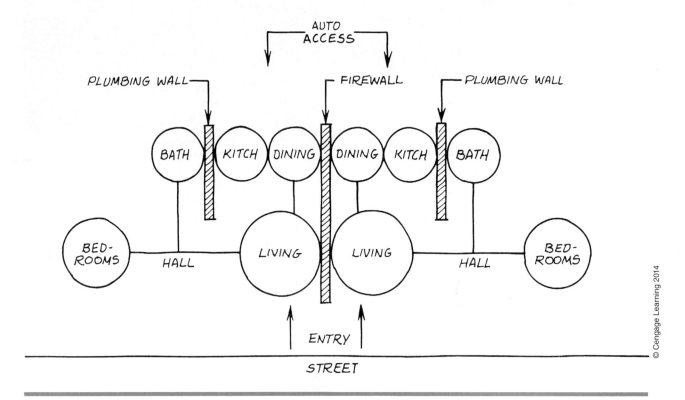

**Figure 1–2.** Balloon sketch of Duplex.

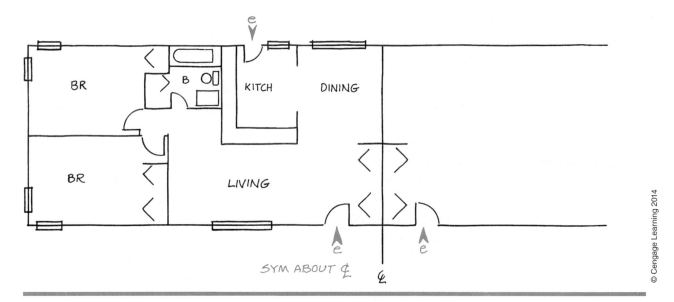

**Figure 1–3.** Straight line sketch of Duplex.

**GREEN NOTE**

*The design and planning for a green home involves not only the owner and designer, but the general contractor and key trade contractors as well. The designer's preliminary house plans are reviewed by those who will build the home, each looking for ways to improve energy efficiency, incorporate durable construction details, and simplify utility systems installation. Group meetings are often conducted in which the designer, owner, contractor, and key trade contractors discuss the plans and examine the impact of each recommendation and how the work will be carried out. The general contractor and trade contractors can also recommend green building materials that best suit the project.*

**Figure 1–4.** 2012 International Residential Code® for One- and Two-Family Dwellings.

Building codes specify requirements to ensure that buildings are safe from fire hazards, earthquakes, termites, surface water, and other concerns of the community. Most building codes are based on a model code. For example, the International Code Council (ICC) publishes several model codes, one of which is the *International Residential Code® for One- and Two-Family Dwellings* (see **Figure 1–4**). This often is called simply the Residential Code. It is a model code, because it is a model that may be used by state and local building authorities as a basis for their own local code. A model code has no authority on its own. The government having jurisdiction in a locale must adopt its own building code. Very often the government body having jurisdiction (called the *Authority Having Jurisdiction*, or AHJ) adopts the model code. Sometimes the AHJ adds specific clauses to the model, and, in rare cases, it writes an all-new code. State building codes allow local governments to adopt a local building code, but they require that the local code be at least as stringent as the state code.

The local building code is administered by a building department of the local government. The building department reviews the architect's plans before construction begins and inspects the construction throughout its progress to ensure that the code is followed.

Most communities also have zoning laws. *A zoning law* divides the community into zones where only certain types of buildings are permitted. Zoning laws prevent such problems as factories and shopping centers being built in the same neighborhood as homes.

Building departments usually require that very specific procedures are followed for each construction project. A building permit is required before construction begins. The building permit notifies the building department about planned construction. Then, the building department can make sure that the building complies with all the local zoning laws and building codes. When the building department approves the completed construction, it issues a *certificate of occupancy*. This certificate is not issued until the building department is satisfied that the construction has been completed according to the local code. The owner is not permitted to move into the new building until the certificate of occupancy has been issued.

If the building is more complex than a home or simple frame building, engineers may be hired to help design the structural, mechanical, electrical, or other aspects of the building. Consulting engineers specialize

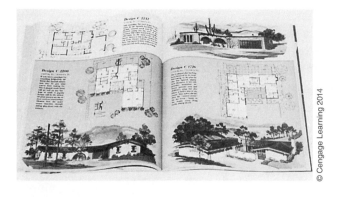

**Figure 1–5.** Stock plans can be ordered from catalogs.

in certain aspects of construction and are employed by architects to provide specific services. Finally, architects and their consultants prepare construction drawings that show all aspects of the building. These drawings tell the contractor specifically what to build.

Many homes are built from stock plans available from catalogs of house designs, building materials dealers, or magazines (see **Figure 1–5**). However, many states require a registered architect to approve the design and supervise the construction.

## Starting Construction

After the architect and the owner decide on a final design, the owner obtains financing. The most common way of financing a home is through a mortgage. A *mortgage* is a guarantee that the loan will be paid in installments. If the loan is not paid, the lender has the right to sell the building in order to recover the money owed. In return for the use of the lender's money, the borrower pays interest—a percentage of the outstanding balance of the loan.

When financing has been arranged (sometimes before it is finalized), a contractor is hired. Usually, a general contractor is hired with overall responsibility for completing the project. The general contractor in turn hires subcontractors to complete certain parts of the project. All stages of construction may be subcontracted. The parts of home construction most often subcontracted are plumbing and heating, electrical, drywall, painting and decorating, and landscaping. The relationships of all the members of the design and construction team are shown in **Figure 1–6**. Utility installers should carefully investigate all the drawings, especially the architectural drawings, in order to determine the installation locations of their equipment.

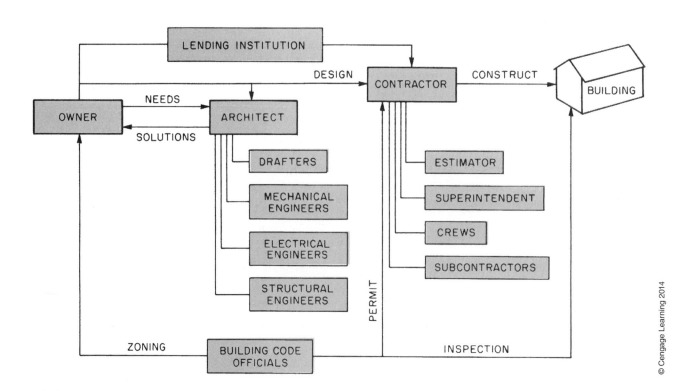

**Figure 1–6.** Design and construction team.

6 UNIT I

**USING WHAT YOU LEARNED**

Everyone involved in the design, construction, and ownership of a building needs to know who the major players are in the process. Only by understanding what role each agency, company, and individual plays in the process can a construction worker know where to go with questions and for information. For example, no work can begin on the site until a building permit has been issued. The owner, contractor, and superintendent all need to know who issues that permit. Building permits are issued by the building department of the city, town, or county where the building is to be constructed.

## Assignment

1. Who acts as the owner's agent while the building is being constructed?
2. Who designs the structural aspects of a commercial building?
3. Who would normally hire an electrical engineer for the design of a store?
4. Who is generally responsible for obtaining financing for a small building?
5. To whom would the general contractor go if there were a problem with the foundation design for a home?
6. If local building codes require specific features for earthquake protection, who is responsible for seeing that they are included in a home design?
7. Whom would the owner inform about last-minute changes in the interior trim when the building is under construction?
8. What regulations specify what parts of the community are to be reserved for single-family homes only?
9. Who issues the building permit?
10. What regulations are intended to ensure that all new construction is safe?

# UNIT 2: Views

## Objectives

After completing this unit, you will be able to perform the following tasks:

○ Recognize oblique, isometric, and orthographic drawings.

○ Draw simple isometric sketches.

○ Identify plan views, elevations, and sections.

## Isometric Drawings

A useful type of pictorial drawing for construction purposes is the **isometric drawing.** In an isometric drawing, vertical lines are drawn vertically, and horizontal lines are drawn at an angle of 30° from horizontal, as shown in **Figure 2–1.** All lines on one of these isometric axes are drawn in proportion to their actual length. Isometric drawings tend to look out of proportion because we are used to seeing the object appear smaller as it gets farther away.

Isometric drawings are often used to show plumbing (see **Figure 2–2**). The ability to draw simple isometric sketches is a useful skill for communicating on the job site. Try sketching a brick in isometric as shown in **Figure 2–3**.

Step 1.  Sketch a Y with the top lines about 30° from horizontal.
Step 2.  Sketch the bottom edges parallel to the top edges.
Step 3.  Mark off the width on the left top and bottom edges. This will be about twice the height.
Step 4.  Mark off the length on the right top and bottom edges. The length will be about twice the width.
Step 5.  Sketch the two remaining vertical lines and the back edges.

Other isometric shapes can be sketched by adding to or subtracting from this basic isometric brick (see **Figure 2–4).** Angled surfaces are sketched by locating their edges and then connecting them.

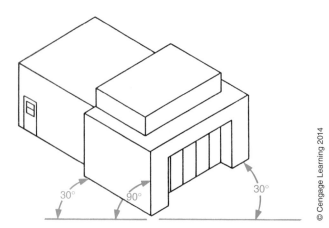

**Figure 2–1.** Isometric of building.

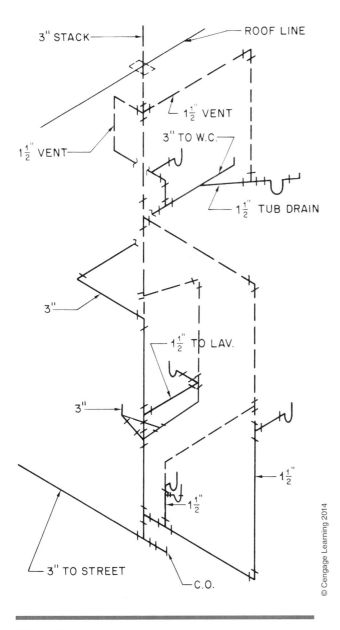

**Figure 2–2.** Single-line plumbing isometric.

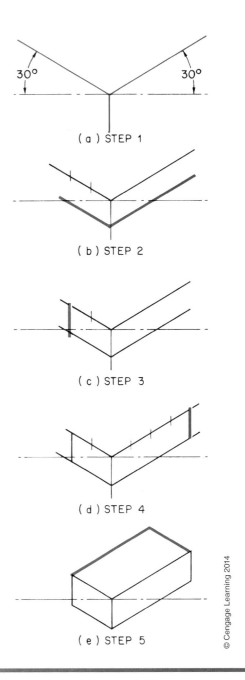

**Figure 2–3.** Sketching an isometric brick.

### GREEN NOTE
*Waste water is classified as either black water or gray water. Black water is that which is discharged from toilets and kitchens, where the water can be very contaminated and must be extensively treated before it can be released into the environment or reused. Gray water is from sinks, laundry areas, and storm water and is less dangerous to the environment. It is becoming increasingly common for green homes to treat gray water and reuse it for irrigation and toilets.*

## Oblique Drawings

When an irregular shape is to be shown in a pictorial drawing, an **oblique drawing** may be best. In oblique drawings, the most irregular surface is drawn in proportion as though it were flat against the drawing surface. Parallel lines are added to show the depth of the drawing as shown in **Figure 2–5**.

## Orthographic Projection

To show all information accurately and to keep all lines and angles in proportion, most construction drawings

Views **9**

(a) GABLE ROOF BUILDING

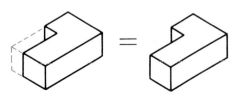

(b) ELL-SHAPED BUILDING

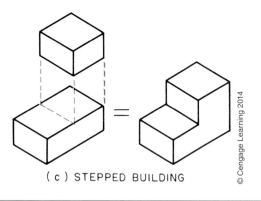

(c) STEPPED BUILDING

**Figure 2–4.** Variations on the isometric brick.

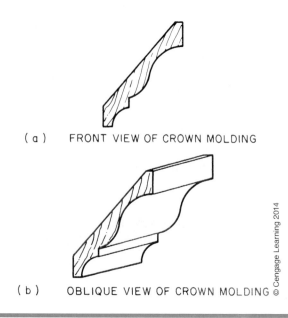

(a) FRONT VIEW OF CROWN MOLDING

(b) OBLIQUE VIEW OF CROWN MOLDING

**Figure 2–5.** Oblique drawing.

are drawn by **orthographic projection**. Orthographic projection is most often explained by imagining the object to be drawn inside a glass box. The corners and the lines representing the edges of the object are then projected onto the sides of the box (see **Figure 2–6**). If the box is unfolded, the images projected onto its sides will be on a single plane, as on a sheet of paper (see **Figure 2–7**). In other words, in orthographic projection, each view of an object shows only one side (or top or bottom) of the object.

All surfaces that are parallel to the plane of projection (the surface of the box) are shown in proportion to their actual size and shape. However, surfaces

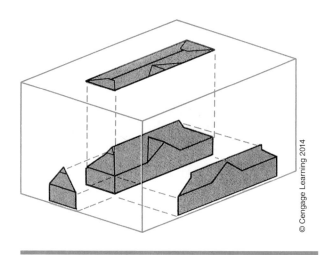

**Figure 2–6.** Duplex inside a glass box; method of orthographic projection of roof, front side, and end.

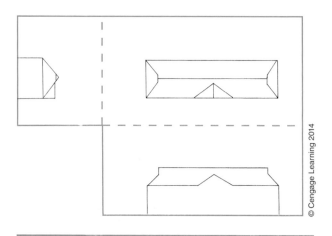

**Figure 2–7.** Orthographic projection unfolded on a flat sheet of paper.

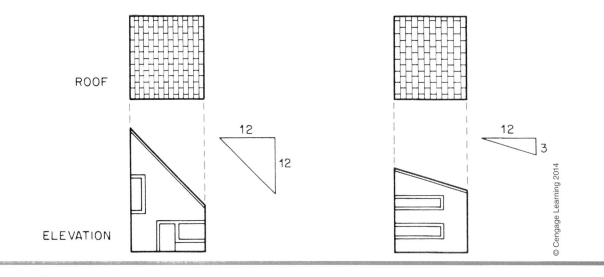

**Figure 2–8.** Views of two shed roofs.

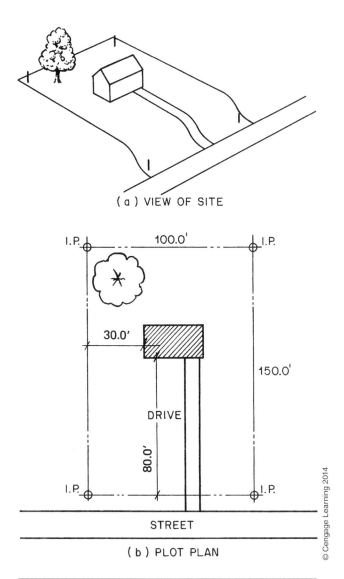

**Figure 2–9.** Plan view.

that are not parallel to the plane of projection are not shown in proportion. For example, both of the roofs in the top views of **Figure 2–8** appear to be the same size and shape, but they are quite different. To find the actual shape of the roof, you must look at the end view.

In construction drawings, the views are called plans and elevations. A *plan view* shows the layout of the object as viewed from above (see **Figure 2–9).** A set of drawings for a building usually includes plan views of the site (lot), the floor layout, and the foundation. **Elevations** are drawings that show height. For example, a drawing that shows what would be seen standing in front of a house is a building elevation (see **Figure 2–10**). Elevations are also used to show cabinets and interior features.

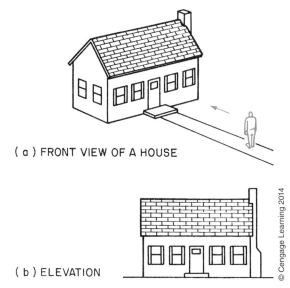

**Figure 2–10.** Building elevation.

Views

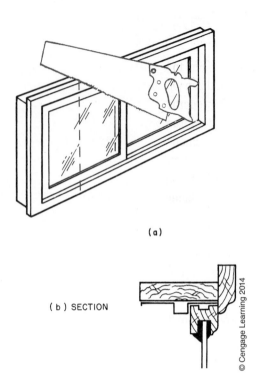

Because not all features of construction can be seen in plan views and elevations from the outside of a building, many construction drawings are **section views**. A section view, usually referred to simply as a *section,* shows what would be exposed if a cut were made through the object (see **Figure 2–11**). Actually, a *floor plan* is a type of section view (see **Figure 2–12**). It is called a *plan* because it is in that position—viewed from above—but it is a type of *section* because it shows what would be exposed if a cut were made through the building. Most section views are called sections, but floor plans are customarily referred to as plans or floor plans.

**Figure 2–11.** Section of a window sash.

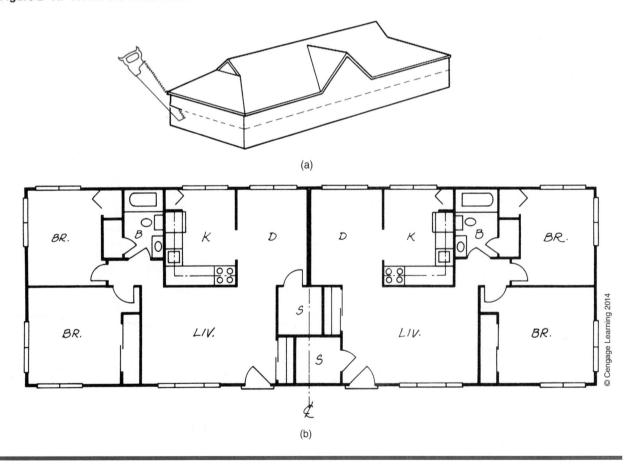

**Figure 2–12.** A floor plan is actually a section view of the building. (a) An imaginary cut is made at a level that passes through all windows and doors. (b) The floor plan shows what is left when the top is removed.

### USING WHAT YOU LEARNED

As you look for specific information on a set of construction prints, it is helpful to know what type of drawing you are looking at. For example, if it is an orthographic projection, the lines you see will be drawn in true proportion to their actual sizes. However, if it is oblique or isometric, they may not be in proportion. The Assignment questions in this unit require you to identify various drawing types. Take a look at the door frame types on Sheet 2 of the Two-Unit Apartment in the drawing packet accompanying this textbook. It is a section view because it shows parts as though a cut were made through the door jamb, revealing the interior construction. It is a plan view because it shows what would be seen looking straight down from above. It is an orthographic projection drawing because what we see is what would have been projected onto the top of a glass box placed over the door jamb.

## Assignment

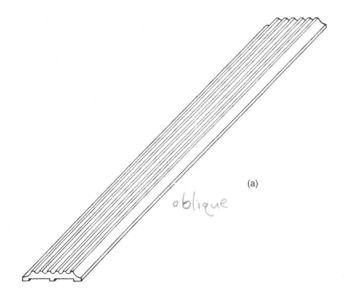

1. Identify each of the drawings in **Figure 2–13** as oblique, isometric, or orthographic.
2. Identify each of the drawings in **Figure 2–14** as elevation, plan, or section.
3. In the view of the house shown in **Figure 2–15**, which lines are true length?
4. What type of pictorial drawing is easiest to draw on the job site? *orthographic*
5. What type of drawing is used for working drawings?

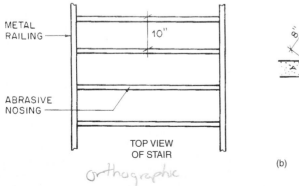
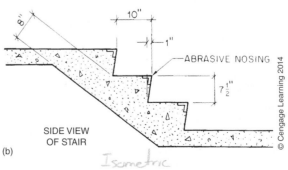

**Figure 2–13.**

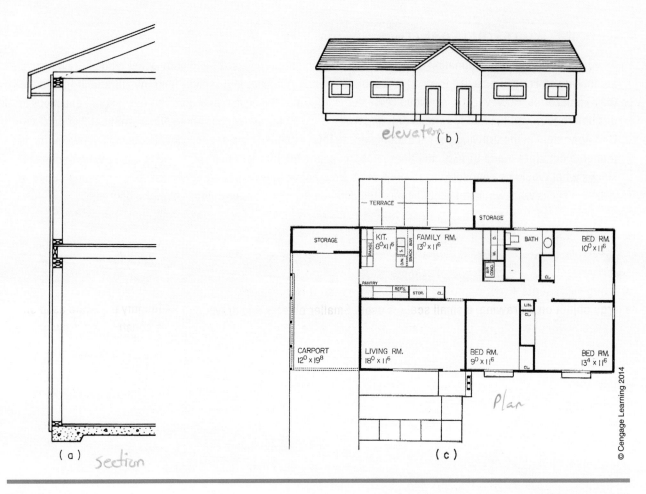

**Figure 2–14.**

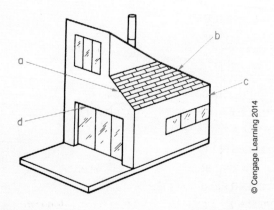

**Figure 2–15.**

# UNIT 3

# Scales

## Scale Drawings

Because construction projects are too large to be drawn full size on a sheet of paper, everything must be drawn proportionately smaller than it really is. For example, **floor plans** for a house are frequently drawn 1/48th of the actual size. This is called *drawing* to scale. At a scale of ¼" = 1'-0", ¼ inch on the drawing represents 1 foot on the actual building. When it is necessary to fit a large object on a drawing, a small scale is used. Smaller objects and drawings that must show more detail are drawn to a larger scale. The floor plan in **Figure 3–1** was drawn to a scale of ¼" = 1'-0". The detail drawing in **Figure 3–2** was drawn to a scale of 3" = 1'-0" to show the construction of one of the walls on the floor plan.

The scale to which a drawing is made is noted on the drawing. The scale is usually indicated alongside or beneath the title of the view.

## Reading an Architect's Scale

All necessary dimensions should be shown on the drawings. The instrument used to make drawings to scale is called an **architect's scale** (see **Figure 3–3**). Measuring a drawing with an architect's or engineer's scale is a poor practice. At small scales it is especially difficult to be precise. The following discussion of how to read an architect's scale is presented only to ensure an understanding of the scales used on drawings. The triangular architect's scale includes eleven scales frequently used on drawings.

Full Scale
| | |
|---|---|
| 3/32" = 1'- 0" | 3/16" = 1'- 0" |
| 1/8" = 1'- 0" | 1/4" = 1'- 0" |
| 3/8" = 1'- 0" | 3/4" = 1'- 0" |
| 1/2" = 1'- 0" | 1" = 1'- 0" |
| 11/2" = 1'- 0" | 3" = 1'- 0" |

Two scales are combined on each face, except for the full-size scale, which is fully divided into sixteenths (see **Figure 3–4**). The combined scales work together because one is twice as large as the other, and their zero points and extra divided units are on opposite ends of the scale.

The fraction, or number, near the zero at each end of the scale indicates the unit length in inches that is used on the drawing to represent 1 foot of the actual building. The extra unit near the zero end of the scale is subdivided into twelfths of a foot (inches) as well as fractions of inches on the larger scales.

## Objectives

After completing this unit, you will be able to perform the following tasks:

○ Identify the scale used on a construction drawing.

○ Read an architect's scale.

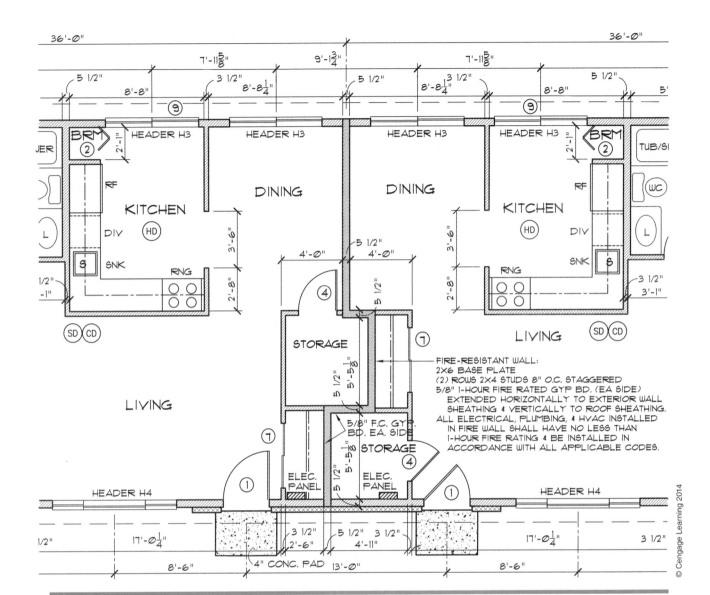

**Figure 3–1.** Portion of a plan view with a firewall.

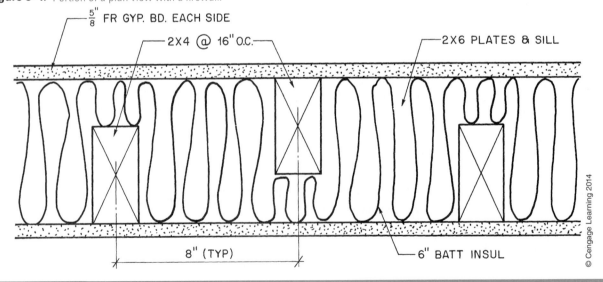

**Figure 3–2.** Detail (plan at firewall).

16 UNIT 3

**Figure 3–3.** Architect's scale.

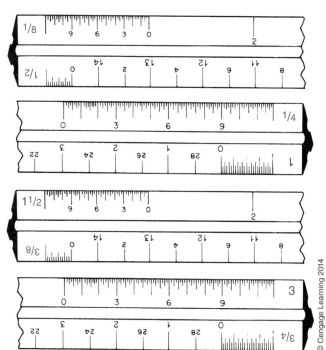

**Figure 3–4.** Architect's triangular scales.

To read the architect's scale, turn it to the ¼-inch scale. The scale is divided on the left from the zero toward the ¼ mark so that each line represents 1 inch. Counting the marks from the zero toward the ¼ mark, there are 12 lines marked on the scale. Each one of these lines is 1 inch on the ¼" = 1'-0" scale.

The fraction 1/8 is on the opposite end of the same scale. This is the 1/8-inch scale and is read in the opposite direction. Notice that the divided unit is only half as large as the one on the ¼-inch end of the scale. Counting the lines from zero toward the 1/8 mark, there are only six lines. This means that each line represents 2 inches at the 1/8-inch scale.

Now look at the 1½-inch scale. The divided unit is broken into twelfths of a foot (inches) and also fractional parts of an inch. Reading from the zero toward the number 1½, notice the figures 3, 6, and 9. These figures represent the measurements of 3 inches, 6 inches, and 9 inches at the 1½" = 1'-0" scale. From the zero to the first long mark, that represents 1 inch (which is the same length as the mark shown at 3) and four lines. This means that each line on the scale is equal to ¼ of an inch. Reading from the zero to the 3, read each line as follows: ¼, ½, ¾, 1, 1¼, 1½, 1¾, 2, 2¼, 2½, 2¾, and 3 inches. Do not confuse the engineer's scale with the architect's scale. The engineer's scale uses feet and decimal parts of a foot.

### GREEN NOTE

*Many homes have green or sustainable features such as solar panels to provide energy or extra insulation to conserve energy, but what makes a home a green home? There are programs to certify homes as green. A green certification program is a set of standards for green building practices that are followed by the building team and verified by an independent third party. These programs usually have a system for awarding points meeting certification standards that, when added up, determine whether the home can be certified as a green home. The two best known national green certification programs are LEED (Leadership in Energy and Environmental Design) and NAHB Green from the National Association of Home Builders Research Center.*

### USING WHAT YOU LEARNED

Construction drawings are rarely drawn the actual size of what they depict. They are almost always smaller than the actual object. Drawings for buildings are drawn to one of the scales found on an architect's scale. For this reason it is important to understand how to read an architects scale. If a drawing is made to a scale of ¼" = 1', what would be the dimension represented by a line 3⅜ inches long? Each ¼ inch represents 1 foot, so 3 inches represents 12 feet. (There are four ¼s in an inch and 3 × 4 = 12.) 3/8 inch is actually 1 and ½ quarters of an inch, so 3/8" represents 1½' or 1' foot 6". 12 feet plus 1 foot six inches is 13 feet 6 inches, normally written as 13'-6".

## Assignment

1. What are the dimensions indicated on the scale in **Figure 3–5?**
2. What scales are used for the following views of the duplex? (Refer to the duplex drawings in your textbook packet.)
   a. Floor plan
   b. Site plan
   c. Front elevation
   d. Typical wall section

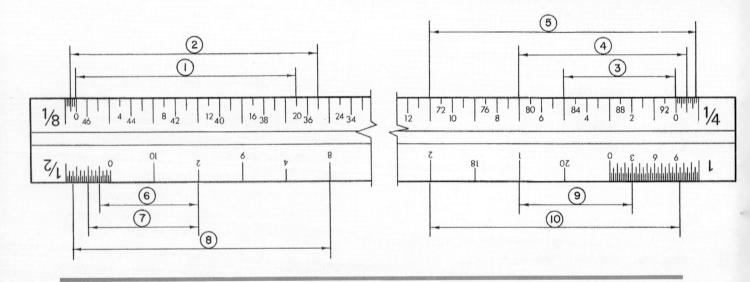

Figure 3–5.

# Alphabet of Lines

## Unit 4

That drawings are used in construction for the communication of information has already been discussed in Unit 2. Indeed, drawings serve as a language for the construction industry. The basis for any language is its alphabet. The English language uses an alphabet made up of twenty-six letters. Construction drawings use an *alphabet of lines* (see **Figure 4–1**).

The weight or thickness of lines is sometimes varied to show their relative importance. For example, in **Figure 4–2** notice that the basic outline of the building is heavier than the windows and doors. This difference in line weight sometimes helps distinguish the basic shape of an object from surface details.

## Object Lines

*Object lines* are used to show the shape of an object. All visible edges are represented by object lines. All the lines in **Figure 4–2** are object lines. Drawings usually include many solid lines that are not object lines, however. Some of these other solid lines are discussed here. Others are discussed later.

### Objectives

After completing this unit, you will be able to identify and understand the meaning of the listed lines:

- Object lines
- Dashed lines (hidden and phantom)
- Extension lines and dimension lines
- Centerlines
- Leaders
- Cutting-plane lines

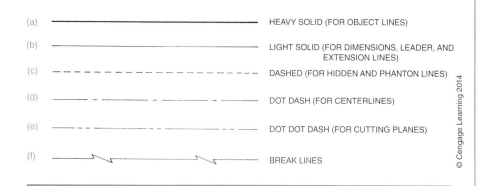

**Figure 4–1.** Alphabet of lines.

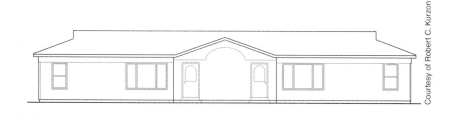

**Figure 4–2.** Elevation outlined.

Alphabet of Lines 19

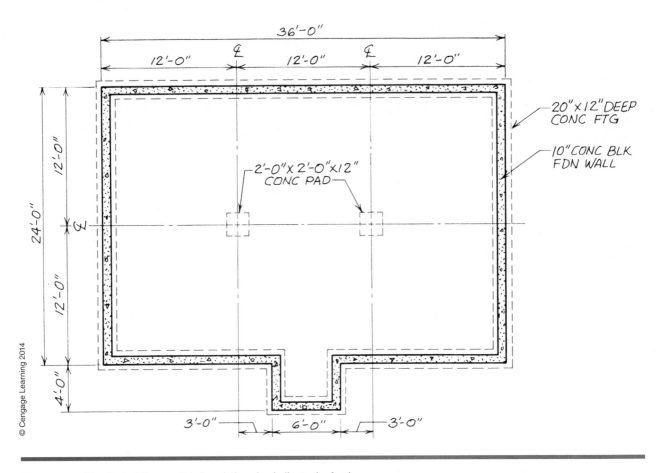

**Figure 4–3.** The dashed lines on this foundation plan indicate the footing.

## Dashed Lines

Dashed lines have more than one purpose in construction drawings. One type of dashed line, the *hidden line,* is used to show the edges of objects that would not otherwise be visible in the view shown. Hidden lines are drawn as a series of evenly sized short dashes (*see* **Figure 4–3**). If a construction drawing were to include hidden lines for all concealed edges, the drawing would be cluttered and hard to read. Therefore, only the most important features are shown by hidden lines.

Another type of dashed line is used to show important overhead construction (*see* **Figure 4–4**). These dashed lines are called *phantom lines.* The objects they show are not hidden in the view—they are simply not in the view. For example, the most practical way to show exposed beams on a living room ceiling may be to show them on the floor plan with phantom lines. Phantom lines are also used to show alternate positions of objects (*see* **Figure 4–5**). To avoid confusion, the dashed lines may be made up of different weights

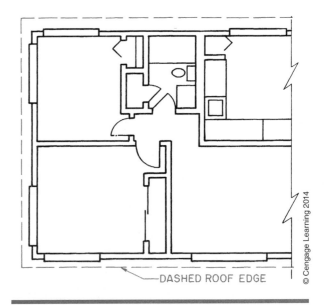

**Figure 4–4.** The dashed lines on this floor plan indicate the edge of the roof overhang.

and different length dashes, depending on the purpose (*see* **Figure 4–6**).

## Extension Lines and Dimension Lines

*Extension lines* are thin, solid lines that project from an object to show the extent or limits of a dimension. Extension lines do not quite touch the object they indicate (*see* **Figure 4–7**).

*Dimension lines* are solid lines of the same weight as extension lines. A dimension line is drawn from one extension line to the next. The dimension (distance between the extension lines) is lettered above the dimension line. On construction drawings, dimensions are expressed in feet and inches. The ends of dimension lines are drawn in one of three ways, as shown in **Figure 4–8**.

Dimensions that can be added together to come up with one overall dimension are called *chain dimensions*.

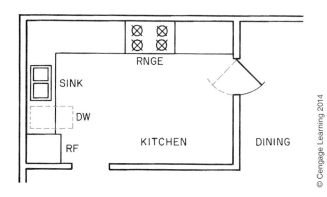

**Figure 4–5.** The dashed lines here are phantom lines to show alternate positions of the double-acting door and the door of the dishwasher.

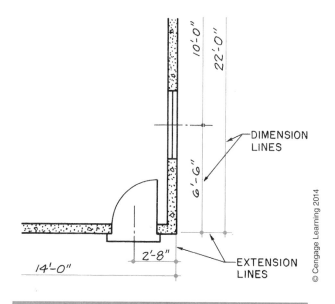

**Figure 4–7.** Dimension and extension lines.

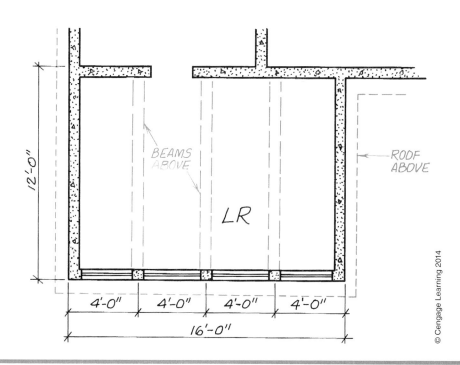

**Figure 4–6.** Different types of dashed lines are used to show different features.

Alphabet of Lines  21

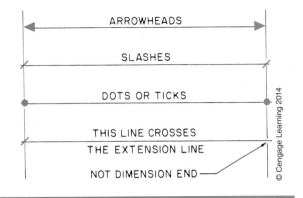

**Figure 4–8.** Dimension line ends.

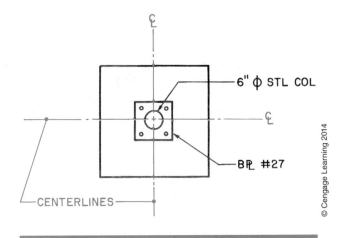

**Figure 4–10.** When centerlines show the center of a round object, the short dashes of two centerlines cross.

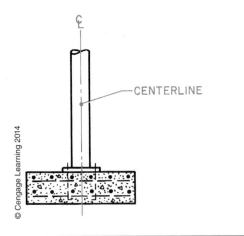

**Figure 4–9.** This centerline indicates that the column is symmetrical, or the same, on both sides of the centerline.

**Figure 4–11.** Method of showing the radius of an arc.

The dimension lines for chain dimensions are kept in line as much as possible. This makes it easier to find the dimensions that must be added to find the overall dimension.

## Centerlines

*Centerlines* are made up of long and short dashes. They are used to show the centers of round or cylindrical objects. Centerlines are also used to indicate that an object is *symmetrical,* or the same on both sides of the center (*see* **Figure 4–9**). To show the center of a round object, two centerlines are used so that the short dashes cross in the center (*see* **Figure 4–10**).

To lay out an arc or part of a circle, the radius must be known. The *radius* of an arc is the distance from the center to the edge of the arc. On construction drawings, the center of an arc is shown by crossing centerlines. The radius is dimensioned on a thin line from the center to the edge of the arc (*see* **Figure 4–11**).

Rather than clutter the drawing with unnecessary lines, only the short, crossing dashes of the centerlines are shown. If the centerlines are needed to dimension the location of the center, only the needed centerlines are extended.

## Leaders

Some construction details are too small to allow enough room for clear dimensioning by the methods described earlier. To overcome this problem, the

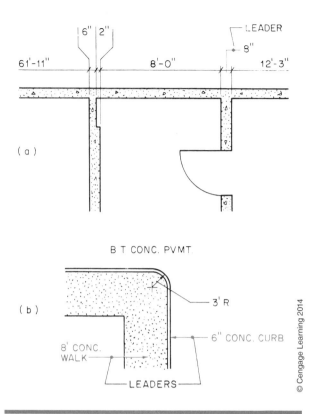

**Figure 4–12.** Leaders used for dimensioning.

dimension is shown in a clear area of the drawing. A thin line called a *leader* shows where the dimension belongs (see **Figure 4–12**).

GREEN NOTE

*There are many ways to wire a house, plumb a bathroom, or frame a house that meet the minimum requirements of building codes. Green building goes beyond the building codes and raises the bar for construction practices to the highest level of quality. Materials that are installed just well enough to meet a building code might not perform efficiently or last as long as those that are installed perfectly. For example, fiberglass insulation that is overstuffed into a wall cavity and has air spaces along the top and bottom may only perform at half the expected insulating value due to the installation defects. Best practices require insulation to be properly installed so that it fits every space snugly with out over packing it.*

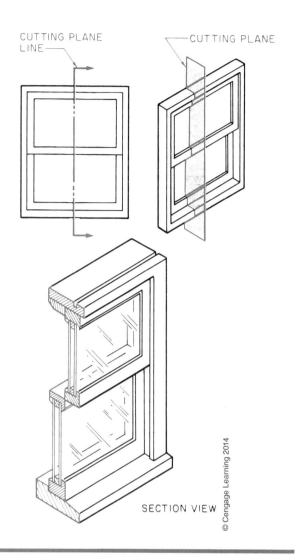

**Figure 4–13.** A cutting-plane line indicates where the imaginary cut is made and how it is viewed.

## Cutting-plane Lines

It was established earlier that section views are needed to show interior detail. In order to show where the imaginary cut was made, a *cutting-plane line* is drawn on the view through which the cut was made (see **Figure 4–13**). A cutting-plane line is usually a heavy line with long dashes and pairs of short dashes. Some drafters, however, use a solid, heavy line. In either case, cutting-plane lines always have some identification at their ends and arrowheads to indicate the direction from which the section is viewed. Cutting-plane-line identification symbols are discussed in the next unit.

Some section views may not be referenced by a cutting-plane line on any other view. These are *typical sections* that would be the same if drawn from an imaginary cut in any part of the building (see **Figure 4–14**).

Alphabet of Lines   23

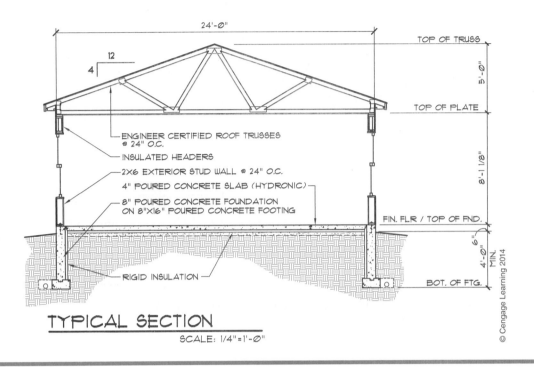

**Figure 4–14.** Building section.

### USING WHAT YOU LEARNED

On very simple drawings, it is usually easy to understand what each line represents, but on complex drawings, conveying a lot of information, there can be many types of lines, each with a different meaning. Look at the Site Plan for the Two-Unit Apartment. Why would some of the lines representing the building be much heavier than the lines across the middle of the building? What kinds of lines are these? The bold lines are object lines showing the basic shape and location of the building as viewed from above. The thinner lines are not part of the basic building outline. However, because this building is two dwelling units, the drafter has used these thinner lines to show how the building is divided. A beter practice might have been to have used dashed hidden lines to show the division, because they cannot be seen from above the building.

## Assignment

Refer to the drawings of the Two-Unit Apartment in your textbook packet. For each of the lines numbered A5.1 through A5.10, identify the kind of line and briefly describe its purpose on these drawings. The broad arrows with A5 numbers are for use in this assignment.

**Example:** A5.E, object line, shows the end of the building.

# Unit 5: Use of Symbols

An alphabet of lines allows for clear communication through drawings; the use of standard symbols makes for even better communication. Many features of construction cannot be drawn exactly as they appear on the building. Therefore, standard symbols are used to show various materials, plumbing fixtures and fittings, electrical devices, windows, doors, and other common objects. Notes are added to drawings to give additional explanations.

It is not important to memorize all the symbols and abbreviations used in construction before you learn to read drawings. There are commonly accepted standards for architectural symbols, but many architects and drafters use their own variations of standard symbols. Even so, with very little practice, you can develop the ability to interpret the symbols that are commonly used on construction drawings, whether standard or variant. Typically, an architectural symbol is a simplified picture of the material or item it represents. In many cases, the material represented by a symbol is also labeled with words or abbreviations. Some of the most common symbols are shown in this chapter and additional symbols are shown in the Appendix.

## Objectives

After completing this unit, you will be able to identify and understand the meaning of the listed symbols:

- Door and window symbols
- Materials symbols
- Electrical and mechanical symbols
- Reference marks for coordinating drawings
- Abbreviations

## Door and Window Symbols

Door and window symbols show the type of door or window used and the direction the door or window opens. There are three basic ways for household doors to open—swing, slide, or fold (see **Figure 5–1**). Within each of these basic types

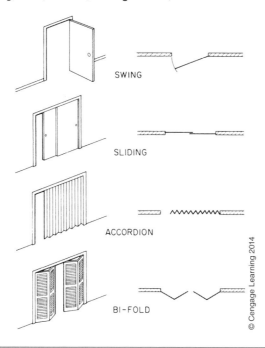

**Figure 5–1.** Types of doors and their plan symbols.

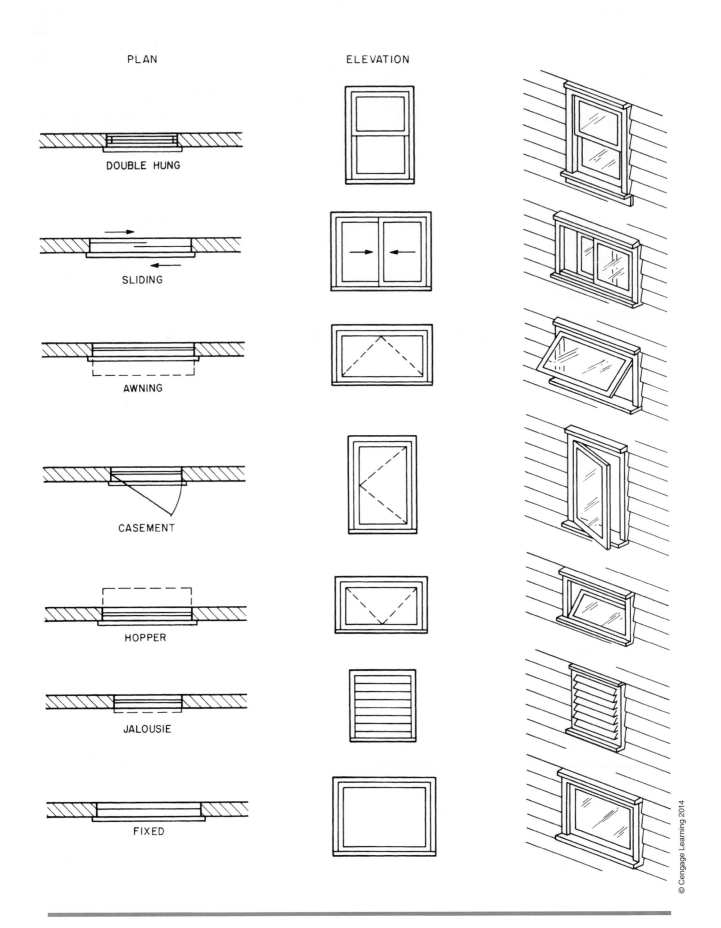

**Figure 5–2.** Window symbols.

there are variations that can be readily understood from their symbols. The direction a swing-type door opens is shown by an arc representing the path of the door.

There are seven basic types of windows. They are named according to how they open (see **Figure 5–2**). The symbols for hinged windows—awning, casement, and hopper—indicate the direction they open. In elevation, the symbols include dashed lines that come to a point at the hinged side, as viewed from the exterior.

The sizes of windows and doors are usually shown on a special window schedule or door schedule, but they might also be indicated by notes on the plans near their symbols. Door and window schedules are explained later. The notations of size show width first and height second. Manufacturers' catalogs usually list several sets of dimensions for every window model (see **Figure 5–3**). The glass size indicates the area that will actually allow light to pass. The rough opening size is important for the carpenter, who will frame the wall into which the window will be installed. The masonry opening is important to masons. The notations on plans and schedules usually indicate nominal dimensions. A *nominal dimension* is an approximate size and may not represent any of the actual dimensions of the unit. Nominal dimensions are usually rounded off to whole inches or feet and inches and are used only as a convenient way to refer to the window or door size. The actual dimensions should be obtained from the manufacturer before construction begins.

## Material Symbols

The drawing of an object shows its shape and location. The outline of the drawing may be filled in with a material symbol to show what the object is made of (see **Figure 5–4**). Many materials are represented by one symbol in elevations and another symbol in sections. Examples of such symbols are concrete block and brick. Other materials look pretty much the same when viewed from any direction, so their symbols are drawn the same in sections and elevations.

> **GREEN NOTE**
>
> *Life Cycle Assessment (LCA), also called* cradle-to-grave assessment, *is a technique to evaluate or assess all of the environmental impacts involved with the harvesting, mining, or manufacture; transportation; use; repair, and maintenance; and eventual disposal of a product. This analysis can be used to help determine which materials are most advantageous for a green home project.*

When a large area is made up of one material, it is common to only draw the symbol in a part of the area (see **Figure 5–5**). Some drafters simplify this even further by using a note to indicate what material is used and omitting the symbol altogether.

## Electrical and Mechanical Symbols

The electrical and mechanical systems in a building include wiring, electrical devices, piping, pipe fittings, plumbing fixtures, registers, and heating and air conditioning ducts. It is not practical to draw these items as they would actually appear, so standard symbols have been devised to indicate them.

The electrical system in a house includes wiring as well as devices such as switches, receptacles, light

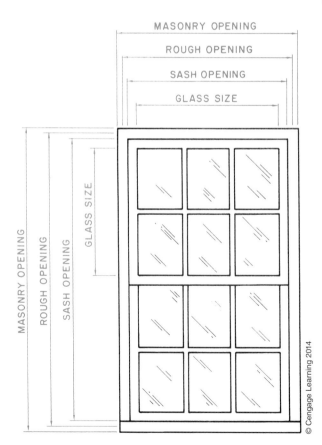

**Figure 5–3.** Windows and doors can be measured in several ways.

Use of Symbols 27

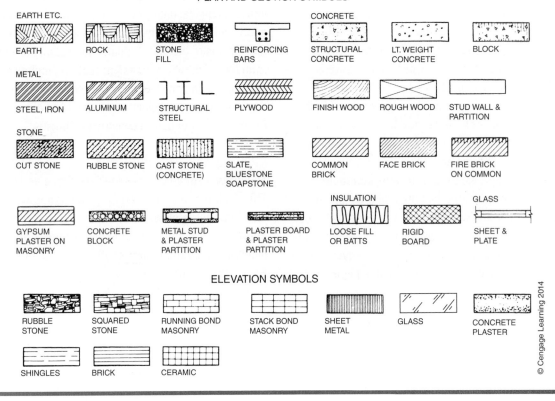

Figure 5–4. Material symbols.

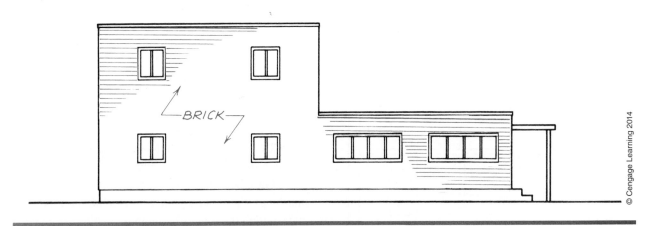

Figure 5–5. Only part of the area is covered by the brick symbol, although the entire building will be brick.

fixtures, and appliances. Wiring is indicated by lines that show how devices are connected. These lines are not shown in their actual position. They simply indicate which switches control which lights, for example. Outlets (receptacles) and switches are usually shown in their approximate positions. Major fixtures and appliances are shown in their actual positions. A few of the most common electrical symbols are shown in **Figure 5–6.**

Mechanical systems—plumbing and HVAC (heating, ventilating, and air conditioning)—are not usually shown in much detail on drawings for single-family homes. However, some of the most important features may be shown. Piping is shown by lines; different types of lines represent different kinds of piping. Symbols for pipe fittings are the same basic shape as the fittings they represent. A short line, or *hash mark,* represents the joint

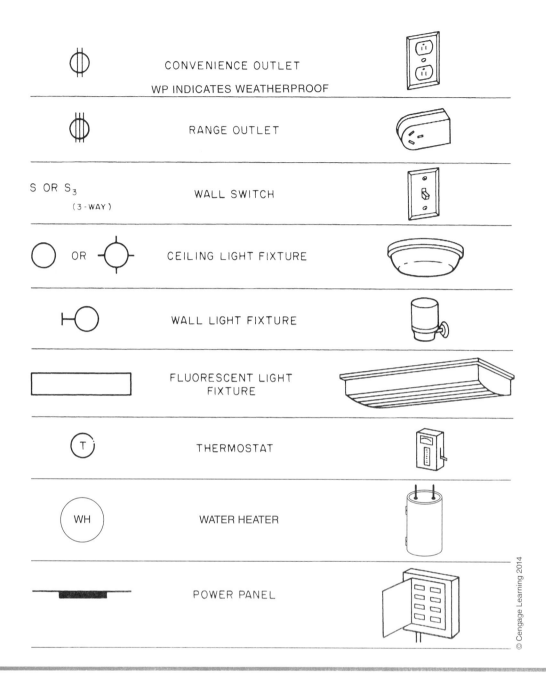

**Figure 5–6.** Some common electrical symbols.

between the pipe and the fitting. Plumbing fixtures are drawn pretty much as the actual fixture appears. A few plumbing symbols are shown in **Figure 5–7**.

## Reference Marks

A set of drawings for a complex building may include several sheets of section and detail drawings. These sections and details do not have much meaning without some way of knowing what part of the building they are meant to show. Callouts, called *reference marks,* on plans and elevations indicate where details or sections of important features have been drawn. To be able to use these reference marks for coordinating drawings, you must first understand the numbering system used on the drawings. The simplest numbering system for drawings consists of numbering the drawing sheets and naming each of the views. For example, Sheet 1 might include a site plan and foundation plan; Sheet 2, floor plans; and Sheet 3, elevations.

On large, complex sets of drawings, the sheets are numbered according to the kind of drawings shown. Architectural drawing sheets are numbered A-1, A-2, and so on for all the sheets. Electrical drawings are

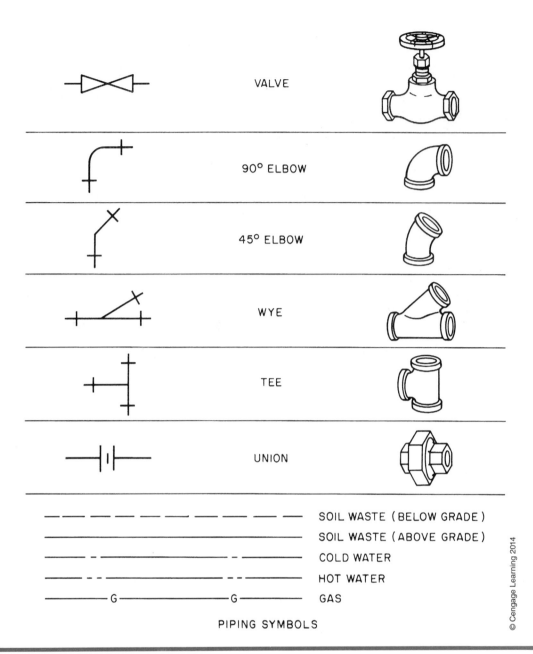

**Figure 5–7.** Some common plumbing symbols.

numbered E-1, E-2, and E-3. A view number identifies each separate drawing or view on the sheet. **Figure 5–8** shows drawing 5 on Sheet A-4.

Because most of the drawings for a house are architectural, and the drawing set is fairly small, letters indicating the type of drawing are not usually included. Instead, the views are numbered, and a second number shows on which sheet it appears. For example, the fourth drawing on the third sheet would be 4/3, 4.3, or 4-3.

Numbering each view and the sheet on which it appears makes it easy to reference a section or detail to another drawing. The identification of a section view is given with the cutting-plane line showing where it is taken from. For example, the section view shown in **Figure 5–9** shows the fireplace at the cutting-plane line in **Figure 5–10**. Notice that the cutting-plane line in **Figure 5–10** indicates that the section is viewed from the top of the page toward the bottom, with the fireplace opening on the right. That is how the section view in **Figure 5–9** is drawn. This numbering system is also used for details that cannot be located by a cutting-plane line. The drawing in **Figure 5–11** is a *typical wall section,* meaning that it is typical of a section view of any outside wall. This typical wall section includes a callout referencing a detail drawing of the roof

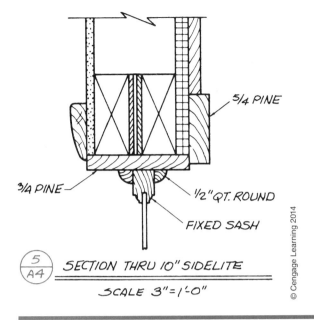

**Figure 5–8.** This is drawing 5 on Sheet A-4.

cornice or eave. The cornice detail is Drawing 4 on Sheet A-4 and is shown in **Figure 5–12**. Notice that the detail faces the opposite direction from the typical wall section. That is because, being a typical section, not a specific section, it represents all exterior walls regardless of the direction they face.

## Abbreviations

Drawings for construction include many notes and labels of parts. These notes and labels are usually abbreviated as much as possible to avoid crowding the drawing. The abbreviations used on drawings are usually a shortened form of the word and are easily understood. For example, BLDG stands for building. The abbreviations used throughout this textbook and on the related drawings are defined in the Appendix.

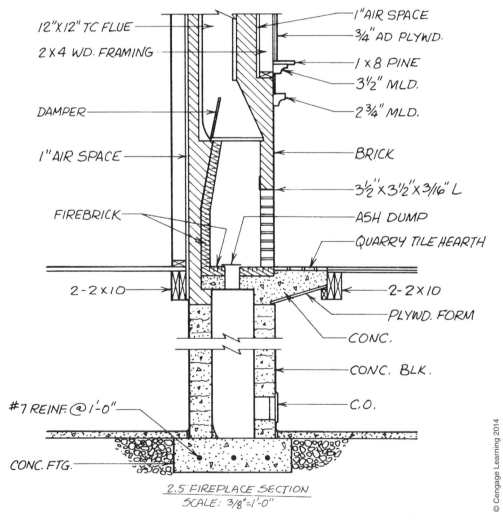

**Figure 5–9.** This section view is drawing 2 on Sheet 5.

Use of Symbols 31

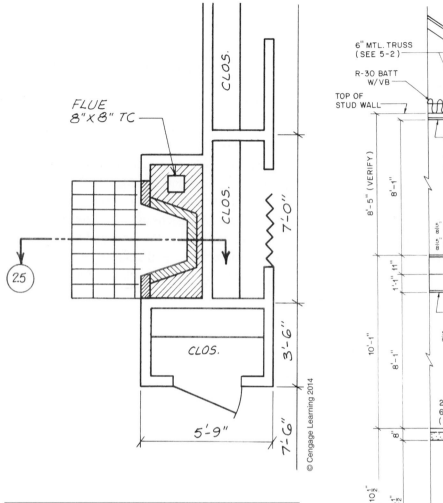

**Figure 5-10.** Plan for fireplace detailed in Figure 5-9.

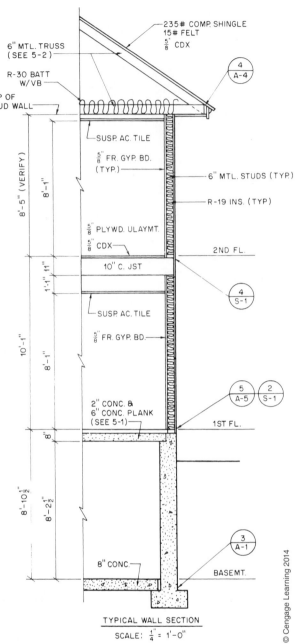

**Figure 5-11.** The detail of this cornice is shown in drawing 4 on Sheet A-4, Figure 5-12.

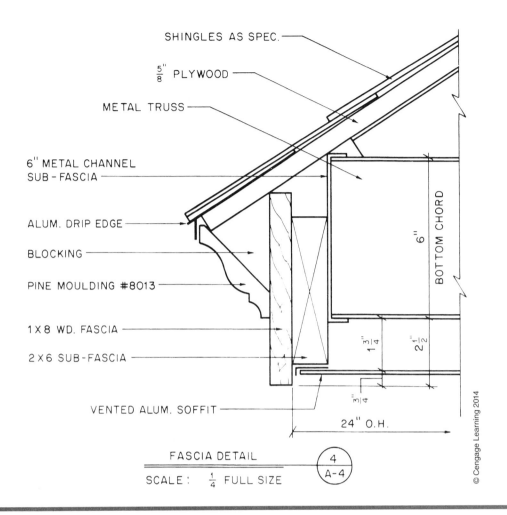

**Figure 5–12.** This is the detail of the cornice in Figure 5–11.

### USING WHAT YOU LEARNED

Construction drawings use many symbols to represent various materials and devices. Usually, one of the first pages in the drawing set includes an index of the symbols used on the drawings, but sometimes the most common symbols are not included in the index. To be able to read all of the information on the drawings, you must be able to interpret these symbols. For example, what is the framing material shown in **Figure 5–8?** The two rectangles with Xs drawn from corner to corner are dimensional lumber. Based on the proportions of width and thickness, they are probably 2×4s. Between the 2×4s is a piece of plywood.

Use of Symbols 33

## Assignment

1. What is represented by each of these symbols?
   a.
   b. 
   c. 
   d. 
   e. 
   f. 
   g. ——G——
   h. S₃
   i. (WH)
   j. 

2. What is meant by each of these abbreviations?
   a. GYP. BD.
   b. FOUND.
   c. FIN. FL.
   d. O.C.
   e. REINF.
   f. EXT.
   g. COL.
   h. DIA.
   i. ELEV.
   j. CONC.

3. Where in a set of drawings would you find a detail numbered 6.4?

4. Where in a set of drawings would you find a detail numbered 5/M–3?

# UNIT 6

# Plan Views

You learned earlier in Unit 2 that plans are drawings that show an object as viewed from above. Many of the detail and section drawings in a set show parts of the building from above. Some of the plan views that show an entire building are discussed here. This brief explanation will help you feel more comfortable with plans, although it does not cover plans in depth. You will use plans frequently throughout your study of the remainder of this textbook. Each of the remaining units helps you understand plan views more thoroughly.

## Site Plans

A **site plan** gives information about the site on which the building is to be constructed. The boundaries of the site (property lines) are shown. The property line is usually a heavy line with one or two short dashes between longer line segments. The lengths of the boundaries are noted next to the line symbol. Property descriptions are usually the result of a survey by a surveyor. Surveyors and engineers usually work with decimal parts of feet, rather than feet and inches. Therefore, site dimensions are usually stated in tenths or hundredths of feet (see **Figure 6–1**).

A symbol or north arrow of some type indicates what compass direction the site faces. Unless this north arrow includes a correction for the difference between true north and magnetic north, it may be only an approximation. However, it is sufficient to show the general direction the site faces.

## Objectives

After completing this unit, you will be able to explain the general kinds of information shown on the listed plans:

○ Site plans

○ Foundation plans

○ Floor plans

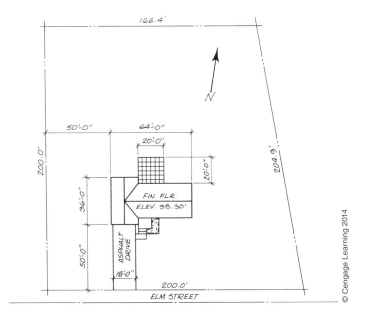

**Figure 6–1.** Minimum information shown on a site plan.

Plan Views 35

The site plan also indicates where the building is positioned on the site. At a minimum, the dimensions to the front and one side of the site are given. The overall dimensions of the building are also included. Anyone reading the site plan will have this basic information without having to refer to the other drawings. If the finished site is to include walks, drives, or patios, these are also described by their overall dimensions.

## Foundation Plans

A foundation plan is like a floor plan, but instead of showing the living spaces, it shows the foundation walls and any other structural work to be done below the living spaces.

Two types of foundations are commonly used in homes and other small buildings. One type has a concrete base, called the **footing,** supporting foundation walls (see **Figure 6–2**). The other is the slab-on-grade type. A *slab-on-grade* foundation consists of a concrete slab placed directly on the soil with little or no other support. Slabs-on-grade are usually thickened at their edges and wherever they must support a heavy load (see **Figure 6–3**).

When the footing-and-wall-type foundation is used, girders are used to provide intermediate support to the structure above (see **Figure 6–4**). The girder is

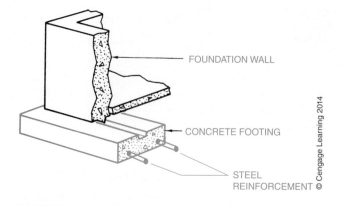

**Figure 6–2.** Footing and foundation wall.

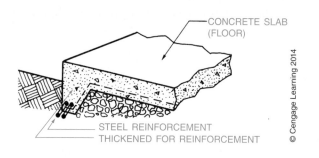

**Figure 6–3.** Slab-on-grade foundation.

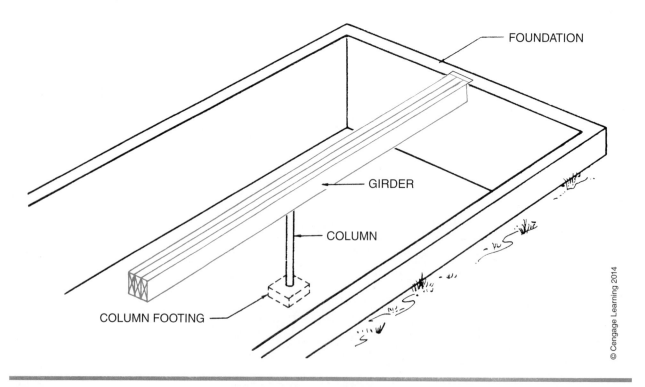

**Figure 6–4.** A girder provides intermediate support between the foundation walls.

shown on the foundation plan by phantom lines and a note describing it.

The foundation plan includes all the dimensions necessary to lay out the footings and foundation walls. The footings follow the walls and may be shown on the plan. If they are shown, it is usually by means of hidden lines to show their outline only. In addition to the layout of the foundation walls, dimensions are given for opening windows, doors, and ventilators. Notes on the plan indicate areas that are not to be excavated, concrete-slab floors, and other important information about the foundation (see **Figure 6–5**).

## Floor Plans

A floor plan is similar to a foundation plan. It is a section view taken at a height that shows the placement of walls, windows, doors, cabinets, and other important features. A separate floor plan is included for each floor of the building. The floor plans provide more information about the building than any of the other drawings.

## Building Layout

The floor plans show the locations of all the walls, doors, and windows. Therefore, the floor plans show how the building is divided into rooms and how to get from one room to another. Before attempting to read any of the specific information on the floor plans, it is wise to familiarize yourself with the general layout of the building.

To quickly familiarize yourself with a floor plan, imagine that you are walking through the house. For example, imagine yourself standing in the front door of the left side of the Duplex—plans for which are included in the drawing packed with this text. You are looking across the living

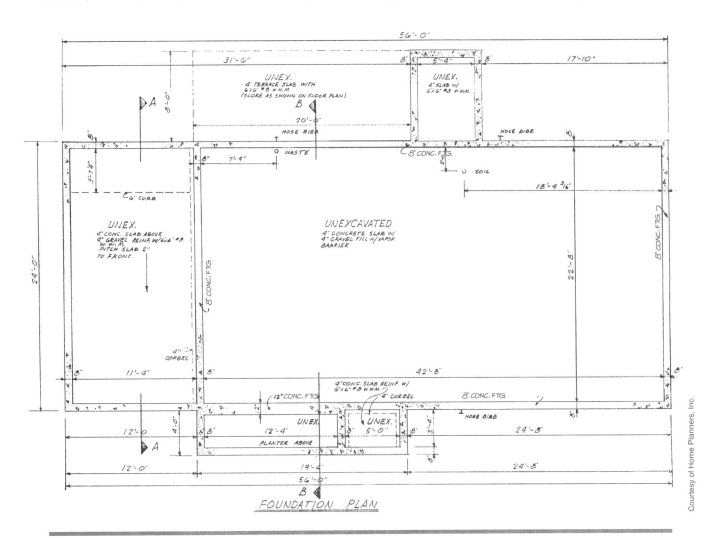

**Figure 6–5.** Foundation plan. *Courtesy of Home Planners, Inc.*

Plan Views 37

room. There is a closet on your right and a large window on your left. Straight ahead is the dining room with doors into a storage room and the kitchen. Looking in the kitchen doorway (notice there is no door in this doorway), you see cabinets, a sink, and a refrigerator on the opposite wall. More cabinets and a range are located on the left. Now, walk out of the kitchen and into the bedroom area. There are three doors; one leads into a large front bedroom with a long closet, another opens into a smaller bedroom, and the third opens into the bathroom. The bathroom includes a linen closet with bifold doors.

## Dimensions

Dimensions are given for the sizes and locations of all walls, partitions, doors, windows, and other important features. On frame construction, exterior walls are usually dimensioned to the outside face of the wall framing. If the walls are to be covered with stucco or masonry veneer, this material is outside the dimensioned face of the wall frame. Interior partitions may be dimensioned to their centerlines or to the face of the studs. (**Studs** are the vertical members in a wall frame.) Windows and doors may be dimensioned about their centerlines (see **Figure 6–6**) or to the edges of the openings.

Solid masonry construction is dimensioned entirely to the face of the masonry (see **Figure 6–7**). Masonry openings for doors and windows are dimensioned to the edge of the openings.

### GREEN NOTE

*It would be difficult to say exactly when green building began. Some practices, such as using local and renewable materials or passive solar design, date back millennia. The modern green building movement arose out of the need and desire for more energy efficient and environmentally friendly building practices, which began with the environmental movement of the 1960s. In the 1970s, drastically rising oil prices also spurred significant research and activity to improve energy efficiency and find renewable energy sources and, as a result, led to the earliest experiments with contemporary green building.*

*A few early milestones in the United States include:*

- *American Institute of Architects (AIA) formed the Committee on the Environment (1989).*
- *Environmental Resource Guide published by AIA, funded by U.S. Environmental Protection Agency (EPA) (1992).*
- *EPA and the U.S. Department of Energy launched the ENERGY STAR program (1992).*
- *First local green building program introduced in Austin, Texas (1992).*
- *U.S. Green Building Council (USGBC) founded (1993).*

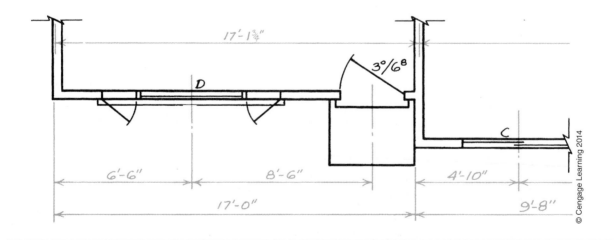

**Figure 6–6.** Frame construction dimensioning.

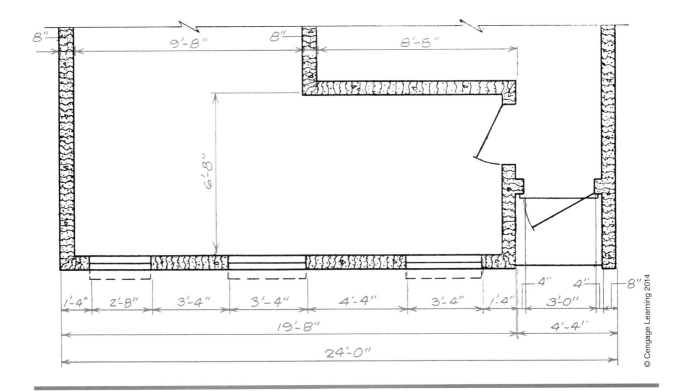

**Figure 6–7.** Masonry construction dimensioning.

## Other Features of Floor Plans

The floor plan includes as much information as possible without making it cluttered and hard to read. Doors and windows are shown by their symbols, as explained in Unit 5. Cabinets are shown in their proper positions. The cabinets are explained further by cabinet elevations and details, which are discussed in Unit 8. If the building includes stairs, these are shown on the floor plan. Important overhead construction is also indicated on the floor plans. If the ceiling is framed with joists, their size, direction, and spacing are shown on the floor plan. Architectural features such as exposed beams, arches in doorways, or unusual roof lines may be shown by phantom lines.

### USING WHAT YOU LEARNED

A lot of information is shown on plan views. For example, carpenters use floor plans extensively to lay out walls, windows, and doors. This way, they can answer such questions as how far the edge of the 3'-6" opening is from the dining room into the kitchen from the inside face of the north wall of the building. Finding the answer involves a few steps.

First, find the north arrow on the site plan to determine that the north wall is the one at the top of the floor plan. Now look at the dimensions at either end of the floor plan to see that the kitchen side of the wall separating the living room and the kitchen is 11'-4¼" from the inside face of the north wall. Now looking in the dining room we see that the north side of the opening is 2'-8" plus 3'-6" from the face of the living room–kitchen wall. Subtract 6'-2" from 11'-4¼" to find that the dimension we are looking for is 5'-2¼".

Plan Views 39

## Assignment

Refer to the drawings for the Two-Unit Apartment (which are included in your textbook packet) to complete this assignment.

1. In what direction does the apartment face?
2. What is the length and width of the apartment site?
3. How far is the front of the apartment from the front property line?
4. What is the overall length and width of the apartment?
5. What are the inside dimensions of the living room?
6. What is the thickness of the partitions between the two bedrooms?
7. What is the thickness of the interior wall between the two dining rooms?
8. With two exceptions, the units in the apartment are exactly reversed. What are the two exceptions?
9. What is the distance from the west end of the apartment to the centerline of the west front entrance?
10. What is indicated by the small rectangle on the floor plan outside each main entrance?
11. What is the distance from the ends of the apartment to the centerlines of the $6^0/6^8$ sliding glass doors?
12. What is indicated by the dashed line just outside the front and back walls on the floor plan of the apartment?

# Unit 7: Elevations

Drawings that show the height of objects are called *elevations*. However, when builders and architects refer to building elevations, they mean the exterior elevation drawings of the building (see **Figure 7–1**). A set of working drawings usually includes an elevation of each of the four sides of the building. If the building is very complex, there may be more than four elevations. If the building is simple, there may be only two elevations—the front and one side.

## Objectives

After completing this unit, you will be able to perform the following tasks:

- Orient building elevations to building plans.
- Explain the kinds of information shown on elevations.

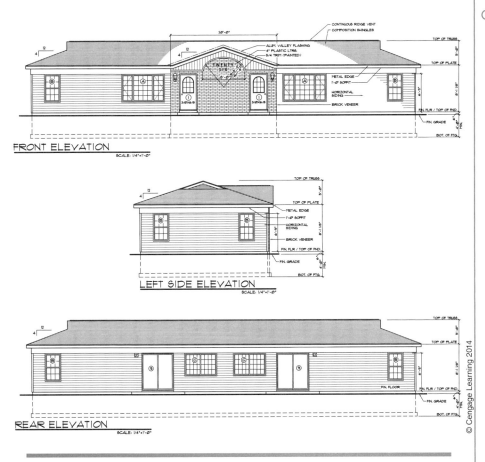

**Figure 7–1.** Building elevations.

## Orienting Elevations

It is important to determine the relationship of one drawing to another. This is called *orienting* the drawings. For example, if you know which elevation is the front, you must be able to picture how it relates to the front of the floor plan.

Elevations are sometimes named according to compass directions (see **Figure 7–2**). The side of the house that faces north is the north elevation, and the side that faces south is the south elevation, for example. When the elevations are named according to compass direction, they can be oriented to the floor plan, foundation plan, and site plan by the north arrow on those plans. It might help to label the edge of the plans according to the north arrow (see **Figure 7–3**).

Labeling elevations according to compass direction is not always possible. When drawings are prepared to be sold through a catalog, or when they are for use on several sites, the compass directions cannot be included. In this case, the elevations are named according to their position as you face the building (see **Figure 7–4**). To orient these elevations to the plans, find the front on the plans. The front is usually at the bottom of the sheet, but it can be checked by the location of the main entrance.

> **GREEN NOTE**
>
> *One of the earliest green practices designers and builders discovered was that by facing the building in the best direction, they could take advantage of natural heating and cooling. If the side with the most window area faces west, the afternoon sun will add considerable heat to the building. This can help heat the building in cold weather—but it can also add to the cooling load during the warm months. Designers have learned to both take advantage of these factors as well as adapt to them when planning buildings that use less energy.*

## Information on Building Elevations

Building elevations are normally quite simple. Although the elevations do not include a lot of detailed

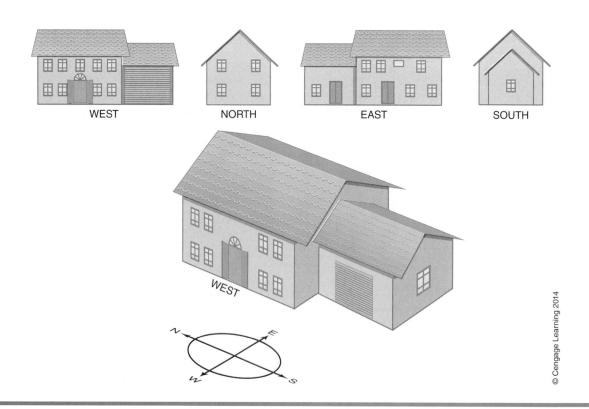

**Figure 7–2.** Elevations are usually named according to their compass directions.

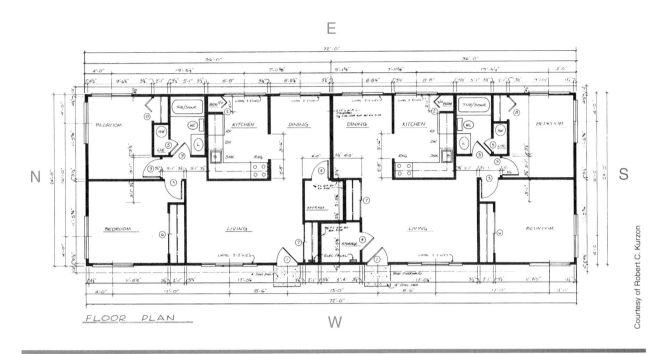

**Figure 7–3.** Plan labeled to help orientation to north arrow.

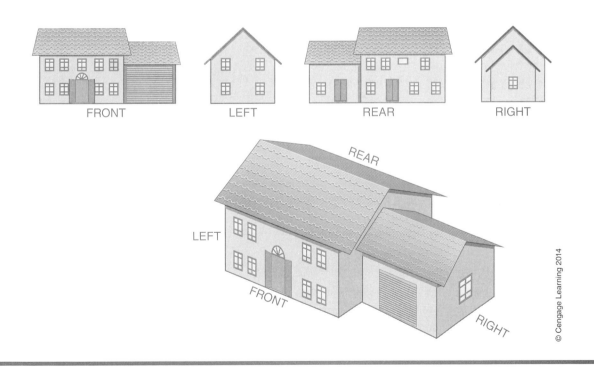

**Figure 7–4.** Elevations can be named according to their relative positions.

Elevations 43

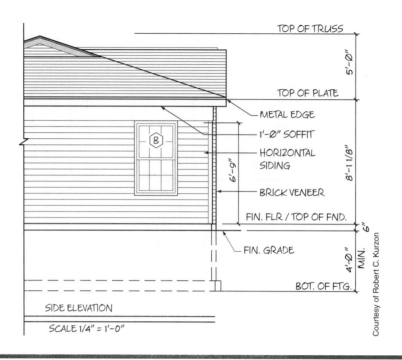

**Figure 7–5.** Underground portion of the building is shown with dashed lines.

dimensions and notes, they show the finished appearance of the building better than other views. Therefore, elevations are a great aid in understanding the rest of the drawing set.

The elevations show most of the building, as it will actually appear, with solid lines. However, the underground portion of the foundation is shown as hidden lines (see **Figure 7–5**). The footing is shown as a rectangle of dashed lines at the bottom of the foundation walls.

The surface of the ground is shown by a heavy solid line, called a grade line. The *grade line* might include one or more notes to indicate the elevation above sea level or another reference point (see **Figure 7–6**). Elevation used in this sense is altitude, or height—not a type of drawing. All references to the height of the ground or the level of key parts of the building are in terms of elevation. Methods for measuring site elevations are discussed in Unit 9.

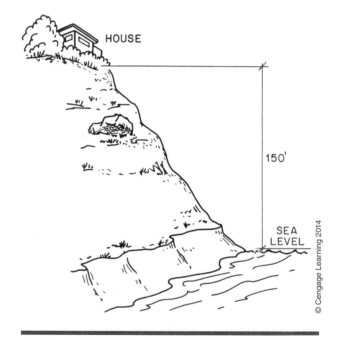

**Figure 7–6.** The elevation of this site is 150 feet.

Some important dimensions are included on the building elevations. Most of them are given in a string at the end of one or more elevations (see **Figure 7–7**). The dimensions most often included are listed here:

- Thickness of footing
- Height of foundation walls
- Top of foundation to finished first floor
- Finished floor to ceiling or top of plate (The **plate** is the uppermost framing member in the wall.)
- Finished floor to bottom of window headers (The **headers** are the framing across the top of a window opening.)
- Roof overhang at eaves

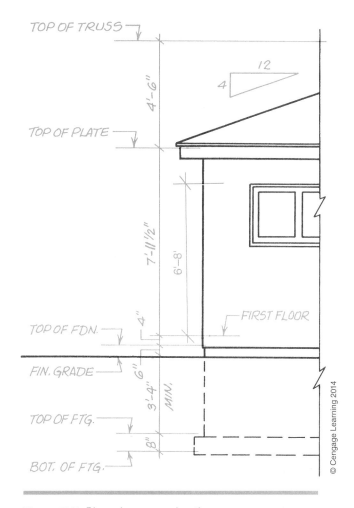

**Figure 7–7.** Dimensions on an elevation.

### USING WHAT YOU LEARNED

The building elevations are often the best drawings for a quick picture of what the finished building will look like. They also convey a lot of useful information in the construction of a building.

One requirement sometimes found in zoning ordinances is a limit on the overall height of a building, What is the height of the ridge of the two-unit apartment above the finished grade? This information is shown at the right end of each of the building elevations. The finished floor is 6 inches above the finished grade. The top of the wall plate is 8'-1 ⅛ above the finished floor. The top of the roof (i.e., top of truss) is 5'-0" above the top of the plate. All of these added together, 13'-7 ⅛ is the overall height of the building.

Elevations 45

## Assignment

Refer to the drawings of the Two-Unit Apartment in your textbook packet to complete this assignment.

1. Which elevation is the north elevation?
2. In what compass direction does the left end of the apartment face?
3. What is the dimension from the surface of the floor to the top of the wall framing?
4. What is the overall height of the building above the finished grade?
5. How far does the foundation project above the ground?
6. How far below the surface of the ground is the bottom of the footing?
7. What is the total height of the foundation walls?
8. What is the minimum depth of the bottom of the footings?

# Sections and Detail

It is not possible to show all the details of construction on foundation plans, floor plans, and building elevations. Those drawings are meant to show the relationships of the major building elements to one another. To show how individual pieces fit together, it is necessary to use larger-scale drawings and section views. These drawings are usually grouped together in the drawing set. They are referred to as **sections** and **details** (see **Figure 8–1**).

## Objectives

After completing this unit, you will be able to perform the following tasks:

- Find and explain information shown on section views.
- Find and explain information shown on large-scale details.
- Orient sections and details to the other plans and elevations.

### GREEN NOTE

*The fill under the concrete slab in the Two-Unit Apartment in your textbook packet is run-of-bank (ROB) gravel. Many slabs are placed on crushed stone, but by using ROB gravel, the apartment reduces the impact on the environment. The gravel does not require running a stone crusher and grader, which consumes energy and emits stone dust into the atmosphere. The gravel may be available more locally than crushed stone, thereby reducing the need to transport it from any great distance.*

## Sections

Nearly all sets of drawings include, at least, a typical wall section. The typical section may be a section view of one wall, or it may be a full section of the building. Full sections are named by the direction in which the imaginary cut is made. **Figure 8–2** shows a transverse section. A *transverse section* is taken from an imaginary cut across the width of the building. Transverse sections are sometimes called *cross sections.* A full section taken from a lengthwise cut through the building is called a *longitudinal section* (see **Figure 8–3**).

Full sections and wall sections normally have only a few dimensions but have many notes with leaders to identify the parts of the wall. The following is a list of the kinds of information that are included on typical wall sections with most sets of drawings:

- Footing size and material (This is specified by building codes.)
- Foundation wall thickness, height, and material
- Insulation, waterproofing, and interior finish for foundation walls
- Fill and waterproofing under concrete floors
- Concrete floor thickness, material, and reinforcement

Sections and Detail 47

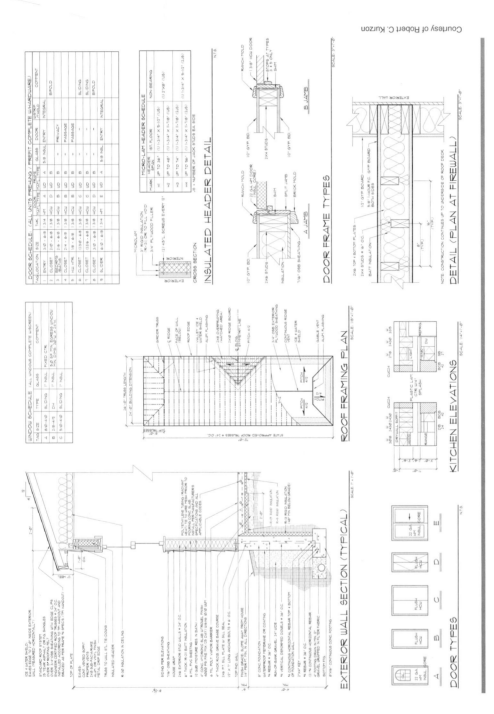

**Figure 8–1.** Typical sheet of sections and details for a small building.

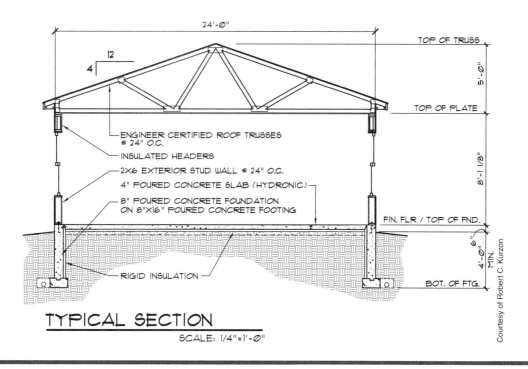

**Figure 8–2.** Transverse section.

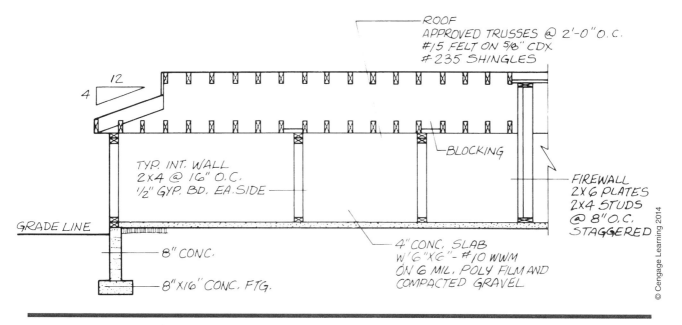

**Figure 8–3.** Longitudinal section.

- Sizes of floor framing materials
- Sizes of wall framing materials
- Wall covering (sheathing, siding, stucco, masonry, and interior wall finish) and insulation
- Cornice construction—materials and sizes (The **cornice** is the construction at the roof eaves.)
- Ceiling construction and insulation

Other section drawings are included as necessary to explain special features of construction. Wherever wall construction varies from the typical wall section, another wall section should be included. Section views are used to show any special construction that cannot be shown on normal plans and elevations. **Figure 8–4** is an example of a special section in elevation. This section view is said

Sections and Detail  49

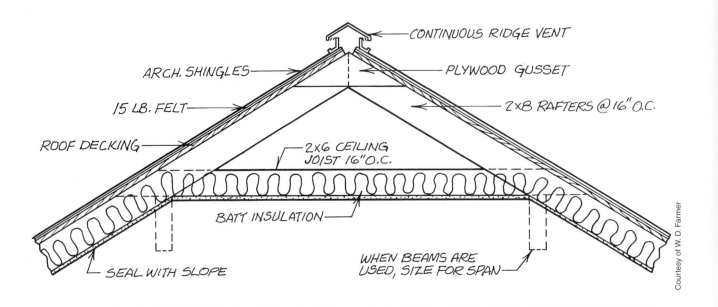

**Figure 8–4.** Special section of ventilated ridge.

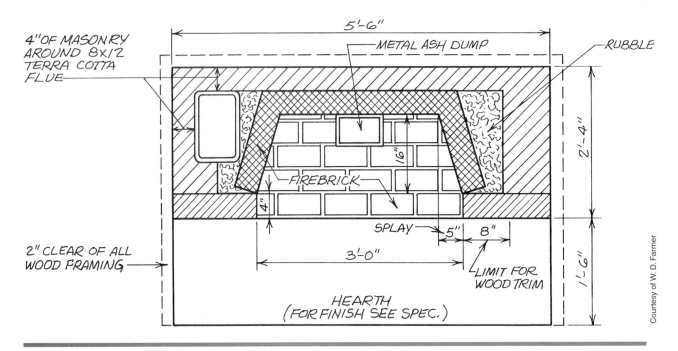

**Figure 8–5.** A section in plan.

to be in elevation because it shows the height of the ridge construction. **Figure 8–5** is in plan because it shows the interior of the fireplace as viewed from above.

## Other Large-Scale Details

Sometimes necessary information can be conveyed without showing the interior construction. A large scale may be all that is needed to show the necessary details. The most common examples of this are on cabinet installation drawings (see **Figure 8–6**). Cabinet elevations show how the cabinets are located, without showing the interior construction.

Many details are best shown by combining elevations and sections or by using isometric drawings. **Figure 8–7** shows an example of an elevation and a section used together to explain the construction of a fireplace. **Figure 8–8** shows an isometric detail drawing that includes sections to show interior construction.

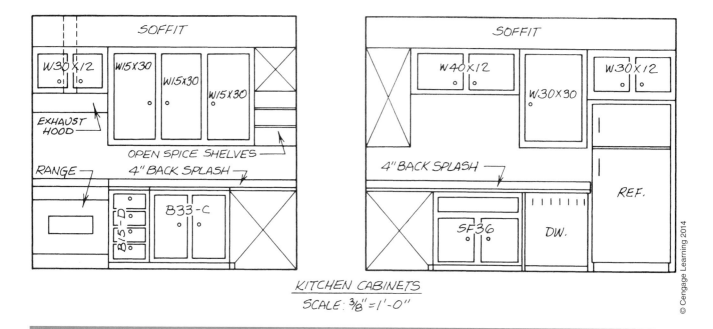

**Figure 8–6.** Cabinet elevations.

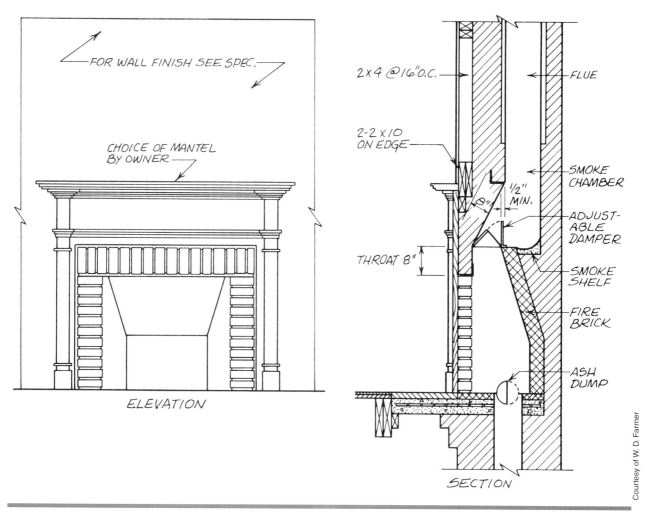

**Figure 8–7.** Fireplace details.

Sections and Detail 51

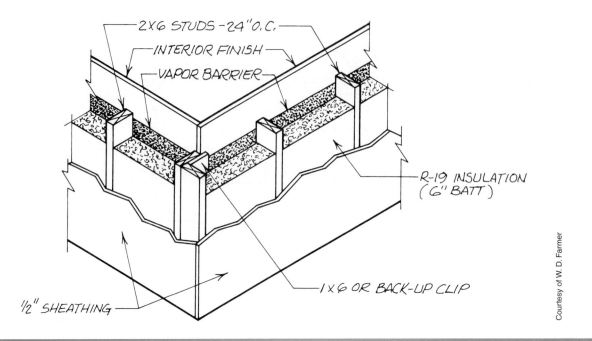

**Figure 8–8.** Isometric section.

GREEN NOTE

*When frame construction was first developed, horizontal or diagonal boards were applied as sheathing. After World War II, plywood replaced board sheathing. Today Oriented Strand Board (OSB) is the most commonly used sheathing material, but other engineered products are becoming increasingly popular. Gypsum-based structural sheathing and magnesium oxide structural sheathing are water resistant, decay resistant, and form a vapor barrier, eliminating the need for a separate moisture barrier.*

## Orienting Sections and Details

As explained earlier, some sections and details are labeled as typical. These drawings describe the construction that is used throughout most of the building.

Sections and details that refer to a specific location in the building include a reference that indicates where the section or detail came from. That larger source drawing has a cutting-plane line to show exactly where the section cut or detail is taken from. The cutting-plane line has an arrowhead or some other indication of the direction from which the detail is viewed. The top drawing in **Figure 8–9** shows that there is a section view or detail of the construction at the skylight. The little flag at the top points to the right, so that is the direction from which the detail at the bottom of the figure is viewed. The label on the bottom drawing, the skylight detail, includes the number of the drawing, corresponding to the number indicated at the cutting-plane line in the top view. A reference mark near the arrow indicates where the detail drawing is shown. The reference marks that are used for orienting details may vary from one set of drawings to another. It is important, although not usually difficult, to study the drawings and learn how the architect references details. Usually a system of sheet numbers and view numbers is used. One such numbering system was explained earlier.

Some basic principles of details and sections have been discussed here. You will gain more practice later in reading details and sections.

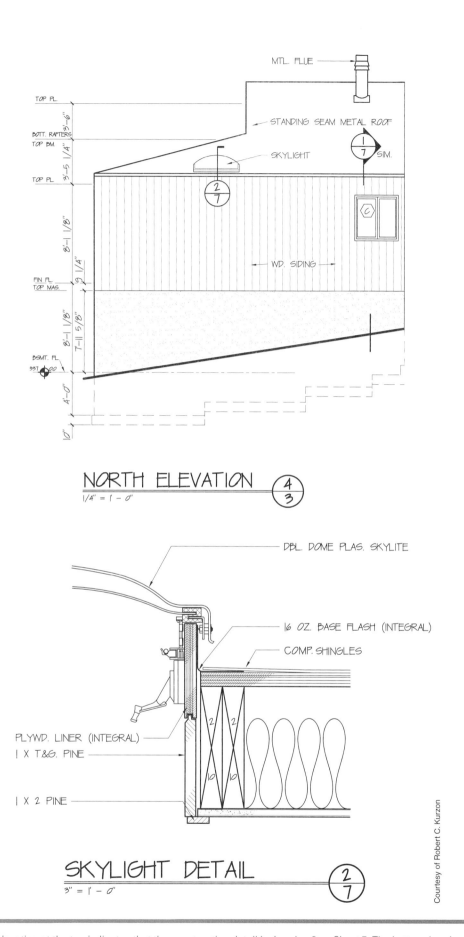

**Figure 8–9.** The elevation at the top indicates that the construction detail is drawing 2 on Sheet 7. The bottom drawing is that detail.

Sections and Detail

### USING WHAT YOU LEARNED

A lot of information can be found on the various section views and detail drawings. Some of this is shown simply by drawing a symbol and adding a note or callout. Any material that is to be placed beneath a concrete slab or imbedded in the slab must be in place before the concrete is delivered. Beneath the concrete slab in the Two-Unit Apartment in your textbook packet there is rigid foam insulation. What is immediately beneath the insulation? The exterior wall section on Sheet 2 has a callout and a leader pointing to a line of long dashes. The callout identifies this as 6 mil poly vapor barrier.

## Assignment

Refer to the drawings of the Two-Unit Apartment in your textbook packet to complete the assignment.

1. What is used to show the detail of a complex design, installation, or product?
2. What kind of section drawing is the Typical Section on Sheet 1?
3. What kind and size material is to be used for the foundation walls?
4. What material is immediately beneath the outer two feet of the concrete slab?
5. What kind and size of insulation is used around the foundation? Is this insulation used on the inside or outside of the foundation?
6. What kind and size of material is to be used on the inside face of the frame walls?
7. Sheet 2 includes a firewall detail. Where in the apartment is this firewall?
8. What is the distance between the centerlines of the studs in the firewall?
9. What is the total thickness of the firewall? (Remember that a 2 × 6 is actually 5½" wide.)
10. Were the cabinet elevations drawn of the kitchen on the east side or the west side of the Two-Unit Apartment?
11. How would the kitchen elevations be different if they were drawn from the other kitchen?
12. What is the distance from the kitchen counter top to the bottom of the wall cabinets?
13. How far does the roof overhang project beyond the exterior walls?
14. Where is the electrical panel located for the west apartment?

# Test

**A.** Identify each of the dimensions indicated on the illustrated scale.

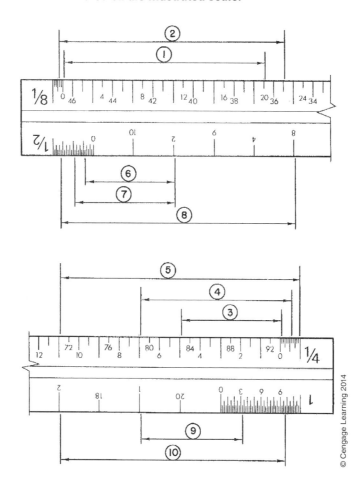

**B.** Which occupation or individual listed in Column II performs the task listed in Column I?

| I | II |
|---|---|
| 1. Obtains a building permit | a. Architect |
| 2. Issues the building permit | b. Building inspector |
| 3. Acts as the owner's representative | c. Owner |
| 4. Issues certificate of occupancy | d. Mechanical engineer |
| 5. Lays out rooms for efficient use | e. Municipal building department |
| 6. Designs plumbing in large buildings | f. General contractor |
| 7. Hires and supervises carpenters | |
| 8. Checks to see that codes are observed | |

**C. Which of the lines shown in Column II is most likely to be used for each purpose in Column I?**

| I | II |
|---|---|
| 1. Outline of a window | a. ─────────── |
| 2. Alternate position of a fold-down countertop | b. ─────────── |
| 3. Centerline of a round post | c. ── ── ── ── |
| 4. Extension line to show extent of a dimension | d. ───── ── ───── |
| 5. Buried footing | e. ───── ── ── ───── |
| 6. Point at which an imaginary cut is made for a section view | f. ── ── ── ── ── ── ── ── |

**D. Which of the symbols shown in Column II is used for each of the objects or materials in Column I?**

I
1. Awning window in elevation
2. Bifold door in plan
3. Earth
4. Rough wood
5. Batt insulation
6. Concrete
7. Ceiling light fixture
8. Finish wood
9. Shutoff valve (plumbing)
10. Hopper window in elevation

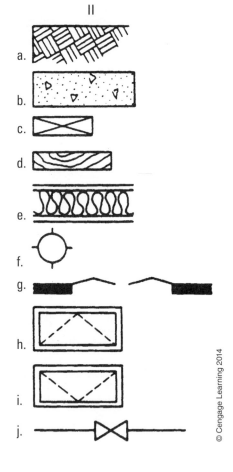

**E. Select the one best answer for each question.**
1. What kind of regulation controls the types of buildings allowed in each part of a community?
   a. building code
   b. zoning law
   c. specification
   d. certificate of occupancy
2. Which of the listed kinds of information can be clearly shown on construction drawings?
   a. size of parts
   b. location of parts
   c. shape of parts
   d. all of these

3. What type of drawing is the 2 × 4 shown in Illustration 3?
   a. isometric
   b. oblique
   c. perspective
   d. none of these

4. What type of drawing is the 2 × 4 shown in Illustration 4?
   a. isometric
   b. oblique
   c. perspective
   d. none of these

5. What type of drawing is the 2 × 4 shown in Illustration 5?
   a. isometric
   b. oblique
   c. perspective
   d. none of these

6. What kind of drawing is shown in Illustration 6?
   a. elevation
   b. detail
   c. rendering
   d. plan

7. What kind of drawing is shown in Illustration 7?
   a. elevation
   b. detail
   c. rendering
   d. plan

8. What kind of drawing is shown in Illustration 8?
   a. elevation
   b. detail
   c. rendering
   d. plan

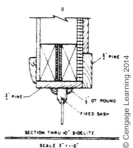

Part I Test 57

9. If an object 12 feet long is drawn at a scale of ¼"= 1'-0", how long is the drawing?
   a. 48 inches
   b. 3 inches
   c. 3 feet
   d. none of these
10. If an object 1'-6" long is drawn at a scale of 1½"= 1'-0", how long is the drawing?
    a. 2¼ inches
    b. 2 inches
    c. 27 inches
    d. none of these
11. Where in a set of drawings would you find detail number 9.6?
    a. sheet 6
    b. sheet 9
    c. ninth sheet in the mechanical section
    d. none of the above
12. In the drawing key 4/A–4, what does the letter A stand for?
    a. architect's initial
    b. first edition of the drawings
    c. architectural
    d. first detail on the sheet
13. On which drawing would you expect to find the height of the foundation wall?
    a. site plan
    b. building elevation
    c. floor plan
    d. foundation plan
14. On which drawing would you expect to find the setback of the building?
    a. site plan
    b. building elevation
    c. floor plan
    d. foundation plan
15. On which drawing would you expect to find the height of the window heads?
    a. site plan
    b. building elevation
    c. floor plan
    d. window detail

F. **Refer to the two-unit apartment drawings in your textbook packet to answer these questions.**

1. How far is the building from the west boundary?
2. What is the dimension from the finished floor to the top of the wall plate?
3. What is the overall length of the building at window height?
4. What is the overall length of the building at the eaves?
5. What is the north-to-south dimension inside the front bedrooms?
6. What is the slope of the roof?
7. What types of windows are used in the bedrooms?
8. How thick is the concrete footing?
9. What material is the foundation wall?
10. What is under the floor at its center?

# READING DRAWINGS FOR TRADE INFORMATION

In Part II, you will examine all the information necessary to build a moderately complex single-family home. The sequence of the units in Part II follows the sequence of actual construction. In some cases, all the information necessary for a particular phase of construction can be found on one sheet of drawings. Other phases require cross-referencing among several drawings. The relationships among the various drawings are discussed as the need to cross-reference them arises.

The assignments in this part refer to the lake house drawings provided in your textbook packet. This lake house was designed as a vacation home on a lake in Virginia. The design is moderately complex, involving several floor levels and some interesting construction techniques.

# UNIT 9

# Clearing and Rough Grading the Site

## Objectives

After completing this unit, you will be able to perform the following tasks:

○ Identify work to be included in clearing a building site according to site plans.

○ Interpret grading indications on a site plan.

○ Interpolate unspecified site elevations.

## Property Boundary Lines

The boundary lines of the building site are shown on the site plan. The direction of a property line is usually expressed as a bearing angle. The *bearing* of a line is the angle between the line and north or south. Bearing angles are measured from north or south depending on which keeps the bearing under 90° (see **Figure 9–1**). Angles are expressed in degrees (°), minutes ('), and seconds ("). There are 360 degrees in a complete circle, 60 minutes in a degree, and 60 seconds in a minute.

The *point of beginning* (POB) may or may not be shown on the site plan. If the point of beginning is not shown on the plan, start at a convenient corner. Corners are usually marked with an *iron pin* (IP) or some permanent feature. The approximate direction of the boundaries can be found with a handheld

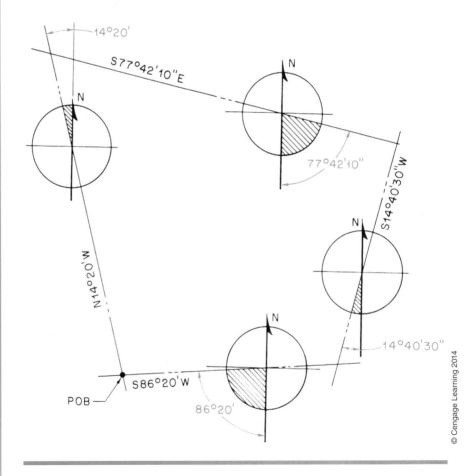

**Figure 9–1.** Bearing angles are always less than 90°.

compass. This approximation should be accurate enough to aid in finding the marker (iron pin, manhole cover, concrete marker, or similar item) at the next corner. Proceed around the perimeter in this manner to find all corners. All construction activity should be kept within the property boundaries unless permission is first obtained from neighboring landowners.

## Clearing the Site

The first step in actual construction is to prepare the site. This means clearing any brush or trees that are not to be part of the finished landscape. The architect's choice of trees to remain is based on consideration of many factors. Trees and other natural features can be an important part of architecture—not only for their natural beauty, but for energy conservation. For example, deciduous trees, which lose their leaves in the winter, can be used to effectively control the solar energy striking a house during warmer months as shown in **Figure 9–2**, where the trees shade the south side of the house. In the winter, the sun shines through the deciduous trees on the south side of a house, thus taking advantage of this source of heat as shown in **Figure 9–3**. The lake house in your textbook package offers a good example of the importance of the selection of trees to remain on a site. This house gets a large part of its heat from its passive-solar features, which are described more fully later.

Trees that are to be saved are shown on the plot plan by a symbol and a note indicating their butt diameter and species (see **Figure 9–4**). Areas that are too densely wooded to show individual trees are outlined and marked "woods" (see **Figure 9–5**). Removal of unwanted trees may require felling and stump removal, or may be accomplished with a bulldozer and dump truck. In either case, care must be exercised not to damage the trees that are to be saved.

## Grading

*Grading* refers to moving earth away from high areas and into low areas. Site grading is necessary to ensure that water drains away from the building properly and does not puddle or run into the building. In some cases, grading may be necessary for access to the site. For example, if the site has a steep grade, it may be necessary to provide a more gradual slope for a driveway.

**Figure 9–2.** The winter sun passes through deciduous trees.

**Figure 9–3.** The summer sun is shaded by deciduous trees.

**Figure 9–4.** Typical note and symbol for individual tree.

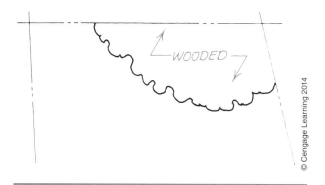

**Figure 9–5.** Typical note and symbol for wooded area.

### GREEN NOTE

*Controlling water is a big part of green construction. Water must be prevented from entering the building envelope with the weather. Water vapor, exhaled by all occupants, even animals, as they breathe, must be vented from the building. Surface water must be controlled as well in order to prevent erosion of the building site or neighboring property. Surface water should not be allowed to collect in puddles or overly damp areas of earth, as these can promote mosquito breeding and harm the growth of desirable vegetation. Proper grading of the site is the most effective way of controlling surface water.*

Grade is measured in vertical feet from sea level or from a fixed object such as a manhole cover. This vertical distance is called **elevation**. The term *elevation* to denote a vertical position should not be confused with elevation drawings that show the height of objects. The elevations of specific points are given as *spot elevations*. Spot elevations are used to establish points in a driveway, a walk, or the slope of a terrace, as shown in **Figure 9–6**. Spot elevations are often given for trees that are to be saved.

The grade of a site is shown by *topographic* **contour lines**. These are lines following a particular elevation. The vertical difference between contour lines is the **vertical contour interval**. For plot plans this is usually 1 or 2 feet. When the land slopes steeply, the contour lines are closely spaced. When the slope is gradual, the contour lines are more widely spaced.

The builder must be concerned with not only the grade or contour of the existing site, but also that of the finished site. To show both contours, two sets of contour lines are included on the plot plan. Broken lines indicate natural grade (NG), and solid lines indicate finished grade (FG) (see **Figure 9–7**).

When the natural-grade elevation is higher than the finished-grade elevation, earth must be removed. This is referred to as *cut*. When the natural grade is at a lower elevation than the finished grade, *fill* is required. To determine the amount of cut or fill required at a given point, find the difference between the natural grade and the finished grade (see **Figure 9–8**).

## Interpolating Elevations

Sometimes it is necessary to find an elevation that falls between two contour lines. This can be done by interpolation. *Interpolation* is a method of finding an unknown value by comparing it with known values.

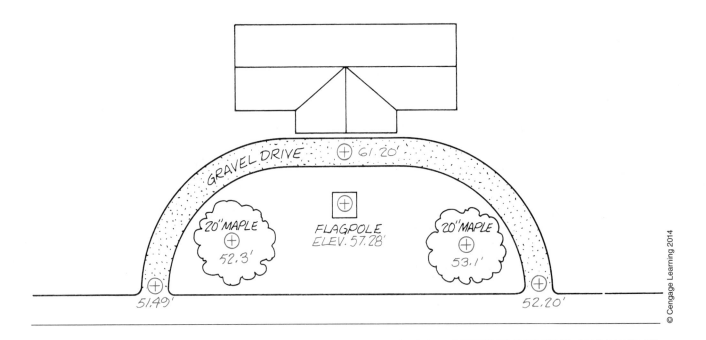

**Figure 9–6.** Spot elevations for specific locations.

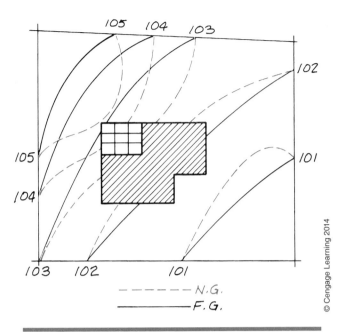

**Figure 9–7.** Two sets of contour lines show that this site will be graded to be more level in the area of the building.

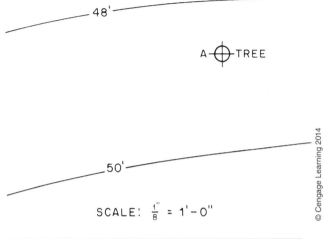

**Figure 9–9.** Interpolate the elevation of the tree.

**Example:** To interpolate the elevation of the tree at point A in **Figure 9–9**, follow the listed steps of the procedure using the information shown in the illustration and the numbers enclosed in parentheses.

Step 1. Scale the distance between the two adjacent contour lines (12 feet).

Step 2. Scale the distance from the unknown point to the nearest contour line (4 feet).

Step 3. Multiply the contour interval by the fraction of the distance between the contour lines to the unknown point. (Contour interval = 2 feet; fraction of distance between contour lines = $4/12 = 1/3$. $2 \times 1/3 = 2/3'$.)

Step 4. If the nearest contour line is below the other one, add this to it. If the nearest contour line is above the other one, subtract this amount. (Nearest contour = 48 feet. This is below the other contour line at 50 feet, so $2/3$ foot is added to 48 feet. $2/3' + 48' = 48.66'$.)

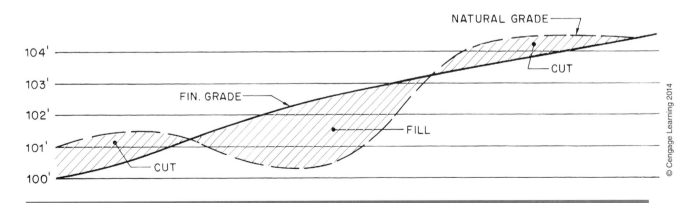

**Figure 9–8.** Cutting is required where NG is above FG. Fill is required where NG is below FG.

Clearing and Rough Grading the Site 63

### USING WHAT YOU LEARNED

One of the first activities on a building site, after laying out the building lines, is excavating for the foundation and any concrete slabs. This requires that you know elevations of the foundations and slabs are compared to the elevation of the natural grade of the site. Let's determine how far the northeast corner of the garage floor is above or below the natural grade at that point.

The garage elevation is clearly marked as 343.0'. The corner fall between the natural-grade contour lines for 342' and 344'. We will have to interpolate to find a close estimate of what that elevation is. Using the inch scale on an architectural scale, measure the distance between the two contour lines at the corner of the garage. The shortest distance between these two lines that crosses the corner of the garage is 1 9/16". The corner is 9/16" from the 344' contour line. These elevations are expressed in decimal parts of a foot, so we will find our elevation to the nearest 1/10th of a foot. Divide 9/16 by 1 9/16 (or 25/16) to find that the corner is 36 percent of the way from 344' to 342'. Thirty-six percent of 2 feet (the contour interval) is 0.72'. Subtract 0.72' from 344' to find that the elevation of the natural grade at the corner is 343.28'. To the nearest 1/10th of a foot this is 343.3 feet, so the garage floor is 0.3 feet above the natural grade at that point.

## Assignment

Refer to the site drawings of the Lake House in your textbook packet to complete the assignment.

1. What are the lengths of the north, east, and west boundaries of the Lake House?
2. Which of the compass points shown in **Figure 9–10** corresponds with the north boundary of the Lake House? Which compass point corresponds with the east boundary?
3. How many trees are indicated for removal?
4. How many trees are to remain on the site? (Do not include wooded areas.)
5. What is the finished-grade elevation at the tree nearest the Lake House?
6. What is the natural-grade elevation of the most easterly tree to be saved?
7. What is the elevation of the tree to be saved nearest the lake?
8. What is the natural-grade elevation at the southwest corner of the Lake House? Do not include the deck as part of the house.
9. What is the finished-grade elevation at the southwest corner of the Lake House?
10. How much cut or fill is required at the entrance of the garage?
11. Is cut or fill required at the southwest corner of the Lake House? How much?
12. What is the elevation at the northeast corner of the site?

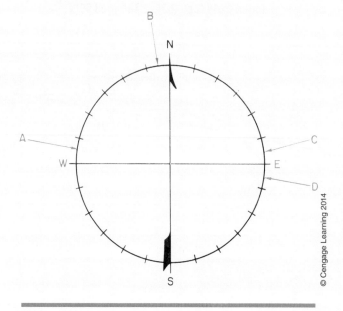

**Figure 9–10.**

# Unit 10: Locating the Building

## Laying Out Building Lines

The position of the building is shown on the site plan. Dimensions show the distance from the street (or lake) to the building and from the side boundaries to the building. The location of one corner can easily be found by measuring with a long (100′ or 200′) steel tape. Refer to **Figure 10–1** as you read the following directions for finding the corners of a rectangular building. The lot boundaries are represented by A, B, C, and D. Along boundary line A-C measure the distance from the front edge of the property, usually a street or body of water. Mark this point with a stake (e). As you drive each stake, drive a nail in the top of the stake to accurately mark the location. Measure the same distance along the other side boundary (B-D) and mark that point with stake f. Stretch a string from stake e to stake f. The front of the building will fall on this string. The distance from the front boundary to the building line is called the front **setback**. Both the front and side setbacks are nearly always regulated by zoning ordinances. Next measure the side setback, which is the distance from the side of the property to the side of the building at stake e to side of the building (line e-G). Drive a stake where the side setback and front setback meet. This stake represents one front corner of the building (corner G). Find the other front corner by measuring the width of the building from the first corner (line G-H). Check the distance from the corners to the front boundary of the lot to make sure it agrees with the setback shown on the site plan.

## Objectives

After completing this unit, you will be able to perform the following tasks:

○ Lay out building lines according to a site plan.

○ Use the 6-8-10 or equal-diagonals method to check the squareness of corners.

○ Use a laser level to measure the depths of excavations.

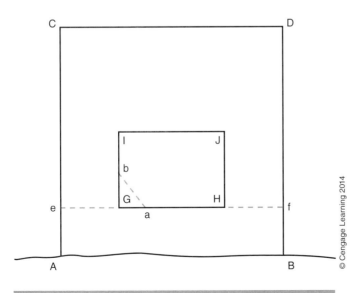

**Figure 10–1.** Laying out building lines.

Now measure the depth of the building from the front corners to the rear of the building (corner I). Use a framing square to make corner G as square as possible. Check the corner for square using the 6-8-10 method (see **Figure 10–2**). Measure 6 feet from the corner along line G-I to find point a. Measure 8 feet from the corner along line G-e to find point b. Line a-b should be exactly 10 feet. (See Math review 24.) Adjust the angle of the side building line as necessary to make a-b 10 feet. Drive a stake at corner I. Repeat the process to find the other rear corner J. Check to see that the rear building line (I-J) is the dimension shown on the plans.

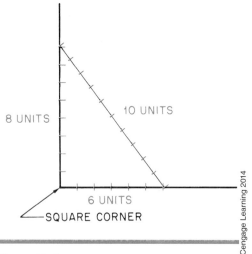

**Figure 10–2.** The 6-8-10 method of checking a square corner.

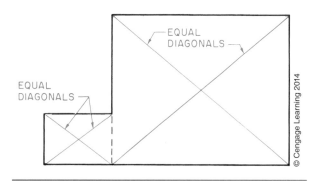

**Figure 10–3.** When the diagonals of a rectangle are the same length, the corners are square.

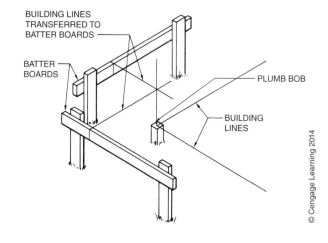

**Figure 10–4.** Batter boards.

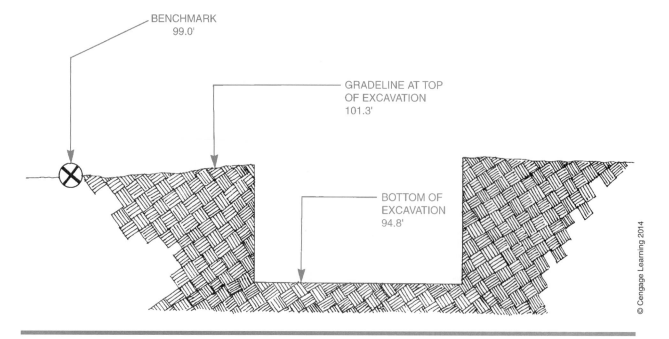

**Figure 10–5.** The top of this excavation is 2.3 feet above the benchmark. Its depth is 6.5 feet (101.3' − 94.8').

Make adjustments as necessary until all corners are square and all dimensions are correct.

Now check the final layout to ensure that rectangles have 90-degree corners by measuring the diagonals (see **Figure 10-3**). When the diagonals are equal, the corners of the rectangle are square.

The **building lines** can be saved, even after the corner stakes are removed for earthwork, by erecting batter boards (see **Figure 10–4**). **Batter boards** are sturdy horizontal boards fastened between 2 × 4 stakes, at least 4 feet outside the building lines. The building lines are extended and marked on the batter boards.

## Excavating

Most buildings require some *excavation* (digging) to prepare the site for a foundation. The depth of the excavation is measured from a fixed benchmark. A benchmark can be any stationary object such as a surveyed point on a street or very large boulder. All elevations (vertical distances) are measured from this benchmark. Only in the case of a real coincidence would the benchmark be at the same elevation as the surface of the ground where the excavation is to be done. The actual depth of the excavation is the difference between the elevation at the surface of the ground and the elevation at the bottom of the excavation (see **Figure 10–5**).

Concrete footings are placed in the bottom of the excavation to support the entire weight of the building (see **Figure 10–6**). These footings are placed on unexcavated earth to reduce the chance of the soil compacting under them. This means that the excavation contractor must measure the depth of the excavation accurately. The footings may be *stepped,* as in **Figure 10–7**,

GREEN NOTE

*The different soil types found at varying depths during an excavation can be surprising. In some areas, there may only be a few inches of top soil to support healthy plant growth. Below that layer can be heavy clay, which contains few nutrients and does not drain well, or gravel that also has few nutrients and does not hold water at all. For this reason, the top soil should be stripped from the area to be excavated and stock piled in a safe place to be used later where landscaping and lawns are planted.*

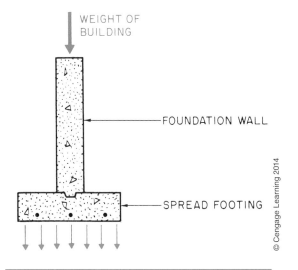

**Figure 10–6.** The footing spreads the weight of the building over a greater soil area.

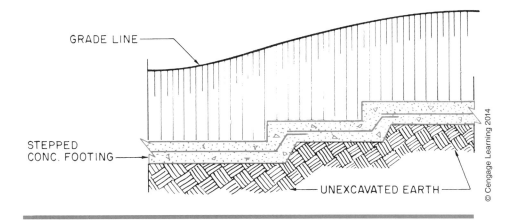

**Figure 10–7.** Footings can be stopped to accommodate a sloping site.

Locating the Building  67

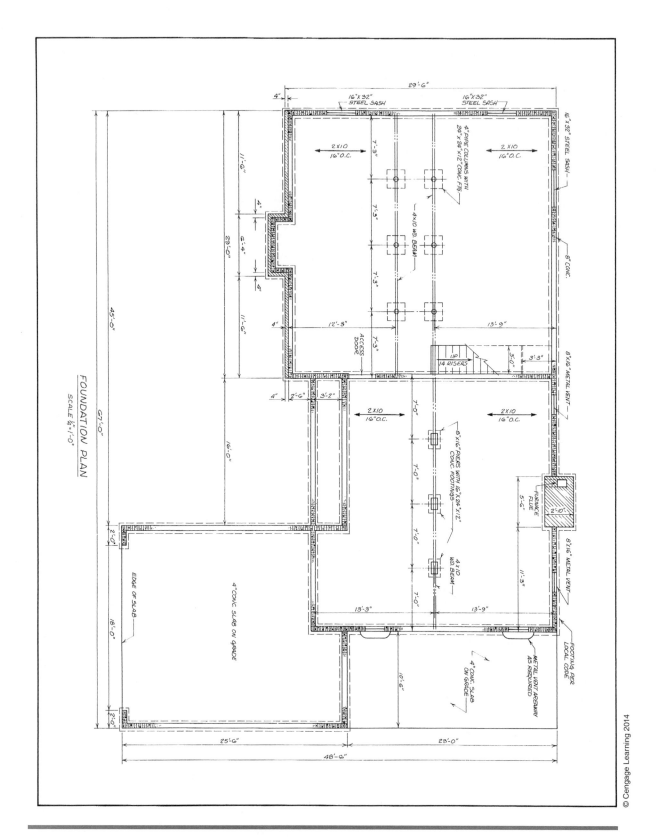

**Figure 10–8.** The foundation plan gives complete dimensions.

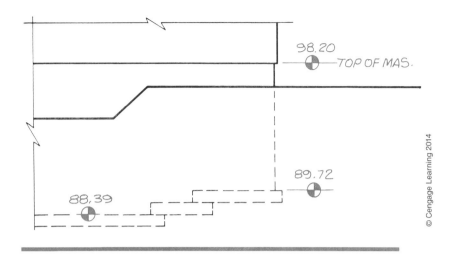

**Figure 10–9.** Spot elevations show key points on the footing and the foundation.

to accommodate a sloping site. This requires measuring the depth at each step of the footing. Information about the footing design is found on the foundation plan and building elevations. The layout of the foundation walls and their footings is shown on the foundation plan. The foundation walls are shown by two solid lines with dimensions to indicate their sizes (see **Figure 10–8**). A dotted line on each side of the foundation indicates the concrete footing. The size of the footing may be omitted when the plan is developed for use in several locations.

The depth of the foundation, including its footing, is shown on the elevations. To simplify calculating excavation depths, many architects indicate the elevations as key points along the footings (see **Figure 10–9**). A section view through all or part of the building may show a typical depth, but it is wise to check all the elevations for steps in the footing. The footings may be shown on the elevation as a double or a single dotted line. In masonry foundations, steps in footings are usually in increments of 8 inches to conform to standard concrete block sizes.

Note: A laser level is often used to measure elevations and depths of excavations. Setting up a laser level is similar to setting up a builder's level. You should follow the manufacturer's instructions and obey the safety precautions provided for the use of the laser level.

Use the following procedure to measure differences in elevation, such as the depth of an excavation:

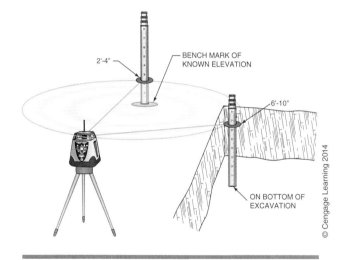

**Figure 10–10.** This excavation is 4'-6' deep—the difference between the two readings on the target rod.

Step 1. Set the instrument up on a tripod, and level it in a convenient location.

Step 2. Hold the target rod on a known elevation, such as a benchmark or ground of known elevation. Note the reading on the target rod where the target registers the laser beam.

Step 3. Take a similar reading with the target rod at the bottom of the excavation. The difference in the two readings is the depth of the excavation, **Figure 10–10**.

### USING WHAT YOU LEARNED

In the old days, before people had much awareness of how site conditions affected the livability of a house, it was common for a builder to begin by removing all of the trees that were near the planned building. It was thought that those trees would just be in the way. Today we are much more conscious of how the site affects the performance of the building. What would be the result in how the house functions if all of the oak and maple trees near the front of the house were removed? Those trees, being on the south side of the house, provide cooling shade in the summer, when they have all of their leaves. They also help hold the soil in place on the steep slope down to the lake.

## Assignment

Refer to the Lake House drawings (in the packet) to complete this assignment.

1. What is the distance from the Lake House to the nearest property boundary? (Do not treat the decks as part of the house for this question.)

2. What is the distance from the Lake House to the lake?

3. What is the distance from the north property line to the garage?

4. What is the area of the basement of the Lake House, including the foundation? Ignore slight irregularities in the shape of the foundation. For ease in calculating, divide the foundation into rectangles, **Figure 10–11.**

5. Find the highest and lowest natural grades meeting the house.

6. How much cut or fill is required for the basement at the northeast corner of the house?

7. Measuring from the natural grade, how deep is the excavation for the footing under the overhead garage door?

8. What is the elevation at the bottom of the deepest excavation for the Lake House? (Do not include the garage.)

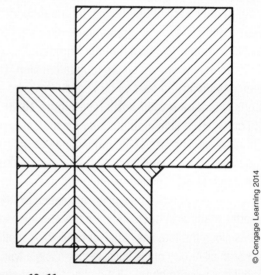

Figure 10–11.

9. Why would a row of large evergreen trees between the Lake House and the lake decrease the energy efficiency of the house?

10. What aspect of the location of a building is most often regulated by local ordinances?

# Site Utilities

## Sewer Drains

The *building sewer* carries the waste to the municipal sewer or septic system (see **Figure 11–1**). Because sewer lines usually rely on gravity flow, they are large in diameter (4 inches, minimum) and are pitched to provide flow. Because water supply lines and gas lines are pressurized, pitch is not important in their installation. Therefore, the sewer is installed first, and other piping is routed around it as necessary. The size, material, and pitch of drains are usually given in a note on the site plan (see **Figure 11–2**). The pitch of a pipe is given in fractions of an inch per foot. A pitch of ¼ inch per foot means that for every horizontal foot, the pipe rises or falls ¼ inch.

In some cases, sewers may have to flow uphill. This is the case with the Lake House. Uphill flow is accomplished by a *grinder pump* (see **Figure 11–3**). A grinder pump grinds solids into small enough particles to be pumped and pumps the sewage at low pressure.

## Objectives

After completing this unit, you will be able to perform the following tasks:

○ Interpret symbols and notes used to describe site utilities.

○ Explain the septic system indicated on a site plan.

○ Determine the pitch of drain lines.

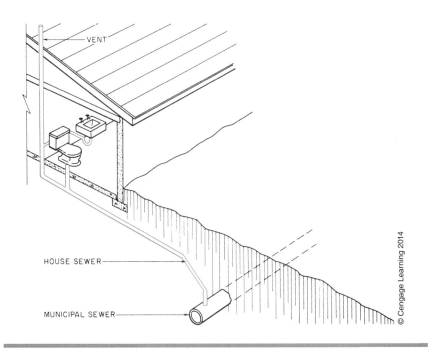

**Figure 11–1.** The house sewer carries waste from the house to the municipal sewer or septic systems.

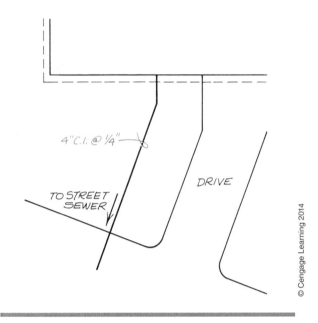

**Figure 11–2.** This note indicates cast iron pipe sloped ¼ inch for every foot of run.

## Building Sewer

Any plumbing that is to be concealed by concrete work must be installed without fixtures or *roughed in* before the concrete is placed. Because the plumbing contractor installs all plumbing inside the building lines, this phase of construction is discussed later with mechanical systems. The sewer, however, is usually installed *after* the building is erected to prevent damage due to machinery on the site during construction. The sewer, too, is generally considered a site utility, which is why it is discussed here.

The workers who install the sewer must be able to determine the elevation at which it passes through the foundation and the pitch of the line outside the building. The sewer line may be shown on plans as a solid or broken line. Although it is usually labeled, this is not always true. When the sewer is not labeled as such, it can still be recognized by its material, pitch, and ending place. In light construction, the sewer is usually the

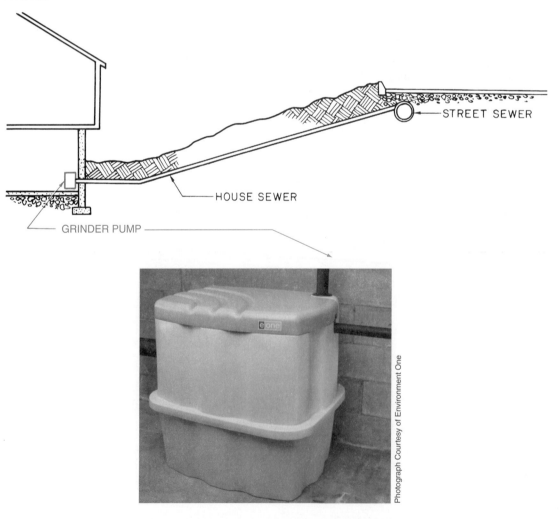

**Figure 11–3.** A grinder is a pump that can move sewage uphill.

only 4-inch pipe to the building. Also, the sewer is the only line with the pitch indicated.

## Municipal Sewers and Septic Systems

In most developed areas, the building sewer empties into a municipal sewer line near the street. The contractor for the new building is responsible for everything from the municipal sewer to the house.

In less developed areas, the sewer carries the sewage to a septic system. The most common type of *septic system* includes a septic tank and drain field (see **Figure 11–4**). The septic tank holds the solid waste while it is decomposed by bacterial action. The liquids pass through the baffles and flow out of the tank to the distribution box. The distribution box (DB) diverts the liquid into *leach lines.* These perforated plastic or loose-fitting tile lines allow the liquid to be absorbed by the surrounding soil. The liquid gradually evaporates from or drains through the soil. The *drain field,* where the leach lines are laid, is usually made of a layer of crushed stone.

The design of septic systems is closely regulated by most health and plumbing codes. These local codes should be checked before designing or installing any septic system. The building code or health department code often requires a *percolation test* before the system can be approved. In a *percolation test,* holes are dug. Then a measured amount of water is poured into each hole. The amount of time required for the water to drain into the soil is an indication of how well the soil

### GREEN NOTE

*There are several materials used to make pipes and fittings. Traditionally sewer pipes were cast iron and supply pipes were galvanized iron or steel. As polyvinyl chloride (PVC) became available, it became the most common material for drainage, waste, and vent piping and for sewer pipes. It requires less energy to manufacture than cast iron or steel, it lasts almost indefinitely, and it requires less time to install. The downside is that it does not withstand heat, so it can't be used for hot water.*

*Copper, with soldered fittings, has been widely used for supply piping, because it is noncorrosive and will last much longer than iron or steel pipe. Well water is sometimes slightly acidic and the acid does corrode copper pipe, but that is not usually a concern because municipal water is treated to control the acidity.*

*In recent years, copper has become very expensive and two kinds of plastic pipe have been developed for use where copper has been the standard. Chlorinated polyvinyl chloride (CPVC) is inexpensive, noncorrosive, and easy to install. Cross-linked polyethylene (PEX) is more flexible, so it requires fewer fitting and is quick to install. Both CPVC and PEX are widely used in home construction today.*

*Manufacture of all plastic piping requires petrochemicals, but as a percentage of the total materials in a home the amount of piping material is very small, plastic pipe is not considered to have much of an environmental impact. There is no one best material for piping in green home construction.*

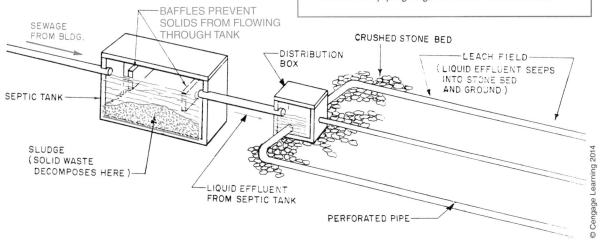

**Figure 11–4.** Septic system.

will accept water from the septic system. This ability to accept water is called *percolation*. The locations of key elements in the system are often shown on the site plan.

To ensure proper drainage for a septic system, the elevation of its drainage piping is critical. For this reason, it is necessary to know the *invert elevation* of a pipe, which is the elevation at the lowest point inside the pipe (see **Figure 11–5**).

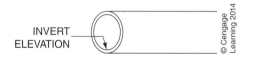

**Figure 11–5.** Invert elevation is the elevation at the lowest point inside the pipe.

## Other Piping

Other utility piping, such as that for water supply or gas, is shown on the site plan. If these lines will pass beneath the concrete footings or be concealed under a concrete slab, they must be roughed in before the concrete is placed. Water supply pipes follow the most direct route from the municipal water main or well to the main shut-off valve or pump. Gas lines run from the main to the gas meter. All supply lines on the plot plan should be labeled according to type and size of piping (see **Figure 11–6**).

## Electrical Service

The electrical service is the wiring that brings electricity to the house. There are two types of electrical service: overhead and underground (or buried).

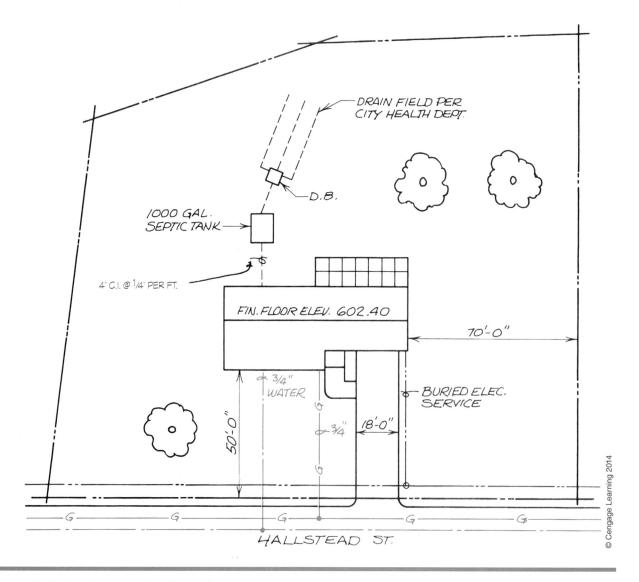

**Figure 11–6.** Partial site plan with utilities indicated.

74 UNIT 11

Overhead service involves a cable from the utility company transformer or pole to a weather-tight fixture called a *service head* or *mast head* on the house (see **Figure 11–7**). The service head is mounted on the top of a pipe, which serves as a conduit to the meter receptacle. In an underground service, the cable is buried (see **Figure 11–8**). Although electrical service is a site utility, it is not usually installed until the building is enclosed.

The electrical service to residential and small commercial buildings is similar, as previously explained and as shown in **Figures 11–7** and **11–8**. This type of service has only one distribution point, the main electrical panel.

Heavy commercial and industrial electrical services require multiconductor feeders, and typically these are underground services from a utility transformer to a service entrance section in the designated electrical or utility room. Additional utility coordination will be required within the building, which is discussed in Unit 14. The placement of these underground electrical utility lines must be coordinated with the locations of other site and building utilities prior to starting installation.

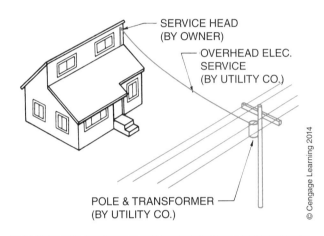

**Figure 11–7.** Overhead electrical service.

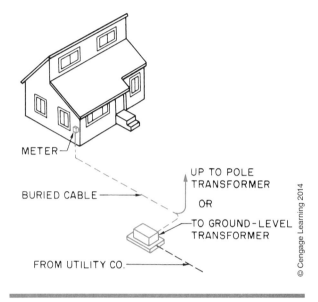

**Figure 11–8.** Underground electrical service.

### USING WHAT YOU LEARNED

All building trades workers need to have an understanding of where the utilities will be located on the site. It is obvious that plumbers need to know where the water supply and sewer enter the building, but masons and carpenters also need this information. For example, if the piping is to run through a sleeve in the concrete, that sleeve must be in place before the concrete is placed. Where does the water supply enter the Lake House? The site plan shows it entering through the north wall of the play room. This is a case where the architect should be consulted to determine if it would be better to run the water supply closer to the sewer pipe, so it enters the utility room.

## Assignment

Refer to the Lake House drawings (in the packet) to complete the assignment.
1. What size is the sewer for the Lake House?
2. How many lineal feet are required from the foundation wall to the septic tank?
3. What is the rise of the sewer from the house to the septic tank?
4. Where does the sewer pass through the foundation?
5. How many lineal feet of perforated pipe are needed for the drain field?
6. How many cubic yards of crushed stone are needed for the drain field?
7. What is the invert elevation of the sewer pipe at the point it enters the house?

Site Utilities

# UNIT 12
# Footings

## Objectives

After completing this unit, you will be able to perform the following tasks:

○ Find all information on a set of drawings pertaining to footing design.

○ Interpret drawings for stepped footings used to accommodate changes in elevation.

○ Discuss applicable building codes pertaining to building design.

All soil changes shape under force. When the tremendous weight of a building is placed on soil, the soil tends to compress under the foundation walls and allow the building to settle. To prevent settling, concrete footings are used to spread the weight of the building over more area. The footings distribute the weight of the building, so that there is less force per square foot of area.

The simplest type of footing used in residential construction is referred to as *slab-on-grade*. In this system the main floor of the building is a single concrete slab, reinforced with steel to prevent cracking. This slab supports the weight of the building (see **Figure 12–1**). Slab-on-grade foundations are common in warm climates. This type of construction is indicated on the floor plan by a note (see **Figure 12–2**), and on section views of the construction (see **Figure 12–3**). If excavation is involved in the construction where a slab-on-grade is to be placed, it is very important to thoroughly compact all loose fill before placing the concrete. Tamping the fill prevents the soil from compacting under the concrete later, causing the concrete to settle or crack.

A thickened slab, sometimes called a haunch, is used to further strengthen the slab where concentrated weight, such as a bearing wall, will be located. A *haunch* is an extra thick portion of the slab that is made by ditching the earth before the concrete is placed (see **Figure 12–4**).

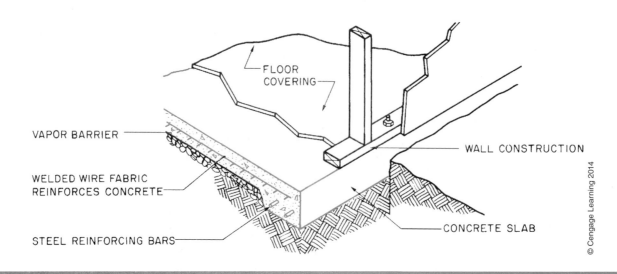

**Figure 12–1.** Slab-on-grade foundation.

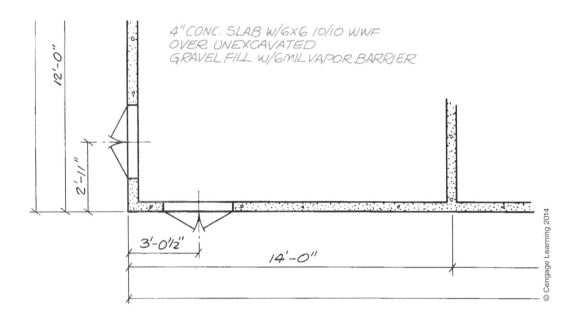

**Figure 12–2.** Note indicating slab-on-grade construction.

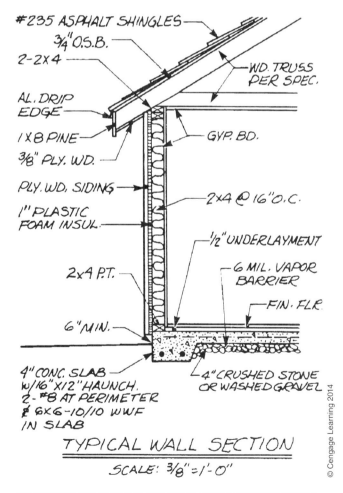

**Figure 12–3.** Typical wall section for slab-on-grade construction.

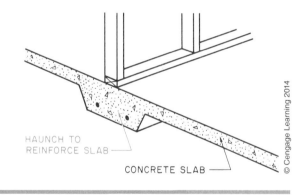

**Figure 12–4.** A haunch is a thickened part of a slab to reinforce it under a load-bearing wall.

## Spread Footings

In most sections of the country, the foundation of the house rests on a footing separate from the concrete floor. This separate concrete footing is called a *spread footing* because it spreads the force of the foundation wall over a wider area (see **Figure 12–5**). Spread footings can be made by placing concrete inside wooden or metal forms (see **Figure 12–6**) or placing the concrete in carefully measured ditches. In either case, the footing is shown on the foundation or basement plan by dotted lines outside the foundation wall lines (see **Figure 12–7**).

Footings 77

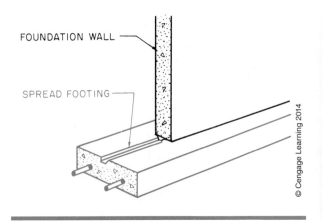

**Figure 12–5.** A spread footing is so named because it "spreads" the downward force of the foundation wall over a greater soil area.

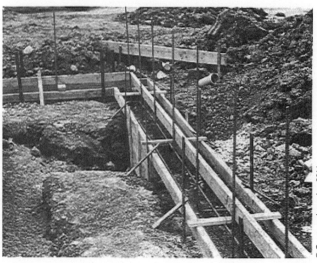

**Figure 12–6.** Footing forms.

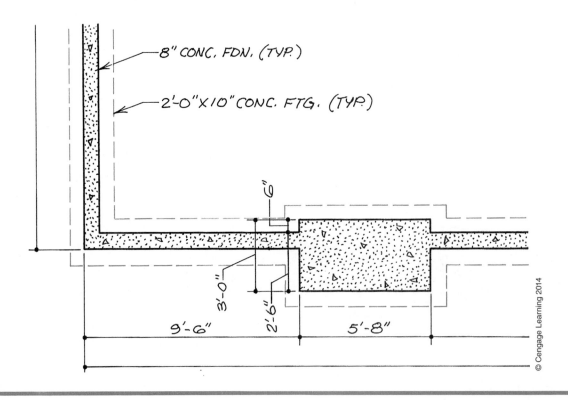

**Figure 12–7.** The footing lines are sometimes shown on the foundation plan by broken lines around the foundation.

The dimensions of the footings can be determined from the dimensions shown for the foundation. The foundation rests on the center of the footing unless otherwise specified, but the dimensions must meet the minimum requirements shown in Table R403.1 of the *International Residential Code®*. Therefore, if an 8-inch foundation rests on a 16-inch footing, the footing projects 4 inches beyond the foundation on each side. To lay out these footing lines, measure 4 inches from the building lines marked on the batter boards. Where footing lines cross to form a corner, suspend a plumb bob. Drive a stake under the plumb bob. Then, drive a nail in the stake to accurately mark the corner. When all corners are located in this manner, stretch a line between the nails to locate the inside of the footing forms.

A complete set of construction drawings also includes sections that show the spread footing in greater detail. However, these drawings are often superseded by local building codes. Building codes for footings include such things as minimum permissible depth of footing, required strength of concrete for footings, the width of footings, and the use of keys or dowels made of reinforcing steel (see **Figure 12–8**). The drawings for the Lake House are drawn to satisfy the building codes that are enforced in the community where it is to be built. If the drawings were done for a plan catalog where the locality is not known in advance, the *International Residential Code®* might be used as a guide, but the builder would be referred to local building codes and many dimensions would be omitted. **Figure 12–9** shows an excerpt from the *International Residential Code®* section on footings.

When reading the foundation plan for footing dimensions, pay particular attention to special features like fireplaces and pilasters. (Pilasters are thickened sections of the foundation wall that add strength to the wall.) The footing in these areas will probably be wider and may be deeper than under straight sections of foundation wall. There are times when a footing or foundation wall must be crossed or penetrated by a utility line. Care must be taken when penetrating a footing or foundation wall. If the penetration is not shown on the drawings, prior approval by an engineer is most often required, and the building code probably regulates how it can be done. Items of concern include expansion and contraction between the utility line and footing or foundation wall, reduction in structural strength, and breaking the moisture barrier.

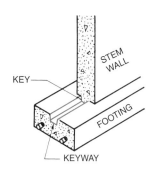
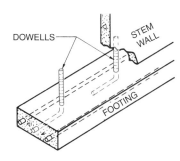
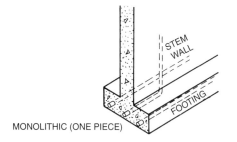

**Figure 12–8.** Anchoring the foundation wall to the footing.

**TABLE R403.1**
**MINIMUM WIDTH OF CONCRETE OR MASONRY FOOTINGS (inches)[a]**

| | LOAD-BEARING VALUE OF SOIL (psf) | | | |
|---|---|---|---|---|
| | 1,500 | 2,000 | 3,000 | ≥ 4,000 |
| **Conventional light-frame construction** | | | | |
| 1-story | 12 | 12 | 12 | 12 |
| 2-story | 15 | 12 | 12 | 12 |
| 3-story | 23 | 17 | 12 | 12 |
| **4-inch brick veneer over light frame or 8-inch hollow concrete masonry** | | | | |
| 1-story | 12 | 12 | 12 | 12 |
| 2-story | 21 | 16 | 12 | 12 |
| 3-story | 32 | 24 | 16 | 12 |
| **8-inch solid or fully grouted masonry** | | | | |
| 1-story | 16 | 12 | 12 | 12 |
| 2-story | 29 | 21 | 14 | 12 |
| 3-story | 42 | 32 | 21 | 16 |

For SI: 1 inch = 25.4 mm, 1 pound per square foot = 0.0479 kN/m$^2$.

a. Where minimum footing width is 12 inches, a single wythe of solid or fully grouted 12-inch nominal concrete masonry units is permitted to be used.

**Figure 12–9.** This is a table from the *International Residential Code®*. The entire section on footings is six pages long.

### GREEN NOTE

*All too often contractors, building with the "margin of safety" in mind, use concrete that is mixed to yield higher strength than is necessary. The extra strength is obtained by increasing the amount of Portland cement in the concrete. Using only what is required to achieve the necessary strength for the job is better for the environment and less expensive. The amount of Portland cement in concrete can also be reduced by replacing a portion of it with some other material. Fly ash, a by-product of burning coal, can be added to concrete to replace some of the cement. The slag from a blast furnace used to produce steel can also be added to concrete to replace some of the Portland cement.*

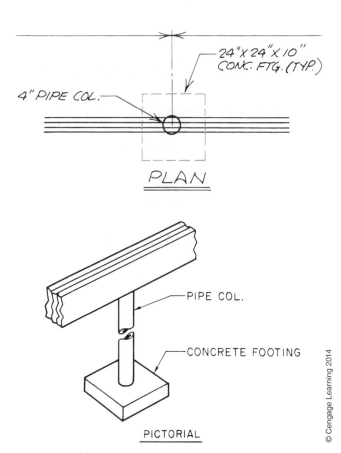

**Figure 12–10.** Column footings appear as a rectangle of broken lines on the plan.

## Column Pads

Where steel columns, masonry piers, and wooden posts are used in the construction, a special concrete pad is indicated on the foundation plan (see **Figure 12–10**). As with other footings, building codes for the area should be consulted for the design of these pads. These pads, as with all other footings, should rest on unexcavated or well-tamped earth.

## Reinforcement

Footings, column pads, and other structural concrete frequently include steel reinforcement. Footing reinforcement is normally in the form of steel reinforcement bars, commonly called *rebars*. Reinforcement bars are designated by their diameters in eighths of an inch (see **Figure 12–11**).

## Depth of Footings

In many sections of North America, the moisture in the surface of the earth freezes in the winter. As this frost forms, it causes the earth to expand. The force of this expansion is so great that if the earth under the footing

| REINFORCEMENT BARS ||
|---|---|
| Size Designation | Diameter in Inches |
| 3 | .375 |
| 4 | .500 |
| 5 | .625 |
| 6 | .750 |
| 7 | .875 |
| 8 | 1.000 |
| 9 | 1.128 |
| 10 | 1.270 |
| 11 | 1.410 |
| 14 | 1.693 |
| 18 | 2.257 |

**Figure 12–11.** Standard sizes of rebars.

of a building is allowed to freeze, it either cracks the footing or moves the building. To eliminate this problem, the footing is always placed below the depth of any possible freezing. This depth is called the **frostline** (see **Figure 12–12**).

The frost-depth map shown in **Figure 12–12** is only approximate and is not generally accurate enough for foundation design. **Figure 12–13** is from the *International Residential Code®* and is to be used by local building departments to specify foundation and footing design more precisely. The building department would fill in each of the spaces on the form for its particular jurisdiction.

Two methods are commonly used to indicate the elevation, or depth, of the bottom of the footings. The easiest to interpret is for spot elevations to be given at key points on the elevation drawings. Where elevations are given in this manner, the tops of the footing forms are leveled with a leveling instrument, using a benchmark for reference.

Another commonly used method is to dimension the bottom of the footing from a point of known elevation. This may be the finished floor or the top of the masonry foundation, for example. These dimensions are given on the building elevations. Footings and other features marked with a reference symbol ⊕, called a **datum** symbol, are to be used as reference points for other dimensions.

For an example, see the South Elevation 3/3 of the Lake House (included in the packet). The left end of this view shows a footing with its bottom at an elevation of 334.83 feet. This is a variation of the usual practice of showing the elevation of the top of the footing. The top of this footing is 10 inches higher, or 335.66 feet. (See Math Review 16.) What room of the Lake House is this footing under?

At the right end of the South Elevation 3/3, the footing is shown to be 5′–4″ below the finished floor and masonry. This is the basement floor, which the Site Plan shows to be at 337.0 feet. Therefore, the top of this footing is at 331′–8″, or 331.67′.

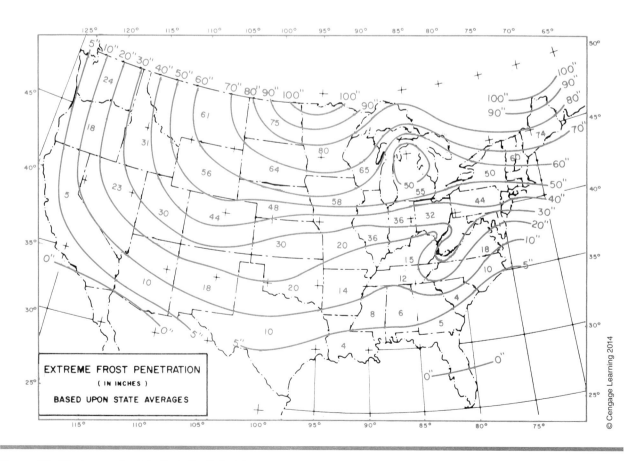

**Figure 12–12.** Average frost depths in the United States.

**TABLE R301.2(1)**
**CLIMATIC AND GEOGRAPHIC DESIGN CRITERIA**

| GROUND SNOW LOAD | WIND DESIGN | | SEISMIC DESIGN CATEGORY[f] | SUBJECT TO DAMAGE FROM | | | WINTER DESIGN TEMP[e] | ICE BARRIER UNDERLAYMENT REQUIRED[h] | FLOOD HAZARDS[g] | AIR FREEZING INDEX[i] | MEAN ANNUAL TEMP[j] |
|---|---|---|---|---|---|---|---|---|---|---|---|
| | Speed[d] (mph) | Topographic effects[k] | | Weathering[a] | Frost line depth[b] | Termite[c] | | | | | |
| | | | | | | | | | | | |

For SI: 1 pound per square foot = 0.0479 kPa, 1 mile per hour = 0.447 m/s.

a. Weathering may require a higher strength concrete or grade of masonry than necessary to satisfy the structural requirements of this code. The weathering column shall be filled in with the weathering index (i.e., "negligible," "moderate" or "severe") for concrete as determined from the Weathering Probability Map [Figure R301.2(3)]. The grade of masonry units shall be determined from ASTM C 34, C 55, C 62, C 73, C 90, C 129, C 145, C 216 or C 652.
b. The frost line depth may require deeper footings than indicated in Figure R403.1(1). The jurisdiction shall fill in the frost line depth column with the minimum depth of footing below finish grade.
c. The jurisdiction shall fill in this part of the table to indicate the need for protection depending on whether there has been a history of local subterranean termite damage.
d. The jurisdiction shall fill in this part of the table with the wind speed from the basic wind speed map [Figure R301.2(4)A]. Wind exposure category shall be determined on a site-specific basis in accordance with Section R301.2.1.4.
e. The outdoor design dry-bulb temperature shall be selected from the columns of 97$\frac{1}{2}$-percent values for winter from Appendix D of the *International Plumbing Code*. Deviations from the Appendix D temperatures shall be permitted to reflect local climates or local weather experience as determined by the building official.
f. The jurisdiction shall fill in this part of the table with the seismic design category determined from Section R301.2.2.1.
g. The jurisdiction shall fill in this part of the table with (a) the date of the jurisdiction's entry into the National Flood Insurance Program (date of adoption of the first code or ordinance for management of flood hazard areas), (b) the date(s) of the Flood Insurance Study and (c) the panel numbers and dates of all currently effective FIRMs and FBFMs or other flood hazard map adopted by the authority having jurisdiction, as amended.
h. In accordance with Sections R905.2.7.1, R905.4.3.1, R905.5.3.1, R905.6.3.1, R905.7.3.1 and R905.8.3.1, where there has been a history of local damage from the effects of ice damming, the jurisdiction shall fill in this part of the table with "YES." Otherwise, the jurisdiction shall fill in this part of the table with "NO."
i. The jurisdiction shall fill in this part of the table with the 100-year return period air freezing index (BF-days) from Figure R403.3(2) or from the 100-year (99 percent) value on the National Climatic Data Center data table "Air Freezing Index-USA Method (Base 32°F)" at www.ncdc.noaa.gov/fpsf.html.
j. The jurisdiction shall fill in this part of the table with the mean annual temperature from the National Climatic Data Center data table "Air Freezing Index-USA Method (Base 32°F)" at www.ncdc.noaa.gov/fpsf.html.
k. In accordance with Section R301.2.1.5, where there is local historical data documenting structural damage to buildings due to topographic wind speed-up effects, the jurisdiction shall fill in this part of the table with "YES." Otherwise, the jurisdiction shall indicate "NO" in this part of the table.

**Figure 12–13.** Climate and geographic design information to be provided by the local building department in the *International Residential Code*®.

## Stepped Footings

On sloping building sites, it is necessary to change the depth of the footings to accommodate the slope. This is done by stepping the footings. When concrete blocks are to be used for the foundation walls, these steps are normally in increments of 8 inches. This allows the concrete blocks to be laid so that the top of each footing step is even with a masonry course. Some buildings require several steps in the footing to accommodate steeply sloping sites. Steps in the footings are shown on the elevation drawings and on the foundation plan by a single line across the footing.

The Lake House has several steps in the footing. For example, see the east side of the garage. This is shown in the East Elevation 2/3. This step is also shown on the Foundation Plan. It is 8′-0″ from the north end of the garage. Notice that the two levels of the footing overlap each other. These are built in one overlapping section (see **Figure 12–14**).

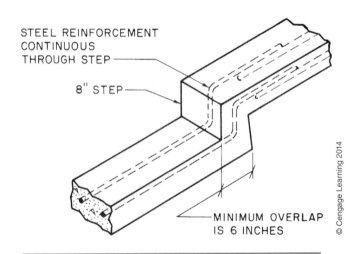

**Figure 12–14.** Stepped footing.

### USING WHAT YOU LEARNED

Every component of a building must be installed or built in the exact location and with the exact dimensions intended by the designer. If a supporting column is only slightly too high or slightly too short, the structure that column supports will not align properly with the rest of the building. What should the elevation be at the top of the columns supporting the deck between the Lake House kitchen and the garage?

Refer to the upper level floor plan to find the elevation of the top surface of the deck. That elevation is 341.67′. Deck detail 3/6 on sheet 6 shows the sizes of materials used in the deck construction. Add all of the dimensions of the materials between the top of the column and the top of the deck: The first thing we encounter is a metal post base. Most post bases raise the wood 1 inch off the concrete, so let's assume that is the dimension in this case. In actual practice, the dimension should be confirmed by measuring the base or consulting the manufacturer's literature before the column is built. On top of that are three 2 × 10s, which would have a depth of 9¼ inches. The joists are 2 × 8s, which would have a depth of 7¼ inches. Finally, we have 1 × 6 composite decking, which is actually 1 inch thick. 1″ + 9¼″ + 7¼″ = 1″ = 18½″. Subtract 18.5′ from the elevation at the top of the deck, 341.67′, to find that the top of the column should be at 340.13′.

## Assignment

Refer to the Lake House drawings in your textbook packet to complete this assignment.

1. What is the typical width and depth of the concrete footings for the Lake House?
2. What is the total length and width (outside dimensions) of the concrete footings for the garage of the Lake House? (Remember to allow for the footings to project beyond the foundation wall.)
3. How many concrete pads are shown for footings under columns or piers in the Lake House?
4. What are the dimensions of these pads?
5. What reinforcement is indicated for these pads?
6. What reinforcement is indicated for the spread footings under the Lake House?
7. What is indicated by the 2-inch dimension between the 12-inch round concrete footings?
8. What is the elevation of the top of the footing under the garage door?
9. What are the elevations of the tops of each section of concrete footing shown on the East Elevation 2/3?
10. How far outside the foundation walls are the typical footings?
11. Refer to the building code in your community (or the model code section shown in **Figure 12–9**), and list the specific differences between the Lake House footings and the minimum code requirements. Assume a soil load-bearing capacity of 2,000 pounds per square foot.

# UNIT 13: Foundation Walls

## Objectives

After completing this unit, you will be able to perform the following tasks:

○ Determine the locations and dimensions of foundation walls indicated on a set of drawings.

○ Describe special features indicated for the foundation on a set of drawings.

## Laying Out the Foundation

When the **concrete** for the footings has hardened and the forms are removed, carpenters can begin erecting forms for concrete foundations, or masons can begin laying blocks or bricks for masonry foundations. Although the material differs, the drawings and their interpretation for each type of foundation are similar.

In Unit 12 you referred to the dimensions on the foundation plan to lay out the footings. The same dimensions are used to lay out the foundation walls. The layout process is also similar. The outside surface of the foundation wall is laid out using previously constructed batter boards. Then the forms are erected

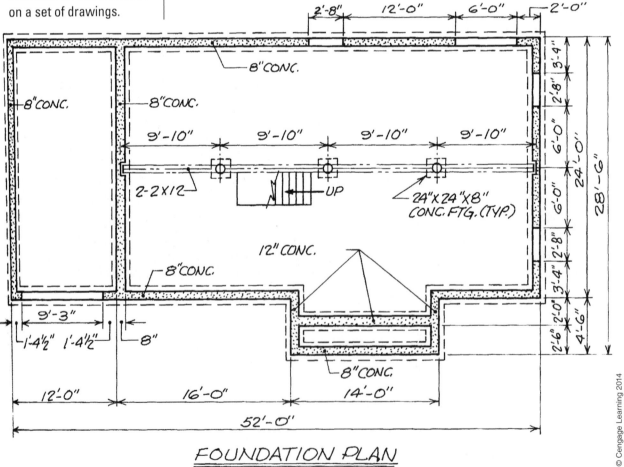

**Figure 13-1.** Dimensioning on a foundation plan.

or the masonry units are laid to these lines. The foundation plan includes overall dimensions, dimensions to interior corners and special constructions, and dimensions of special smaller features. It is customary to place the smallest dimensions closest to the drawing. The overall dimensions are placed around the outside of the drawing (see **Figure 13–1**).

All drawing sets include, at least, a wall section showing how the foundation is built, how it is secured to the footings, and any special construction at the top of the foundation wall (see **Figure 13–2**). Although a typical wall section may indicate the thickness of the foundation wall, you should carefully check around the entire wall on the foundation plan to find any notes that indicate varying thicknesses of the foundation wall. For example, the wall may be 12 inches thick where it has to support brick veneer above, while it is only 8 inches thick on the back of the building where there is no brick veneer. A careful check of the foundation plan for the Lake House shows that the house foundation is 10-inch thick concrete, except the portion under the fireplace, which is 12 inches thick. Some portions of the foundation wall have 6-inch concrete masonry units at the top to create a ledge for special construction and the garage foundation calls for 8-inch concrete.

Structural concrete is the most common foundation material because it is strong and concrete foundations can be built quickly. Some buildings have masonry (usually concrete block) foundations. There are some additional topics to be discussed for masonry (concrete block) foundations.

GREEN NOTE

*Concrete and concrete masonry units are the traditional materials for building foundations. These materials do not provide particularly good thermal insulation, and the processes for the manufacture of concrete and concrete products are not environmentally friendly. In recent years, many new developments in the construction of foundations are taking place. One such innovation that is gaining popularity is the use of insulated concrete forms (ICFs). ICFs are typically made of rigid foam, such as Styrofoam®, but they can also be made of other materials, such as mineralized wood chips.*

*ICFs are made in a variety of sizes that are small enough to be easy to handle. The individual ICF units lock together, so that the result is a wall form into which concrete can be placed, resulting in a wall that is insulated with several inches of foam yet has the structural properties of concrete.*

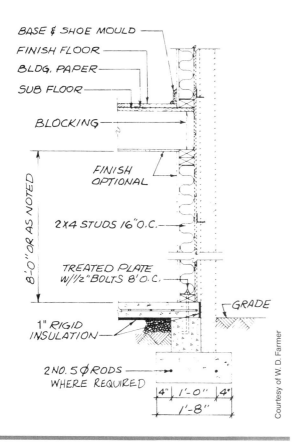

**Figure 13–2.** Section through foundation.

Foundation Walls  85

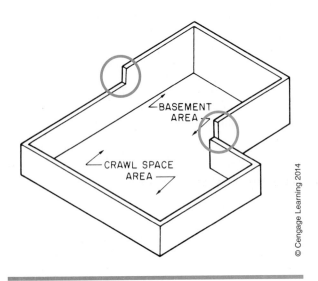

**Figure 13–3.** The top of the foundation may be stepped to allow for a partial basement or varying floor levels.

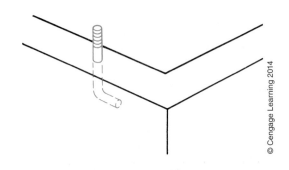

**Figure 13–4.** Anchor bolt.

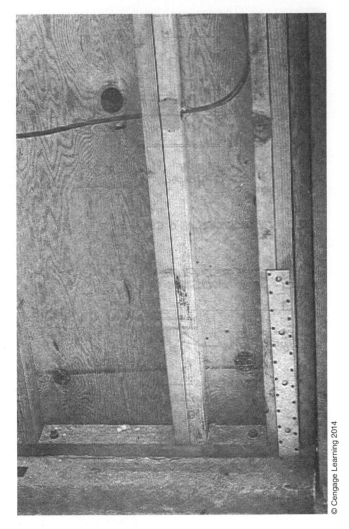

**Figure 13–5.** Anchor bolts and hold-down strap are used in an earthquake zone.

The details for a masonry foundation may call for horizontal reinforcement in every second or third course. This is usually prefabricated wire reinforcement to be embedded in the mortar joints. Prefabricated wire reinforcement is available in varying sizes for different sizes of concrete blocks.

The height of the foundation wall is dimensioned on the building elevations. These are the same dimensions as those used to determine the depth of the footings in the preceding unit. Just as the footing was stepped to accommodate a sloping building site, the top of the foundation wall may be stepped to accommodate varying floor levels in the *superstructure* (construction above the foundation) as shown in **Figure 13–3.**

The top of a foundation may be built with smaller concrete blocks to form a ledge upon which later brickwork will be built. It is also common practice to use one course of 4-inch solid block as the top course of a masonry foundation wall.

In concrete foundations, **anchor bolts** are placed in the top of the foundation (see **Figure 13–4**). These bolts are left protruding out of the top of the foundation so that the wood superstructure can be fastened in place later. Anchor bolts are not normally shown on the foundation plan, but a note on the wall section indicates their spacing. On masonry walls, anchor bolts can be placed in the hollow cores of the concrete blocks. They are held in place by filling the core with mortar grout. **Grout** is a **Portland cement** mixture that has high strength.

In areas where there is a threat of extremely high winds or earthquakes, additional hold-down straps may be called for (see **Figure 13–5**). These hold-downs are normally only used with concrete foundations.

## Special Features

Many foundations include steel or wooden beams, which act as girders to support the floor framing over long spans (see **Figure 13–6**). When the girder is steel, it is indicated by a single line with a note specifying the size and type of structural steel. A wood girder is usually indicated by two or more lines and a note specifying the number of pieces of wood and their sizes in a built-up girder (see **Figure 13–7**).

If the top of the girder is to be flush with the top of the foundation beam, pockets must be provided in the

**Figure 13–6.** The girder supports the floor framing.

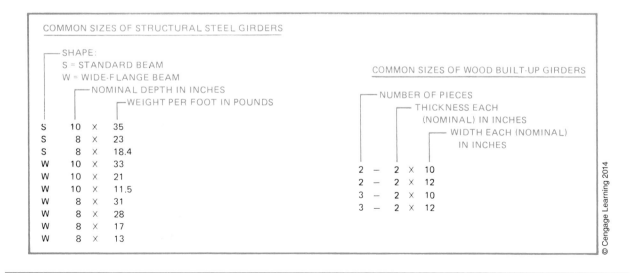

**Figure 13–7.** Typical specifications for structural steel and wood built-up girders—other types are described in the project specifications.

Foundation Walls 87

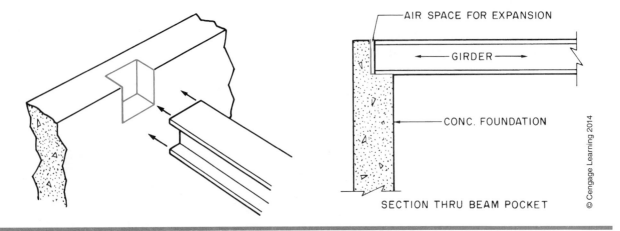

**Figure 13–8.** A beam pocket is a recess in the wall to hold the girder.

foundation (see **Figure 13–8**). **Beam pockets** are shown on the details and sections of the construction drawings. The locations of these beam pockets are dimensioned on the foundation plans.

If windows are to be included in the foundation, the form carpenter or mason must provide rough openings of the proper size. The locations of windows should be dimensioned on the foundation plan. The sizes of the windows may be shown by a note or given on the window schedule. Window sizes and window schedules are discussed in Unit 25. It is important, however, to get the masonry opening size from the window manufacturer before forming the opening in the foundation wall. The *masonry opening* is the size of the opening required in the foundation wall to accommodate the window. This size may be different from the **nominal size** given in a note on the foundation plan. The foundation may include pilasters for extra support. A *pilaster* is a thickened section of the foundation, which helps it resist the pressure exerted by the earth on the outside. The location and size of pilasters are shown on the foundation plan.

The Lake House drawing includes a special feature not commonly found on foundation plans for houses. There are four notes that read 3½″ STD. WT. STL. COL. W/8 × 8 × ½ B.PL. These notes indicate a 3½-inch square, standard-weight, steel column with 8-inch-by-8-inch-by-½-inch-thick base plates. This structural steelwork is explained in more detail later, but to completely understand the foundation plan, it is necessary to know that the steel will be erected.

## Permanent Wood Foundation

Foundations are usually constructed of concrete or concrete block. However, a type of specially treated wood foundation is sometimes used (see **Figure 13–9**). These *permanent wood foundations* do not use concrete footings.

Instead, they are built on 2 × 8s or laid on gravel fill below the frostline. The foundation walls are framed with lumber that has been pressure treated to make it rot resistant and insect resistant. The framing is covered with a plywood skin, and the plywood is covered with polyethylene film for complete moisture proofing (see **Figure 13–10**). All the nails in a wood foundation are stainless steel to prevent rusting.

**Figure 13–9.** Permanent wood foundation.

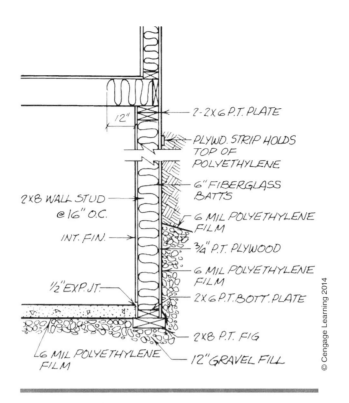

**Figure 13-10.** Section of wood foundation.

### USING WHAT YOU LEARNED

Before any concrete form construction begins, it is necessary to have a good picture in your mind of what the finished product will be. The Lake House foundation has many variations from the typical wall shown in the Typical Wall Section on sheet 4. For example, near the southeast 3½" ⌑ steel column the Foundation Plan has a note that says "6" CONC. BLK. – TOP OF WALL TO BELOW STAIR". What does this note mean and why is that detail necessary? The floor plans show stairs from the living room down to the playroom at this point. The stairs are also shown by the hidden lines on the South Elevation 3/3 and that elevation includes a note that says "6" CMU FDN WALL (BELOW STAIR). The typical foundation wall is 10 inches thick, so the 6-inch concrete masonry units create a 4-inch ledge to support the stairs.

## Assignment

Refer to the Lake House drawings in your textbook packet to complete the assignment.

1. What is the typical thickness of the concrete block foundation for the Lake House?
2. Approximately how many lineal feet of concrete wall are included in the foundations of the Lake House?
3. How thick is the south foundation of the fireplace?
4. What is the elevation of the top of the north end of the east foundation wall of the garage?
5. What is the highest elevation on the entire foundation?
6. How high is the foundation wall at the highest elevation of the foundation?
7. What size anchor bolts are indicated at the top of the Lake House foundation?
8. What spacing is indicated for the anchor bolts?
9. What secures the foundation wall to the footing?
10. What is the elevation of the top of the concrete block wall at the southwest 3½" ⌑ steel column?
11. In how many places are the concrete block walls to be filled with grout?

Foundation Walls 89

# UNIT 14

# Drainage, Insulation, and Concrete Slabs

## Objectives

After completing this unit, you will be able to perform the following tasks:

- Locate and explain information for control of groundwater as shown on a set of drawings.
- Locate and describe subsurface insulation.
- Determine the dimensions of concrete slabs and the reinforcement to be used in concrete slabs.

## Drainage

After the foundation walls are erected and before the excavation outside the walls is **backfilled** (filled with earth to the finished grade line), footing drains, if indicated, must be installed. **Footing drains** are usually perforated plastic pipe placed around the footings in a bed of crushed stone (see **Figure 14–1**). If the site has a natural slope, the footing drains can be run around the foundation wall to the lowest point, then away from the building to drain by gravity. In areas where there is no natural drainage, the drain is run to a dry well or municipal storm drain.

At one time, clay drain tile was the most common type of pipe for this purpose. However, perforated plastic pipe is used in most new construction. Plastic drain pipe is manufactured in 10-foot lengths of rigid pipe and in 250-foot rolls of flexible pipe (see **Figure 14–2**). An assortment of plastic fittings is available for joining rigid plastic pipe. When footing drains are to be included, they are shown on a wall section or footing detail (see **Figure 14–3**). A note on the drawing indicates the size and material of the pipe.

If the floor drains are to be included in concrete-slab floors, they are indicated by a symbol on the appropriate floor plan (see **Figure 14–4**). If these floor

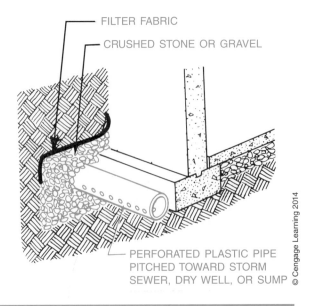

**Figure 14–1.** Footing drain.

90 UNIT 14

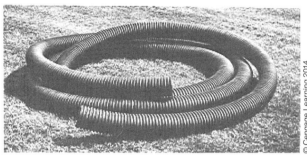

**Figure 14–2.** Plastic drain pipe.

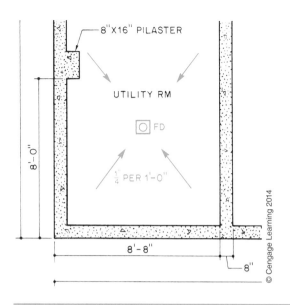

**Figure 14–4.** The floor drain is shown by a symbol. (Notice that the floor is pitched toward the drain.)

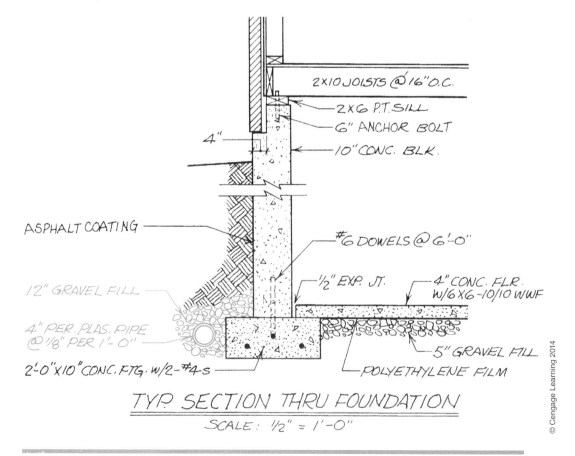

**Figure 14–3.** Footing drains are shown outside the footing.

Drainage, Insulation, and Concrete Slabs 91

drains run under a concrete footing, the piping had to be installed before the footing was placed (see Unit 11). The **riser** (vertical part through the floor) and drain basin are usually set at the proper elevation just prior to placing the concrete floor.

The floor plan may include a spot elevation for the finished drain, or it may be necessary to calculate it from information given for the pitch of the concrete slab. Pitch of concrete slabs is discussed later in this unit.

## Vapor Barriers

Another technique often used to prevent groundwater from seeping through the foundation is coating the foundation wall with asphalt foundation coating.

At this point, subsurface work outside the foundation wall is completed, but backfilling should not be done until the superstructure is framed. The weight and rigidity of the floor on the foundation wall help the wall resist the pressure of the backfill. If the backfilling must be done before the framing, the foundation walls should be braced. To retard the flow of moisture from the earth through the concrete-slab floor, the drawings may call for a layer of gravel over the entire area before the concrete is placed. A polyethylene **vapor barrier** is laid over the gravel underfill. The thickness of polyethylene sheeting is measured in **mils**. One mil equals 1/1,000 of an inch. For vapor barriers, 6-mil polyethylene is generally used.

### GREEN NOTE

*One of the most basic principles of green home construction is the control of groundwater to keep it from penetrating the foundation and causing structural damage as well as interior finish damage and mold. Groundwater is kept away from the foundation by proper grading of the site, careful damp-proofing, and providing a path for groundwater to escape without penetrating the foundation. This path involves placing perforated drain pipes next to or below the footings and pitching it, so all water flows downhill to a daylight exit point. Poor workmanship on any of these aspects of groundwater control can result in devastating water damage and serious implications for the health of the occupants.*

## Insulation

In cold climates, it is desirable to insulate the foundation and concrete slab. This insulation is usually rigid plastic foam board placed against the foundation wall or laid over the gravel underfill (see **Figure 14–5**).

Like all materials, concrete and masonry expand and contract slightly with changes in temperature. To allow for this slight expansion and contraction, the joint between the concrete slab and foundation

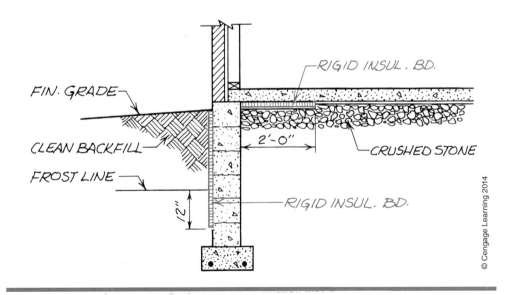

**Figure 14–5.** Rigid plastic foam insulation may be laid under the perimeter of the floor or against the foundation wall.

wall should include a compressible expansion joint material. Expansion joints can be made from any compressible material such as neoprene or composition sheathing material. This expansion joint filler is as wide as the slab is thick and is simply placed against the foundation wall before the concrete is placed.

## Concrete Slabs

When the house has a basement, the floor is a concrete slab-on-grade. The areas to be covered with concrete are indicated on the foundation plan or basement floor plan. This is usually done by an area not giving the thickness of the concrete slab and any reinforcing steel to be used. To help the concrete resist minor stresses, it is usually reinforced with welded wire mesh. Welded wire mesh is sometimes abbreviated WWM and sometimes WWF for welded wire fabric. The specifications for welded wire mesh are explained in **Figure 14–6**. Where the slab must support bearing walls or masonry partitions, it may be haunched, as discussed in Unit 12.

When floor drains are included or where water must be allowed to run off, the slab is *pitched* (sloped slightly). A note on the drawings indicates the amount of pitch. One-quarter inch per foot is common. When there is any possibility of confusion about which way the slab is to be pitched, bold arrows are drawn to show the direction the water will run (see **Figure 14–7**).

When floor drains or forms are set for pitched floors, it is necessary to find the total pitch of the slab. This is done by multiplying the pitch per foot by the number of feet over which the slab is pitched. (See Math Review 8.) For example, if the note on a concrete **apron** in front of a garage door indicates a pitch of ½ inch per foot and the apron is 4 feet wide, the total pitch is 2 inches. The proper elevation for the form at the outer edge of the apron is 2 inches less than the finished floor elevation.

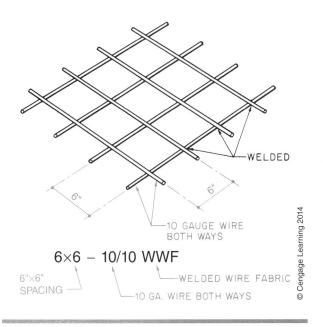

**Figure 14–6.** The callout for welded wire fabric explains the size and spacing of the wires.

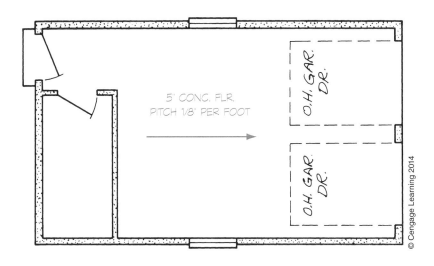

**Figure 14–7.** A bold arrow indicates the way that water will run off a pitched surface.

Drainage, Insulation, and Concrete Slabs

Slab-on-grade installations may require electrical raceways to be installed in or just below the concrete slab in the crushed or gravel fill. These installations must be made so that they do not reduce the structural integrity of the concrete slab. An oversized raceway in the concrete slab may cause the slab to crack and settle unevenly. An electrician should be present during concrete placement to observe and correct any damage to these electrical raceways.

All underground utility systems must be coordinated prior to starting installation, especially in commercial and industrial buildings, as discussed in Unit 11. The service entrance section may have up to six subdistribution sections or panels located throughout the building. Electrical feeders are required from the electrical service entrance section to these subdistribution sections or panels. This requires detailed coordination between the electrical installer and the other utility installers prior to starting the installation of the utility systems. The electrical installation must be laid out around the other utilities, with the other utility layouts normally having priority. The other utilities to be coordinated may include plumbing, fire sprinkler, heating and air conditioning, and specialty systems. This coordination must be done for both the underground and aboveground systems.

Some of the concrete work in the Lake House is of particular interest, because it is part of the passive solar heating system the Lake House uses. Section 1/4 and the lower-level floor plan 1/2 indicate that the area under the living room and dining room floors is a heat sink. A *heat sink* is a mass of dense material that absorbs the energy of the sun during the day and radiates it at night. The living room and dining room are on the south side of the Lake House. In the winter, when the leaves are off the deciduous trees, the sun shines in the large areas of glass in these rooms and warms the heat sink. At night, this heat is radiated into the house to provide additional heat when it is needed most. The floor over the heat sink is a concrete slab similar to that used in the playroom. Detail 2/5 helps explain this concrete slab.

### USING WHAT YOU LEARNED

All of the reinforcement for the concrete slabs must be in place before the concrete is delivered to the site. In the Lake House how many rebars of what size are required for the haunch in the concrete slab between the two north 3½ ◻ steel columns? The haunch is shown on Foundation Plan 4/1 and Detail 3/1 shows the reinforcement. The first callout on the detail is "2 -- #4 CONTINUOUS" with a leader pointing to the symbols for reinforcing bars.

## Assignment

Refer to the Lake House drawings in your textbook packet to complete this assignment.

1. What is the thickness of the concrete slab over the heat sink in the Lake House?
2. Describe the reinforcement used in the concrete slab in the playroom of the Lake House.
3. How many square feet of 2-inch rigid insulation are needed for the Lake House heat sink?
4. What prevents moisture from seeping through the concrete slab floor in the Lake House?
5. What is the finished floor elevation of the Lake House garage?
6. What is the elevation of the floor drain in the utility room of the Lake House?
7. What is the purpose of the 8-inch-thick concrete haunch in the middle of the Lake House slab?
8. How many cubic yards of concrete are required for the garage floor? The basement floor including the crawl space?

# UNIT 15

# Framing Systems

## Platform Framing

**Platform framing**, also called **western framing**, is the type of framing used in most houses built in the last 60 years (see **Figure 15–1**). It is called platform framing because as the rough floor is built at each level, it forms a platform on which to work while erecting the next level (see **Figure 15–2**).

A characteristic of platform framing is that all wall *studs,* the main framing members in walls, extend only the height of one story. Interior walls, called *partitions,* are the same as exterior walls. The bottoms of the studs are held in position by a *bottom* (or *sole*) plate. The tops of the studs are held in position by a *top plate.* Usually, a *double top plate* is overlapped at the corners to tie intersecting walls and partitions together (see **Figure 15–3**). In some construction, the second top plate is not used. Instead, metal framing clips are used to tie

## Objectives

After completing this unit, you will be able to identify each of the following types of framing on construction drawings:

○ Platform
○ Balloon
○ Post-and-beam
○ Energy-saving

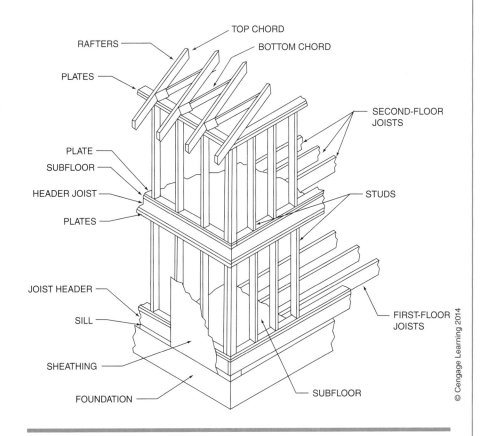

**Figure 15–1.** Platform, or western, framing.

Framing Systems 95

**Figure 15–2.** Platform, or western, framing provides a convenient work surface during construction.

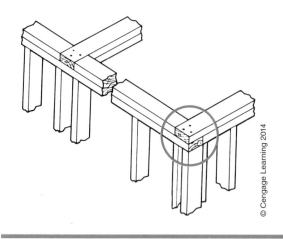

**Figure 15–3.** The double top plate overlaps at the corners.

intersecting walls together. Upper floors rest on the top plate of the walls beneath. The framing members of the upper floors or roof are positioned over the studs of the wall that supports them.

Platform construction can be recognized on wall sections (see **Figure 15–4**). Notice that the studs extend only from one floor to the next.

## Balloon Framing

In **balloon framing**, the exterior wall studs are continuous from the foundation to the top of the wall (see **Figure 15–5**). Floor framing at intermediate levels is supported by *let-in ribbon boards.* This is a board that fits into a notch in each joist and forms a support for the joists. Although balloon framing is not as widely used as it once was, some balloon-framing techniques are still used for special framing situations.

In both platform-frame and balloon-frame construction, the structural frame of the walls is covered with sheathing. **Sheathing** encloses the structure and, if a structural grade is used, prevents wracking of the wall. *Wracking* is the tendency of all the studs to move, as in a parallelogram, allowing the wall to collapse to the side (see **Figure 15–6**). There are two ways to prevent wracking. Plywood or other structural sheathing at the corners of the building prevents this movement. Also, diagonal braces can be attached to the wall framing at the corners to prevent wracking (see **Figure 15–7**).

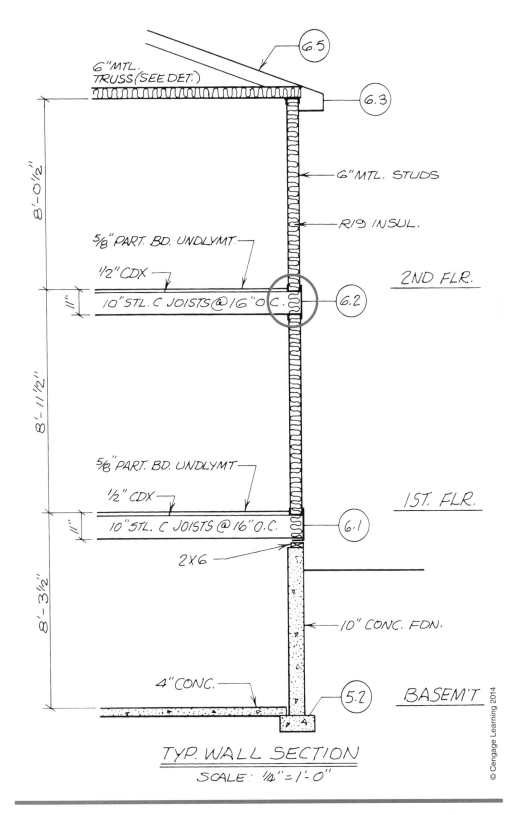

**Figure 15–4.** This can be recognized as platform construction because the studs extend only from one floor to the next.

Framing Systems

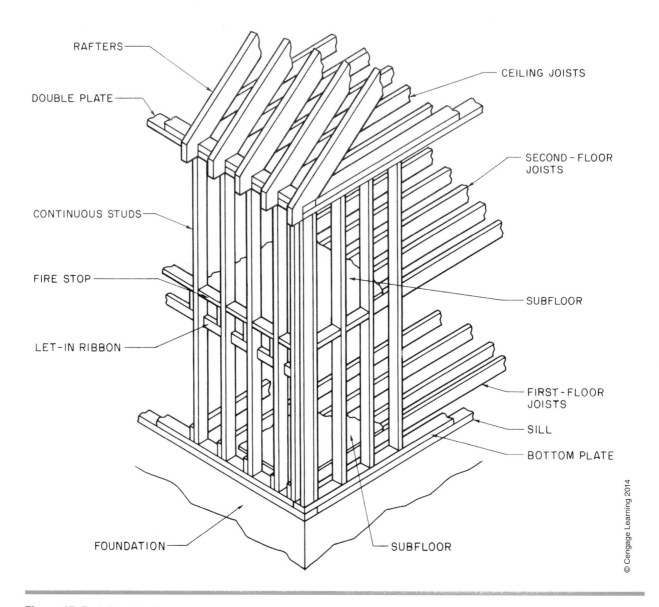

**Figure 15–5.** Balloon framing

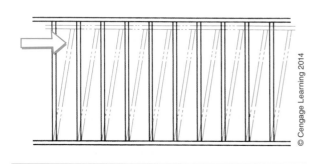

**Figure 15–6.** Wracking.

## Post-and-Beam Framing

*Platform framing* and *balloon framing* are characterized by closely spaced, lightweight framing members (see **Figure 15–8**). *Post-and-beam* framing uses heavier framing members spaced farther apart (see **Figure 15–9**). These heavy timbers are joined or fastened with special hardware (see **Figure 15–10**). Because post-and-beam framing uses fewer pieces of material, it can be erected more quickly. Also, although the framing members are large—they can range from 3 inches by 6 inches to 5 inches by 8 inches—their wider spacing results in a savings of material. However, to span this wider spacing, floor and roof decking must be heavier.

### GREEN NOTE

*Structural integrated panels (SIPs) are factory-built panels with an insulating core to which OSB is adhered to the exterior surface and usually gypsum board is adhered to the interior surface. Both the exterior and interior surfaces can be altered to suit the design of the building. Wiring and piping can be imbedded in the foam core as the panel is manufactured. SIPs are custom designed for each particular application. SIPS are joined together with screws, splines, and glue to create a weather-tight, well-insulated wall, roof, or floor. The advantages of SIPs are that they do not include the number of interior framing members that frame construction does, so they do a better job of thermal insulation; they result in a more weather-tight building envelope; there is virtually no waste created at the job site; and construction is faster.*

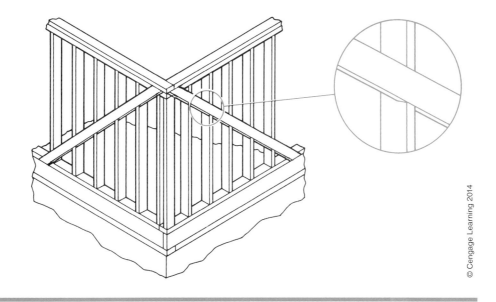

**Figure 15–7.** Let-in bracing prevents wracking.

Post-and-beam framing is sometimes left exposed to create special architectural effects (see **Figure 15–11**).

The structural core of the Lake House in your textbook packet uses posts and beams as shown in detail 6/6. However, this is not purely post-and-beam construction. The posts are 3½-inch square steel tubing, and the beams are plywood box beams. These are properly called **beams** because they carry a load without continuous support from below. These beams are supported only in the beam pockets on the steel posts (see **Figure 15–12**). In the Lake House, the post-and-beam construction does not include exterior walls. In pure post-and-beam construction, the exterior walls have widely spaced posts, and the space between is filled in with nonload-bearing curtain walls. These curtain walls are merely panels that fill in the space between the structural elements—the posts and beams.

Framing Systems 99

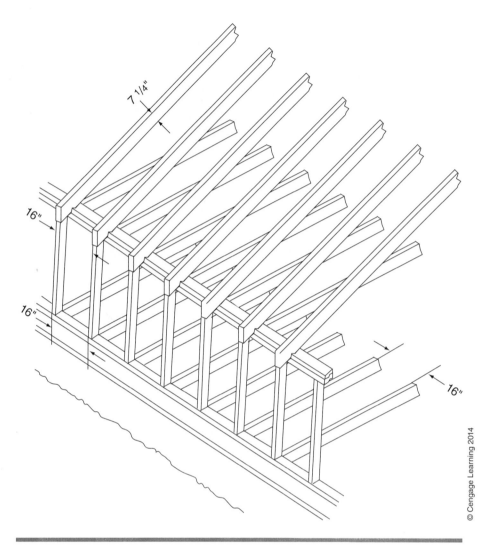

**Figure 15–8.** Conventional framing. 16" OC.

In areas where hurricanes and high tides are a threat, some houses are built as pole structures. Pole buildings are a variation of post-and-beam construction. Poles, which are treated to be insect and rot resistant, are set several feet in the ground and 8 to 12 feet apart. A *band joist*, or *header*, is then bolted to these poles (see **Figure 15–13**). The floors, walls, and roof are framed within the pole structure (see **Figure 15–14**). Pole buildings are strong; they resist severe winds. Pole construction allows buildings to be kept above damaging floodwaters.

## Metal Framing

Light-gauge metal framing is used extensively in commercial construction where the strength of structural steel or reinforced concrete is not required. Metal framing is also very popular in home building where termites and water damage are known to be problems. Light-gauge steel framing uses studs and joists, the same as wood framing. Unlike wood framing, steel framing uses tracks and runners in place of bottom plates and top plates (see **Figure 15–15**). The lighter, less expensive grades of metal framing are not as strong as wood, so they are used primarily for interior partitions or in conjunction with other structural materials. Metal framing can be preassembled just as you would preassemble wood framing. The runner is fastened to the floor with powder-actuated fasteners. The wallboard material is fastened to the metal studs using an electric screw gun with sheet metal screws. Subflooring material is fastened to the metal floor joists similarly with power-driven screws. The construction

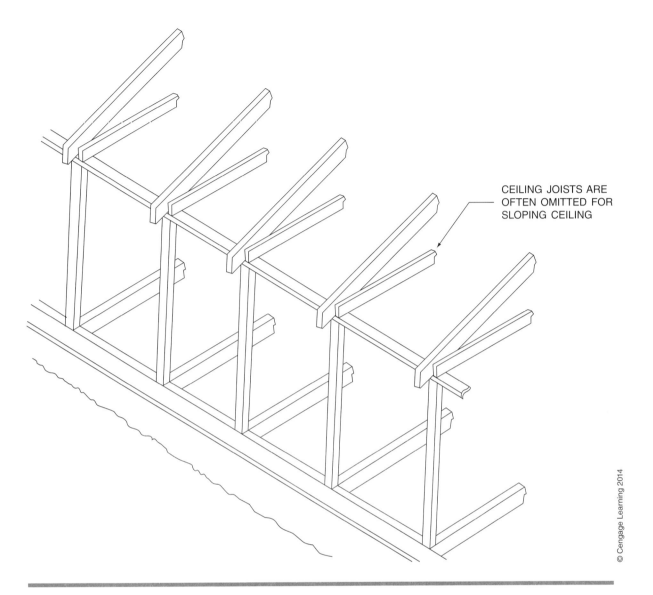

**Figure 15–9.** Post-and-beam framing. 4' OC.

drawings and specifications will indicate the type of metal framing to be used. The gauge must be taken into consideration before using certain tools (stud punches, powder-actuated tools, etc.) and special mounting devices (box hangers, conduit brackets, etc.).

Where electrical nonmetallic-sheathed cable is installed through metal framing, a bushing or grommet must be installed prior to installing the cable (see **Figure 15–16**).

## Energy-Saving Techniques

When wall framing is done with 2 × 4s spaced 16 inches on centers, 9.4 percent of the wall is solid wood. Wood conducts heat out of the building. Only the space between the solid wood framing can be filled with insulation. By using 2 × 6 studs spaced 24 inches on centers, the area of solid wood is reduced to 6 ¼ percent of the wall. The amount of framing material is the same. Not only does this reduce the amount of wood exposed to the surface of the wall, but it also allows for more insulation.

The area of exposed wood is further reduced by special corner construction. In conventional framing, three pieces are used to frame the corner of a wall (see **Figure 15–17**). To reduce heat loss through the wall, only two pieces are used (see **Figure 15–18**). The third piece, which normally provides a nailing surface for the interior **drywall**, is replaced by metal clips.

Examination of the first floor plan and the detail drawings of the Lake House shows that the walls are

Framing Systems 101

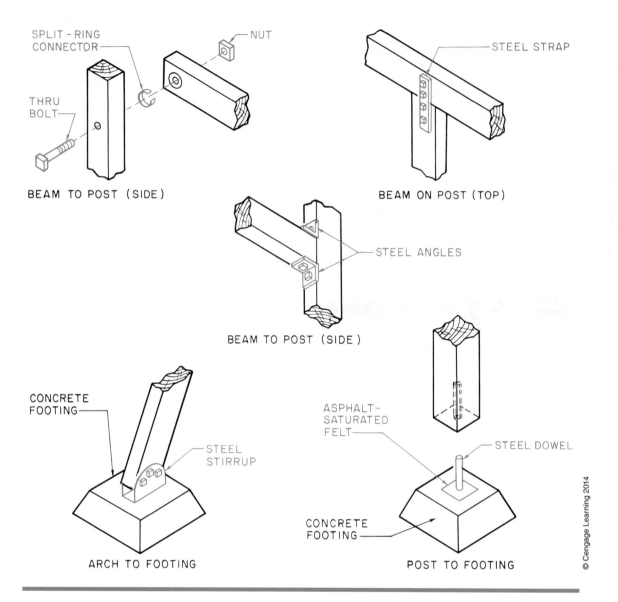

**Figure 15–10.** Common timber fastenings.

framed for maximum efficiency. Several details indicate that the studs are 2 × 6s @ 24 OC. This allows room for more insulation in the wall. Also notice that the house is sheathed with 3/4-inch insulation sheathing. Interior partitions do not need insulation, and so they are framed with 2 × 4s.

When installing nonmetallic-sheathed cable through wood wall studs and floor joists, the cable must be installed at least 1¼ inches from the nearest edge. Where this 1¼ inches cannot be met, a 1/16-inch-thick metal plate must be installed to protect the cable from nails and screws (see **Figure 15–19**).

**Figure 15–11.** The exposed timber framing in this house is an important part of its design.

**Figure 15–12.** Beam pocket for the lake house.

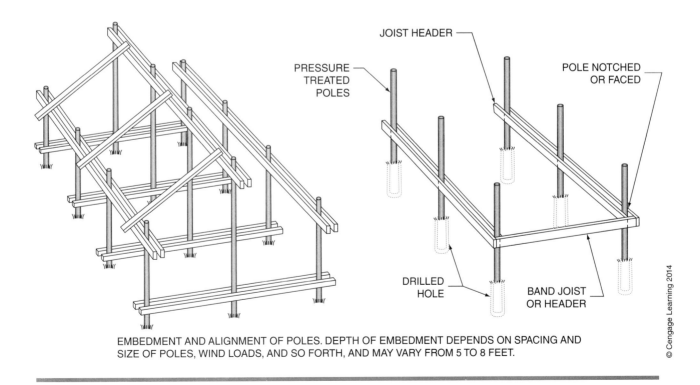

**Figure 15–13.** Basic elements of a pole building.

Framing Systems 103

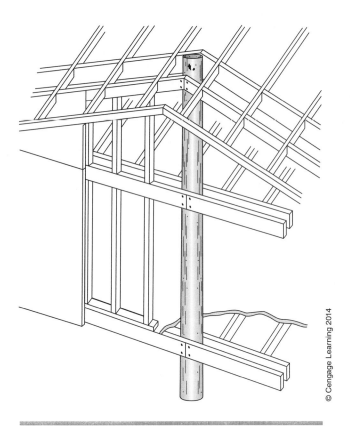

**Figure 15-14.** Framing in a pole building.

**Figure 15-15.** Metal framing.

**Figure 15–16.** Nonmetallic-sheathed cable installed through metal framing.

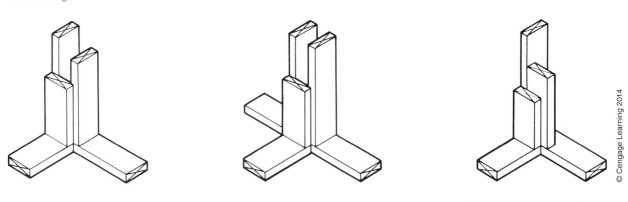

**Figure 15–17.** Conventional corner posts for 2 × 4 framing.

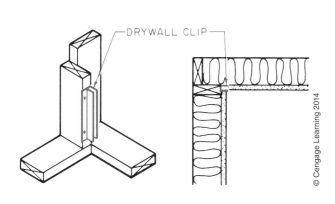

**Figure 15–18.** Corner construction for Arkansas Energy Saving System.

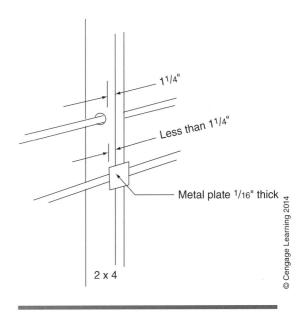

**Figure 15–19.** Nonmetallic-sheathed cable installed in wood framing.

Framing Systems 105

**USING WHAT YOU LEARNED**

A good familiarity of the framing system for a house will be a big help in understanding all of the details necessary to frame the building. One aspect of the framing system is the method for framing corners. In a typical outside corner of the Lake House, what size and how many studs are used? This information is clearly shown in detail 4/4 on Sheet 4 in your textbook packet, the first sheet showing framing details. There are three 2 × 6 studs in a typical corner.

## Assignment

1. Identify *a* through *g* in **Figure 15–20**.
2. What kind of framing is shown in **Figure 15–20**?
3. Identify *a* through *c* in **Figure 15–21**.
4. Identify *a* through *c* in **Figure 15–22**.
5. Sketch a plan view of a conventional corner detail. Include drywall and sheathing.
6. Sketch a plan view of an energy-efficient corner detail. Include drywall and sheathing.
7. What two materials are most often used for framing homes?
8. What framing material requires bushings or grommets for nonmetallic-sheathed cable installations?

Note: Refer to the Lake House drawings (in the packet) to complete the rest of the assignment.

9. Which of the types of framing discussed in this topic is used for the Lake House?
10. What supports the west ends of the floor joists in bedroom #1?
11. What supports the east ends of the kitchen rafters?
12. How are the box beams fastened to the 3½-inch square posts?
13. How are the steel posts anchored?
14. What does the northwest square steel post rest on?
15. What supports the north edge of the living room floor?
16. List the dimensions (thickness × depth × length) of all box beams.

Note: The length can be found by subtracting the outside dimension of the posts from the centerline spacing shown on the plan views.

17. List the length of each piece of 3½" ⬜ steel.

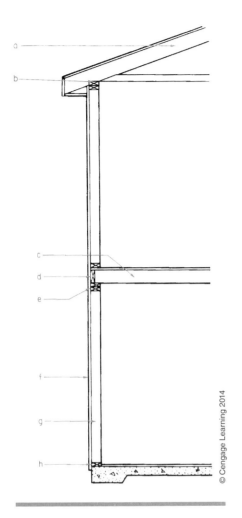

**Figure 15–20.**

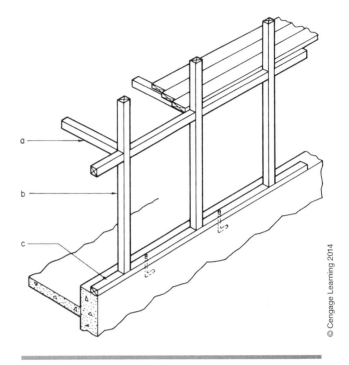

**Figure 15–22.**

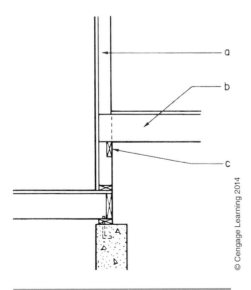

**Figure 15–21.**

Framing Systems 107

# UNIT 16
# Columns, Piers, and Girders

## Objectives

After completing this unit, you will be able to perform the following tasks:

○ Locate columns and piers, and describe each from drawings.

○ Locate and describe the girders that support floor framing.

○ Determine the lengths of columns and the heights of piers.

The most common system of floor framing in light construction involves the use of joists and girders. **Joists** are parallel beams used in the floor framing (see **Figure 16–1**). Usually buildings are too wide for continuous joists to span the full width. In this case, the joists are supported by one or more **girders** (beams) running the length of the building. The girder is supported at regular intervals by wood or metal posts or by masonry or concrete piers.

## Columns and Piers

Metal posts called *pipe columns* are the most common supports for girders. However, masonry or concrete piers may be specified. The locations of columns, posts, or piers are given by dimensions to their centerlines. When metal or wooden posts are indicated, the only description may be a note on the foundation plan (see **Figure 16–2**). This note may give the size and material of the posts only, or it may also specify the kind of bearing plates to be used at the top and bottom of the post (see **Figure 16–3**). A *bearing plate* is a steel plate that provides a flat surface at the top or bottom of the column or post. If the girder is supported by masonry or concrete piers, a special detail may be included to give dimensions and reinforcement details.

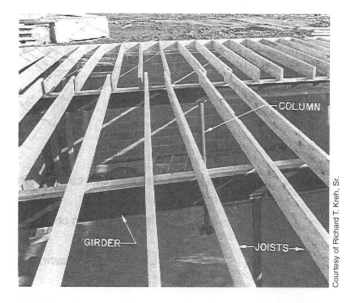

**Figure 16–1.** Joist-and-girder floor framing.

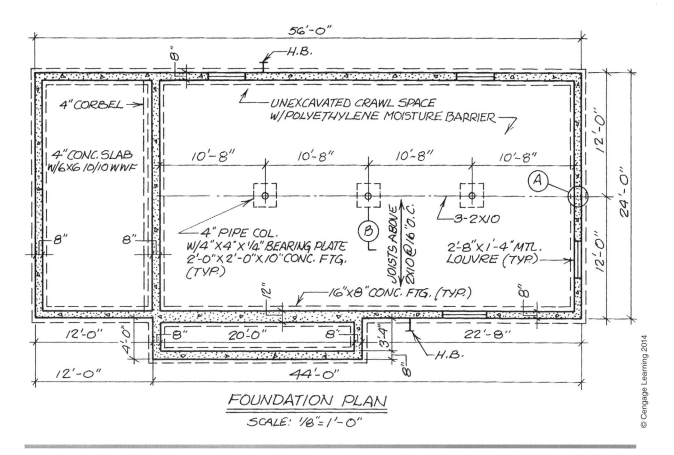

**Figure 16–2.** Foundation plan with note for column and footing.

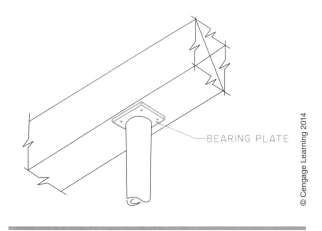

**Figure 16–3.** Bearing plate.

## Girders

Before the total length of the columns or height of the piers can be calculated, it is necessary to determine the size of the girder and its relationship to the floor joist. Three types of girders are commonly used in residential construction. Steel beams are often used where strength is a critical factor. Built-up wood girders are constructed on the site. These consist of three or more pieces of 2-inch lumber nailed together with staggered joints to form larger beams. The sizes and specifications for built-up wood girders and structural steel girders are given in Table R502.5(1) of the *International Residential Code®* and shown on the foundation plan. *Laminated veneer lumber* (LVL) (see **Figure 16–4**), is also popular for use as girders. LVL is an engineered lumber product made by gluing layers of veneer together. These beams can be manufactured in almost any size. They are very strong, because lumber defects can be eliminated and all the grain runs in the best possible direction. The notes commonly found on a foundation plan to indicate the type of girder to be used are shown in **Figure 16–1**.

## Determining the Heights of Columns and Piers

The length of the columns or height of the piers depends on how the joists will be attached to the girder. The floor joists may rest directly on top of the girder or

Columns, Piers, and Girders 109

**Figure 16–4.** LVL (laminated veneer lumber) beams are popular, because they are strong and can be manufactured in any length and depth.

### GREEN NOTE

*Designers, especially those designing with an eye toward green homes, have to weigh the strengths and weaknesses of every material choice they make. In choosing material for a girder, the green designer may see some benefits in using laminated veneer lumber (LVL) instead of built-up lumber girders or steel girders. LVL can be produced in any length and a wide range of thicknesses and depths, yet even the longest LVL girder is made up of smaller pieces of material. This means that high strength can be achieved while using material that might otherwise have been destined for the scrap pile.*

may be butted against the girder so that the top surface of the floor joist is flush with the top surface of the girder (see **Figure 16–5**).

To find the height of the columns or piers, first determine the dimension from the basement floor to the finished first floor. Then subtract from this dimension the depth of the first floor, including all of the framing and the girder. Then add the distance from the top of the basement floor to the bottom of the column. The result equals the height of the column or pier. (See Math Reviews 5 and 6.) For example, the following shows the calculation of the height of the steel column in **Figure 16–6**:

- Dimension from finished basement floor to finished first floor = 8″-10½′
- Allowance for finished floor = 1′
- Nominal 2 × 8 joists = 7¼′
  2 × 4 bearing surface on girder = 1½′
  W8 × 31 = 8′
- Total floor framing = 17¾′ or 1″-5¾′
- Subtract total floor framing
  8″-10½′ minus 1″-5¾′ = 7″-4¾′
- Add thickness of concrete slab
  7″-4¾′ plus 4′ = 7″-8¾′

110 UNIT 16

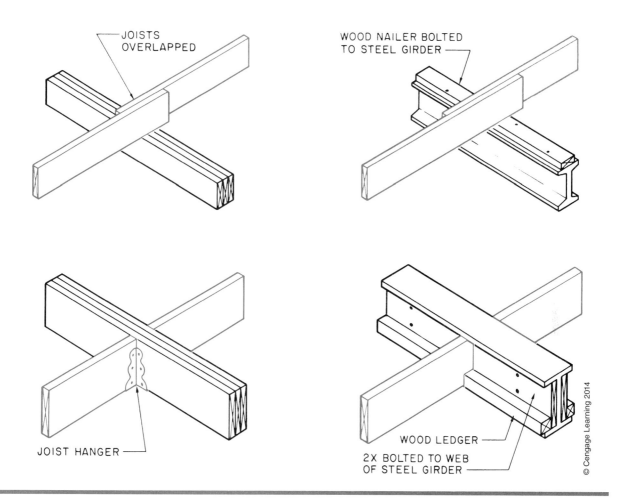

**Figure 16–5.** Several methods of attaching joists to girders.

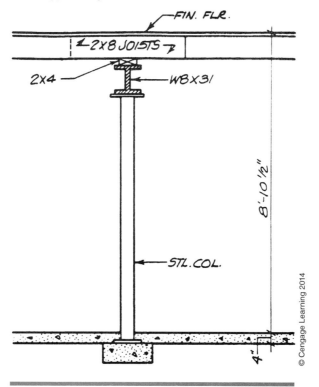

**Figure 16–6.** Calculate the height of the steel column.

Columns, Piers, and Girders 111

### USING WHAT YOU LEARNED

Laminated veneer lumber (LVL) is a popular material for girders in houses because it is strong and available in any length desired. Describe three ways floor joists made of dimensional lumber might be supported by an LVL girder.

1. The joists can rest on top of the girder.
2. A ledger piece can be nailed along the bottom of the LVL to support the joists.
3. Metal joist hangers can be used.

If *I-joists* (often called oTJIs, which are discussed in unit 17) are used for the joists, they would most likely span the full width of the floor and rest on top of the girder because I-joists are available in greater lengths than dimensional lumber.

## Assignment

Questions 1 through 5 refer to **Figure 16–7**.

1. What is the length of the girder?
2. Describe the material used to build the girder, including the size of the material.
3. What supports the girder? (Include material and cross-sectional size.)
4. How many posts, columns, or piers support the girder?
5. What is the height of the columns or piers supporting the girder, including bearing plates?

Questions 6 through 10 refer to **Figure 16–8**.

6. What is the length of the girder?
7. Describe the material used to build the girder, including the size of the material.
8. What supports the girder? (Include material and cross-sectional size.)
9. How many posts, columns, or piers support the girder?
10. What is the dimension from the top of the footing under the pier to the top of the steel beam?

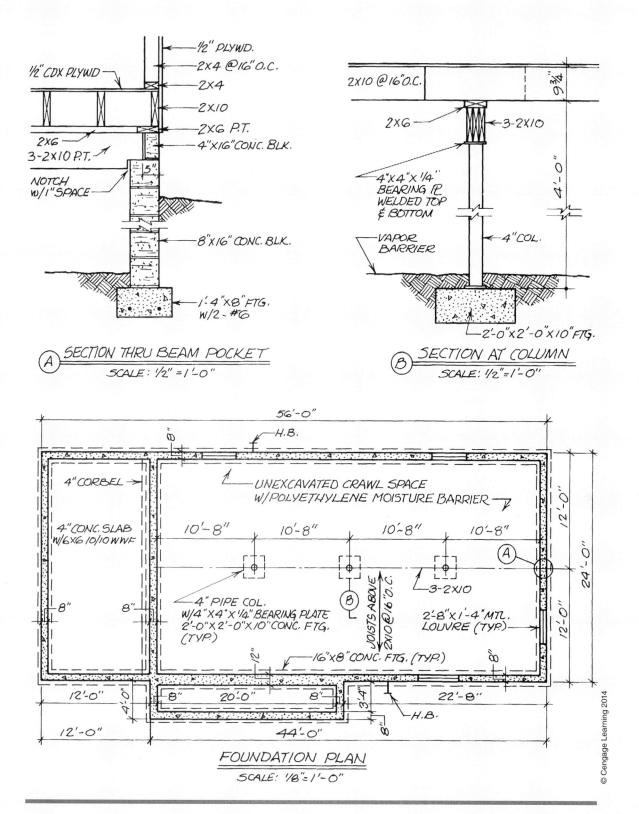

**Figure 16–7.**

Columns, Piers, and Girders

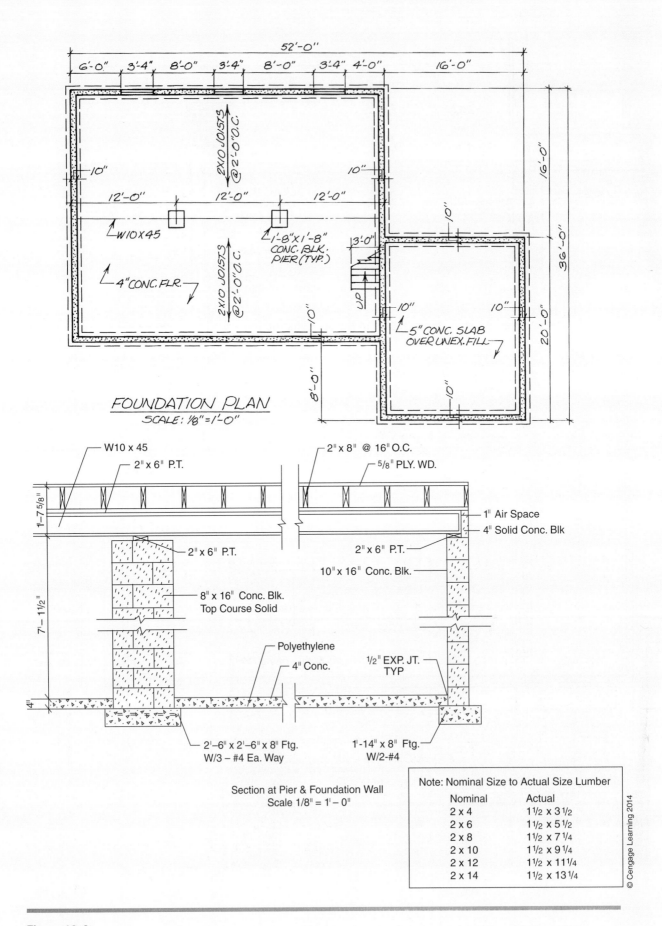

**Figure 16–8.**

# UNIT 17

# Floor Framing

## Sill Construction

Where the framing rests on concrete or masonry foundation walls, the piece in contact with the foundation is called the *sill plate* (see **Figure 17–1**). The sill plate is the piece through which the anchor bolts pass to secure the floor in place. To prevent the sill plate from coming in direct contact with the foundation, and to seal any small gaps, a sill sealer is often included. This is a compressible, fibrous material that acts like a gasket in the sill construction. All lumber in contact with the foundation, including the sill plate, must be *pressure treated* (PT) with an approved chemical to prevent decay.

The entire construction of the floor frame at the top of the foundation is called *sill construction* or the box sill. The box sill is made up of the sill sealer, sill plate, joist, and joist header (see **Figure 17–2**). The sizes of materials are given on a wall section or sill detail. For areas where termites are present, a termite shield is included in the sill construction (see **Figure 17–3**). A termite shield is a continuous metal shield that prevents termites from getting to the wood superstructure.

## Objectives

After completing this unit, you will be able to perform the following tasks:

○ Describe the sill construction shown on a set of drawings.

○ Identify the size, direction, and spacing of floor joists according to a set of drawings.

○ Describe the floor framing around openings in a floor.

**Figure 17–1.** Sill plate.

**GREEN NOTE**

As our sources of lumber have become more depleted and the world's desire to protect our planet has become greater, engineers have developed an increasing array of engineered wood products. These are usually building materials that use wood fibers, wood chips, veneer, and short pieces of wood as their basis, but the wood materials are bound together with synthetic resins. One such material is rim board, a specially engineered product designed for use as the joist header for I-joists. Rim board is made of bonded wood particles much like OSB.

## Floor Joists

*Floor joists* are the parallel framing members that make up most of the floor framing. Until recently, joists in residential construction were 2-inch framing lumber. However, recent advances in the use of materials have produced several types of engineered joists (see **Figure 17–4**). A common type of engineered wood joists is the I-joist, consisting of a top and bottom rib of solid wood and a thinner plywood web between the ribs. I-joists are often called TJIs because the first I-joists, introduced over 40 years ago, were named after the Trus Joist Institute, which developed the first I-joists. Although the materials in each type are different, their use is essentially the same.

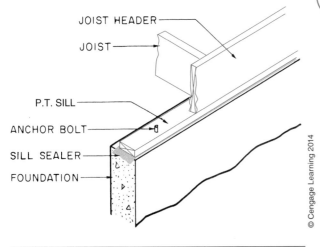

**Figure 17–2.** Box sill.

**GREEN NOTE**

I-joists are becoming increasingly popular for use as floor and ceiling joists because they are lighter than dimensional lumber, are available in greater lengths, and are dimensionally stable. They do not expand and contract or warp nearly as much as conventional lumber with changes in moisture content. They are lighter than dimensional lumber because they use far less material. Weighing less, they require less fuel to transport. Being more dimensionally stable, they are less apt to cause weaknesses in the structure or openings for air infiltration.

Metal floor joists are often used in commercial and industrial construction. Where nonmetallic-sheathed cable is installed through metal studs or floor joists, a bushing or grommet must be installed prior to installing the cable (see **Figure 15–17**).

Notes on the floor plans indicate the size, direction, and spacing of the joists in the floor above (see **Figure 17–5**). For example, notes on the foundation plan give the information for the first floor framing. When the arrangement of framing members is complicated, a framing plan may be included (see **Figure 17–6**). On most framing plans, each member is represented by a single line.

In the simplest building, all joists run in the same direction and are supported between the foundation walls by a single girder. However, irregularities in

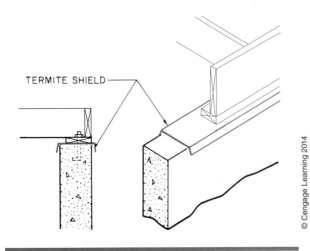

**Figure 17–3.** Termite shield.

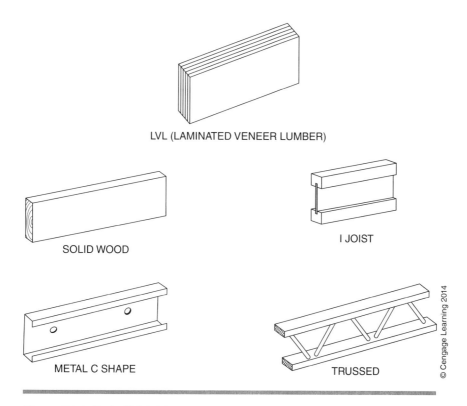

**Figure 17–4.** Typical joist styles.

building shapes require that joists run in different directions (see **Figure 17–7**). As the building design becomes more complex, more variations in floor framing are necessary. In a building such as the Lake House, which has floors at varying levels, the joists are supported by a combination of girders, beams, and load-bearing walls.

## Lake House Floor Framing

The floor framing for the Lake House is shown on Framing Plan 1/6 in your textbook packet. Notice that the framing plan is made up of a simplified floor plan and single lines to represent floor joists, beams, and joist headers. A more elaborate type of framing plan uses double lines to represent the thickness of each member (see **Figure 17–8**). The framing plan shows the location and direction of framing members, but for more detail it will be necessary to refer to Floor Plan 2/2 also.

The floor framing in the Lake House can be studied most easily if it is viewed as having four parts: the kitchen, bedroom #1, bedroom #2 and bathrooms, and the loft. There is no floor framing for the living room and dining room. These two areas form the heat sink for the passive-solar features discussed in earlier units. The floors in these rooms are concrete to absorb and radiate solar heat. As each area is framed, the carpenter must identify the following:

- Joist headers (locate the outer ends of joists)
- Bearing for inner ends of joists (beams, walls, etc.)
- Size and type of framing materials
- Length of joists
- Spacing
- Framing at openings

The joist headers are easily identified on the framing plan, but their exact position should be checked on the detail drawings. They may be set back to create a brick ledge, to accommodate wall finish, or to be flush with the foundation (see **Figure 17–9**).

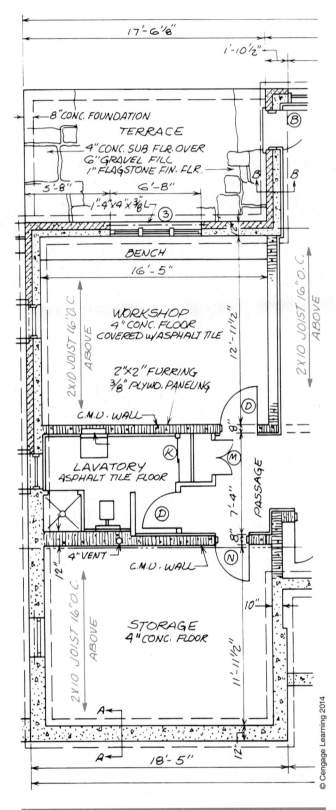

**Figure 17–5.** Joist callouts for the first floor are shown on the foundation plan.

The *bearing* (support) for the inner ends of some of the joists in the Lake House is the structural-steel-and-LVL-beam core, detail 5/6. This core consists of four 3½-inch square steel posts with LVL beams and steel-channel beams. The positions of the posts are shown on all plan views of the house. The C8 × 11.5 steel acts as a beam to support the floor in bedroom #2. The doubled smaller LVLs support the loft. The LVL beams form part of the roof framing. All of the LVL beams rest in beam pockets that are welded to the posts. The structural steel channel (C8 × 11.5) is bolted directly to the posts.

The inner ends of the remaining floor joists in the Lake House are supported by bearing walls in the lower level. For example, some of the kitchen floor joists are supported by the west wall of the playroom, near the fireplace; and some, by the 2 × 10 header that spans the distance from the playroom wall to the foundation in the crawl space.

The size and spacing of material to be used are given on the framing plan by a note. They can also be found on the wall section. Lengths of framing members are usually not included on framing plans. However, these lengths can be found easily by referring to the floor plan that shows the location of the walls or beams on which the joists rest. For example, refer to **Figures 17–10** and **17–11** and find the length of the floor joists in bedroom #2 as follows:

1. The dimension from the outside of the north wall to the centerline of the 3½"□ steel post is 14'-1".
2. The dimension from the centerline of the 3½"□ post is 2", so the overall dimension of the bedroom floor is 14'-3".
3. According to Wall Section 3/4, the joist header is flush with the north foundation wall, so subtract 1½" (the thickness of the joist header) from each end.
4. 14'-3" minus 3" (1½" at each end) equals 14'-0" (the length of the joists).

Some floor framing is cantilevered to create a seemingly unsupported deck.

*Cantilevered* framing consists of joists that project beyond the bearing surface to create a wide overhang. This technique is used extensively for balconies, under bay windows, and for garrison-style houses (see **Figure 17–12**).

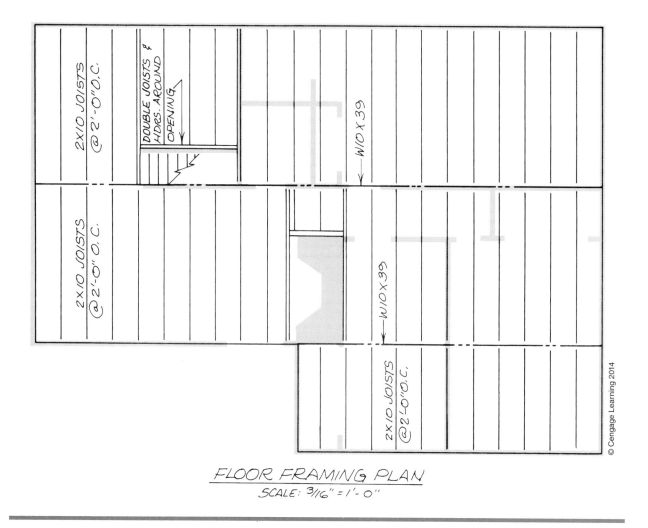

**Figure 17-6.** A simple floor framing plan.

## Framing at Openings

Where stairs and chimneys pass through the floor frame, some of the joists must be cut out to form an opening. The ends of these joists are supported by headers made of two or more members. The full joists at the sides of the opening have to carry the extra load of the shortened joists and headers, so they are also doubled or tripled (see **Figure 17–13**). The number of joists and headers required around openings of various sizes is spelled out in building codes and shown on framing plans.

Floor Framing 119

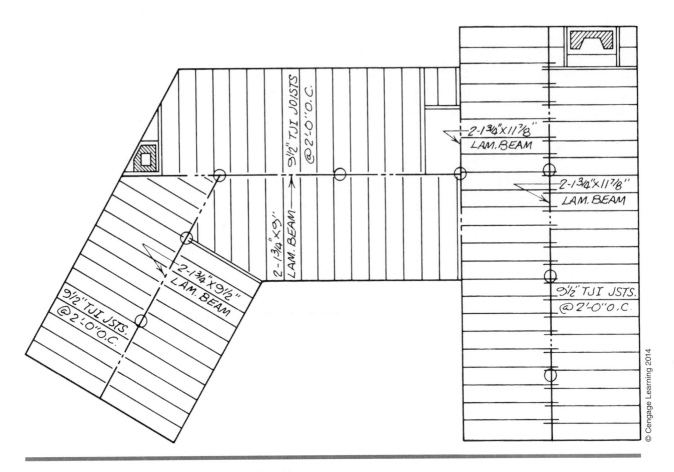

**Figure 17–7.** Floor framing plan for an irregular-shaped house.

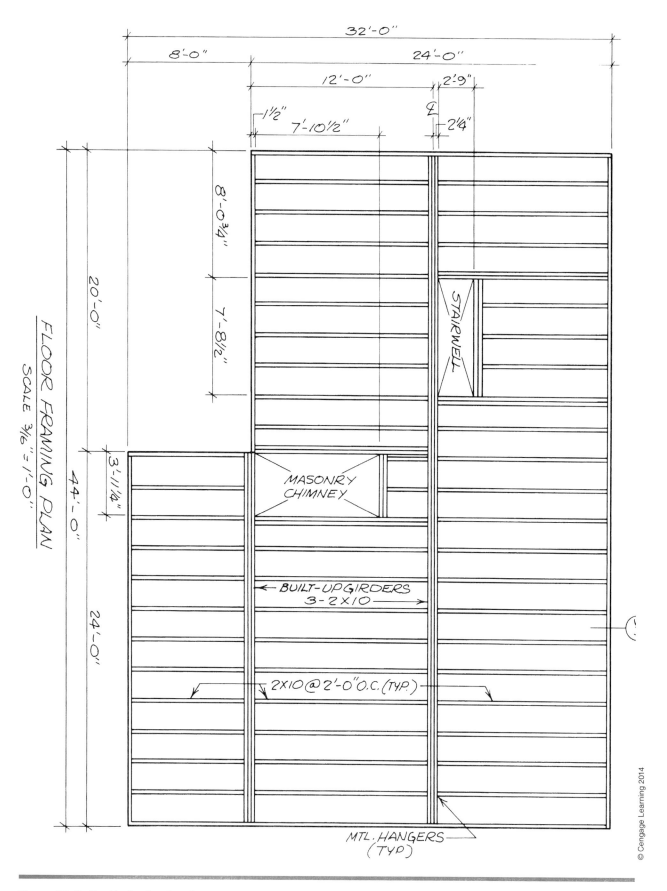

**Figure 17–8.** Double-line framing plan.

Floor Framing 121

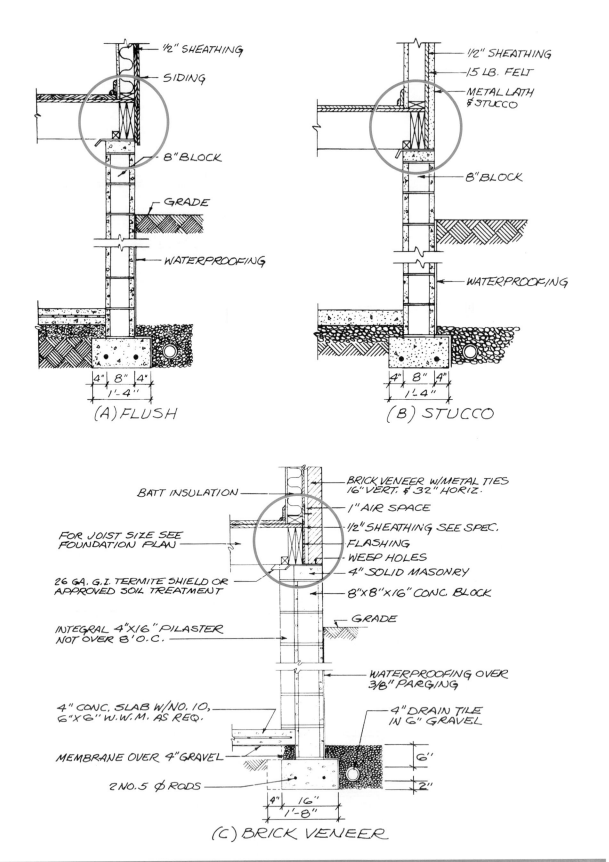

**Figure 17–9.** The joist headers are positioned according to the exterior finish to be used.

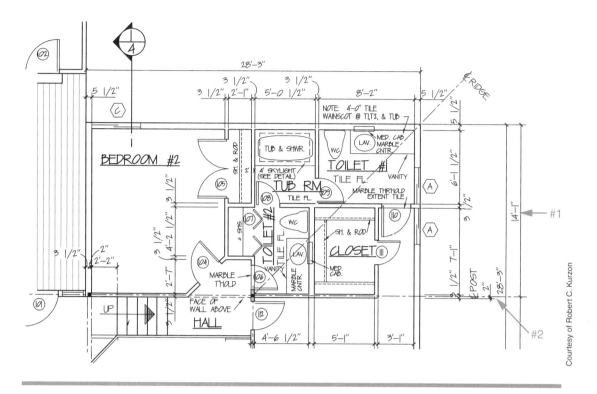

**Figure 17-10.** Part of the Lake House Floor Plan.

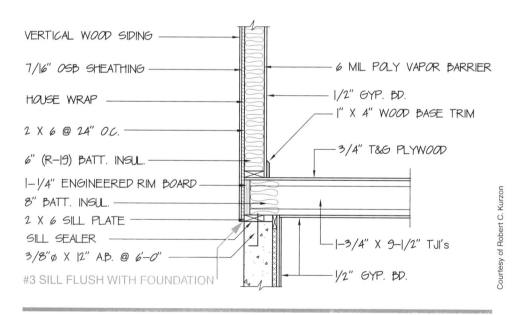

**Figure 17-11.** Part of the Lake House wall section.

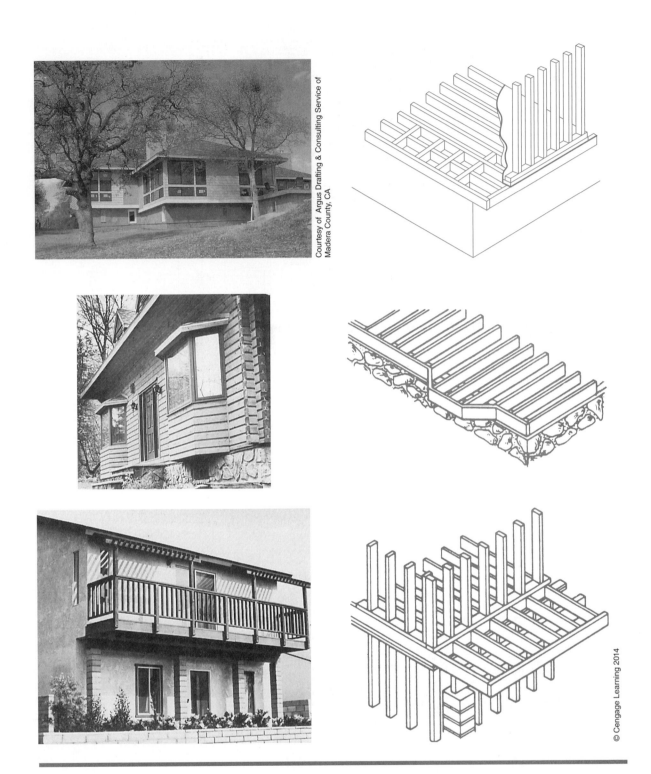

**Figure 17-12.** Typical uses of cantilevered framing.

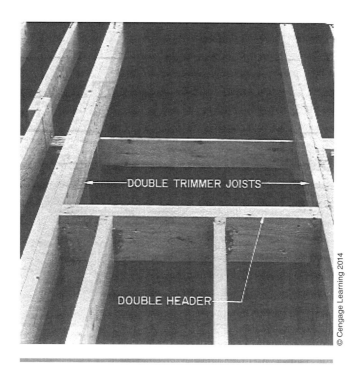

**Figure 17–13.** Framing around an opening in the floor.

### USING WHAT YOU LEARNED

Whether you are actually building the house or estimating the cost of building the house, you need to determine the exact length of all of the floor joists. Some times this requires using more than one of the many sheets of drawings. In this practice exercise, we will find the length of the floor joists in the Lake House kitchen floor.

The Floor Framing Plan, Drawing 1/6, on Sheet 6 tells us that the kitchen floor is framed with 1¾" × 9½" TJIs spaced 16" OC. If we look at the Upper Level Floor Plan, where the kitchen is shown, we see a symbol indicating a section view of the floor construction on Drawing 2/4. That is drawing number 2 on Sheet 4. Section 2/4 shows us where each end of the joist is supported. The West or left end rests on the foundation and ends at a 1¼" engineered rim board, according to the typical Wall Section 3/4. That rim board is flush with the outside of the foundation wall. The east or right end rests on a frame wall beside the fire place in the play room, shown on the Lower Level Floor Plan. The east end has another 1¼" engineered rim board, which is flush with the Playroom side of the frame wall. The face of the playroom wall is 2 inches east of the centerline of the steel posts. The Foundation Plan on Sheet 1 shows that the centerline of the steel posts is 14'-1" from the face of the foundation, so the face of the playroom wall is 14'-3" from the face of the foundation wall. Subtract 1¼" from each end for the rim boards to find the length of the kitchen floor joists as 14'-0½".

# Assignment

Refer to the Lake House drawings in your textbook packet to complete this assignment.

1. What is used for the floor framing in bedroom #2?
2. What size lumber is used for the floor joists in the loft?
3. What size lumber is used for the joist headers in the loft?
4. How long are the joists in the loft?
5. How long are the floor joists in bedroom #1?
6. What supports the west end of the floor joists in bedroom #1?
7. What supports the south end of the floor joists in bedroom #2?
8. How long are the floor joists in bedroom #2?
9. How many floor joists are needed for bedroom #2 and the adjacent closets?
10. What size material is used for the sill?
11. What does the double 1¾" × 9½" LVL in the structural core of the Lake House support?
12. How long are the headers that support the loft floor joists?
13. Is the box sill of the Lake House flush with the foundation wall or set back?
14. When an opening in a floor is framed, why are the joists at the sides of the opening doubled?

# Laying Out Walls and Partitions

# UNIT 18

When the deck (framing and subfloor or concrete slab) is completed, the framing carpenter lays out the location of walls and partitions. The size and location of each wall are indicated on the floor plans. Drawings 1/2 and 2/2 are the floor plans for the Lake House.

Of all the sheets in a set of construction drawings, the floor plans often contain the most information. Before looking for specific details on floor plans, it may help to mentally walk through the house. Start at the main entrance to the main floor, and visualize each room as if you were walking through the house.

## Visualizing the Layout of Walls and Partitions

The lowest floor with frame walls in the Lake House is the basement. Start at the $6^0 \times 6^8$ sliding glass doors (H) in the southeast corner of the playroom. This large L-shaped room covers most of the basement floor. Some plans list overall dimensions for each room. The plans for the Lake House give this information by conventional dimension lines only. Many of the dimensions of the Lake House framing are referenced to the columns of the structural steel core. The faces of the some of the walls are an inch or two from the centerlines of the columns, but for the sake of visualizing the layout of the house, these centerline dimensions are close enough to develop a good mental picture of the layout.

The north–south dimension of the playroom is 28′-3″. The east–west dimension of the north part of the playroom is 14′-1″. The section of the playroom with the fireplace is 13′-8″ by 13′-10″. It may be helpful to notice the overall dimensions of each room as you visualize its shape and relationship to other rooms. It will be necessary to refer to these dimensions many more times. At the north end of the playroom is a small closet. South of the closet is a hall leading to the bathroom and utility room.

The bathroom has a shower, water closet (toilet), and lavatory. Notice that although most interior partitions are 3½ inches thick, the wall behind the water closet and lavatory is 5½ inches thick. This thicker wall is called a *plumbing wall*. Its extra thickness allows room for plumbing (see **Figure 18–1**).

The purpose of the utility room beyond the bathroom is to house mechanical equipment such as the water heater, furnace or boiler, and water pump. In the southwest corner of the utility room is a small opening into the crawl space on the west side of the house. Notice that the concrete ends here.

## Objectives

After completing this unit, you will be able to perform the following tasks:

○ Describe the layout of a house from its floor plans.

○ Find specific dimensions given on floor plans.

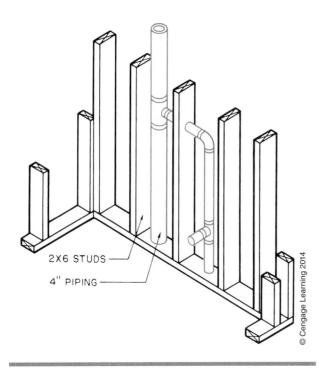

**Figure 18–1.** 6-inch plumbing wall.

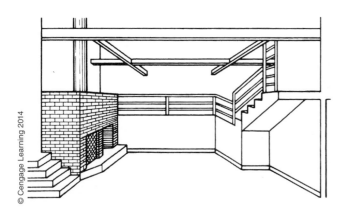

**Figure 18–2.** Lake House playroom.

clearly in the sections on Sheet 4 of the Lake House in your textbook packet. The L-shaped stairs lead up to the living room (see **Figure 18–3**). Against the west and south walls of the living room is a plywood platform or built-in bench.

Another set of stairs in the northwest corner of the living room leads to the kitchen and dining room. These rooms are separated only by a peninsula of kitchen cabinets. The kitchen is separated from a hall by a free-standing closet and enclosure for the refrigerator and oven (see **Figure 18–4**). This closet and enclosure cannot be recognized as free-standing (meaning they do not reach the ceiling) on the floor plan. However, this can be seen on Section 2/4. On the north side of the kitchen is a door to the deck outside. The east side of

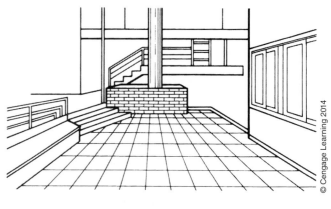

**Figure 18–3.** Lake House living room.

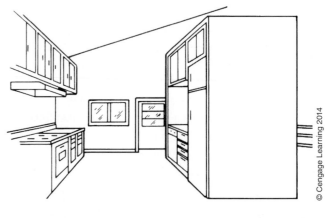

**Figure 18–4.** Lake House kitchen.

Walking back through the playroom to the stairs, you will see some interesting features (see **Figure 18–2**). A broken line in this area indicates the edge of a floor above. These floors can be seen more easily on the first floor plan. Also notice the location of the fireplace, which will extend up through the upper floor. The L-shaped stairs lead up to the first floor. It is obvious that this is not a full story higher because a callout indicates that there are only five risers. This can be seen more

128  UNIT 18

the hall has a railing that continues up another set of stairs to the upper hall. Beyond this railing the floor is open to the playroom below.

Above the open area of the playroom is a loft that provides storage or extra sleeping space (see **Figure 18–5**). Access to the loft is by a ladder in the upper hall. The loft is shown on a separate plan on Sheet 4. The loft is suspended on two beams of 1¾″ × 9½″ LVL. The north and south sides of the loft are enclosed by a wall 2′-6″ high. The east and west ends have a railing.

A door at the end of the upper hall opens into bedroom #1. On the east side of this bedroom is a hall that leads past a large (7′-1″ × 5′-1″) closet to toilet #1 in the northeast corner of the house. This toilet room has only a lavatory and water closet. A tub in the next room also serves toilet #2 to the east. Toilet #2 has a closet with shelves for linen storage. This toilet room can also be entered from the upper hall. The remaining room is bedroom #2.

The garage is attached to the house only by the wood deck and the roof. The garage is a rectangular building with 4-inch walls, an overhead door, and a walk-through door.

## Finding Dimensions

When you understand the relationships of the rooms to one another, you are ready to look for more detailed information. Frame walls are dimensioned in one of three ways (see **Figure 18–6**). Exterior walls are usually dimensioned to the face of the studs or the face of the

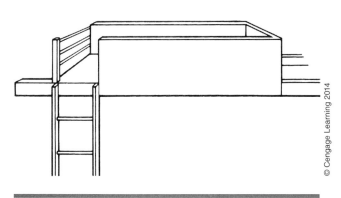

**Figure 18–5.** Lake House loft.

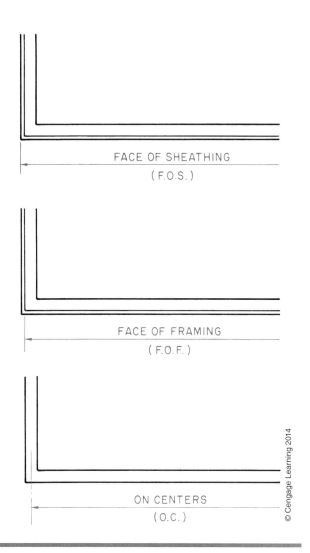

**Figure 18–6.** Three types of dimensioning.

sheathing. Interior walls may be dimensioned either to the face of the studs or to their centerlines.

When walls are dimensioned to their centerlines, one-half of the wall thickness must be subtracted to find the face of the studs. For example, in **Figure 18–7**, the end walls are 12′-4″ on centers. However, the plates for the side walls are 12′-0½″ long (12′-4″-3½″, the width of the studs, equals 12′-0½″).

Dimensions are usually given in a continuous string where practical (see **Figure 18–8**). Overall dimensions and major wall locations are given outside the view. Minor partitions and more detailed features are dimensioned either on or off the view, whichever is most practical.

Laying Out Walls and Partitions 129

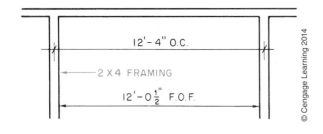

**Figure 18–7.** It is important to consider the true size of the framing material when working with OC dimensions.

The Lake House includes an angled wall. The length of such walls can be accurately found by trigonometry. However, the accuracy obtained by measuring with an architect's scale should be adequate for normal estimating.

As each wall is laid out, the plates are cut to length, and the positions of all openings and intersecting walls are marked. The wall frame is usually assembled flat on the deck. Its position is marked on the deck with a chalk line; then the assembly is tipped up and slid into place.

Commercial buildings may require a thicker wall, where the electrical service panel is flush mounted and the panel box is deeper than the standard 3½-inch wall.

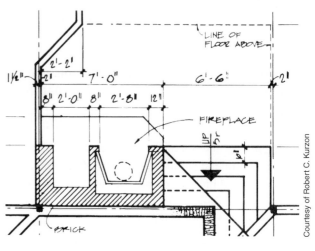

**Figure 18–8.** Dimensions are placed in neat strings as much as possible.

Many times this thicker wall requirement is not shown on the floor plan or drawing details.

The utilities that are to be stubbed up from a concrete floor slab into a wall are installed prior to floor slab-concrete placement and, therefore, prior to the wall locations being determined by the framing contractor. The utility installer must be able to accurately predetermine the specific wall locations.

### USING WHAT YOU LEARNED

Each piece of framing material must be precisely positioned according to the construction drawings. What is the dimension from the inside of the framing of the west wall of the kitchen to the outside of the west end of the loft?

All of the information necessary to determine this dimension is shown on the Upper Level Floor Plan 2/2. Above the loft, in bedroom #2, there is a string of dimensions that show the edge of the loft to be 2'-2" from the inside face of the wall and another 2" to the centerline of the 3½" ☐ steel post. At the bottom of the Floor Plan is a dimension of 14'-1" from the centerline of the posts to the outside of the west wall and another dimension that shows the wall framing to be 5 ½" thick: (2'-2") + 2" + (14'-1") − 5½" = 15'-11½".

## Assignments

Refer to the Lake House drawings (in the packet) to complete the following assignment.

1. What is the inside dimension of each room? Disregard slight irregularities in room shape, but remember to allow for wall thicknesses. Do not include the kitchen and dining room.

   | Room | N-S dimension | E-W dimension |
   |---|---|---|
   | Utility | | X |
   | Basement Bath | | X |
   | Living Room | | X |
   | Bedroom #1 | | X |
   | Bedroom #2 | | X |
   | Toilet #1 | | X |
   | Tub | | X |
   | Toilet #2 | | X |
   | Loft | | X |

2. How many lineal feet of 2 × 4 frame wall are there in the basement?
3. How many lineal feet of 2 × 6 frame wall are there in the basement?
4. What is between the dining room and the living room?
5. What is directly below the loft?
6. What is directly below the tub room?
7. How many lineal feet of 2 × 6 frame wall are there on the Upper Level Floor Plan?
8. What utility must be roughed in prior to concrete slab placement?
9. Where would you find the utility concrete slab rough-in requirements?

Laying Out Walls and Partitions

# UNIT 19
# Framing Openings in Walls

## Objectives

After completing this unit, you will be able to perform the following tasks:

○ Describe typical rough openings.

○ Locate and interpret specific information for framing openings.

Two types of dimensions must be known before window and door openings can be framed: location and size. Opening locations are given by dimensioning to their centerlines in a string of dimensions outside the floor plan (see **Figure 19–1**). Such dimensions are usually given from the face of the studs. One-half the rough opening is then allowed on each side of the centerline.

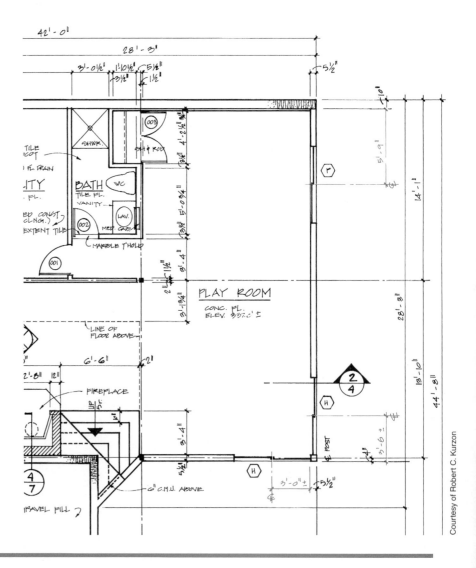

**Figure 19–1.** The locations of openings in framed walls are usually given to their centerlines. The ± on the sliding door dimensions allows the builder to place the doors next to the corner post.

## Dimensions of Rough Openings

The size of the *rough opening* (RO), or opening in the framing, is listed on the door and window manufacturers' specifications. Door and window schedules are lists of all doors and windows in the house, usually with nominal dimensions (see **Figure 19–2**). Doors and windows are identified on the floor plan by a *mark*—a letter or number. All doors or windows of a certain size and type have the same mark. Each mark is listed on the schedule with the information for the doors or windows of that type and size.

### SCHEDULE OF WINDOWS

| | SIZE | TYPE | GLASS | COMMENTS |
|---|---|---|---|---|
| A | 3'-0" X 3'-0" | THERMAL BREAK ALUMINUM | 5/8" INSULATED | |
| B | 3'-0" X 5'-0" | D.O. | D.O. | |
| C | 4'-0" X 4'-0" | D.O. | D.O. | |
| D | 4'-0" X 5'-0" | D.O. | D.O. | |
| E | 5'-0" X 5'-0" | D.O. | D.O. | |
| F | 6'-0" X 4'-0" | D.O. | 1" INSULATED | |
| G | 6'-0" X 5'-0" | D.O. | 1" INSULATED | |
| H | 6'-0" X 6'-8" | D.O. | 1" TEMP. INSUL. | SLIDING GLASS DOORS |

### SCHEDULE OF DOORS

| | SIZE | THK | DOOR MAT'L | DOOR TYPE | FRAME MAT'L | FRAME TYPE | GLASS | LOCKSET | COMMENTS |
|---|---|---|---|---|---|---|---|---|---|
| 001 | 2'-8" X 6'-8" | 1 3/8 | WD. | C | WD. | A | | CLASSROOM | |
| 002 | 2'-0" X 6'-8" | 1 3/8 | WD. | C | WD. | A | | PRIVACY | |
| 003 | PR 2'-0" X 6'-8" | 1 3/8 | WD. | D | WD. | A | | PASSAGE R.H. DUMMY TRIM L.H. | |
| 101 | 3'-0" X 6'-8" | 1 3/4 | H.M. | A | WD. | A | 1" INSUL. | ENTRY | 1'-8" SIDE LIGHT |
| 102 | 3'-0" X 6'-8" | 1 3/4 | H.M. | B | WD. | A | | ENTRY | |
| 103 | PR 2'-4" X 6'-8" | 1 3/8 | WD. | D | WD. | B | | NONE | BI-PASSING HDWARE |
| 104 | 2'-6" X 6'-8" | 1 3/8 | WD. | C | WD. | B | | PRIVACY | |
| 105 | PR 2'-6" X 6'-8" | 1 3/8 | WD. | D | WD. | B | | PASSAGE R.H. DUMMY TRIM L.H. | |
| 106 | 2'-0" X 6'-8" | 1 3/8 | WD. | C | WD. | B | | PRIVACY | |
| 107 | (4) 1'-0" X 6'-8" | 1 3/8 | WD. | E | WD. | B | | NONE | BI-FOLDING HDWARE |
| 108 | 2'-0" X 6'-8" | 1 3/8 | WD. | C | WD. | B | | PRIVACY | |
| 109 | 2'-0" X 6'-8" | 1 3/8 | WD. | C | WD. | B | | PRIVACY | |
| 110 | 2'-6" X 6'-8" | 1 3/8 | WD. | C | WD. | B | | PRIVACY | |
| 111 | 2'-0" X 6'-8" | 1 3/8 | WD. | C | WD. | B | | PASSAGE | |
| 112 | 2'-6" X 6'-8" | 1 3/8 | WD. | C | WD. | B | | PRIVACY | |

Courtesy of Robert C. Kurzon

**Figure 19–2.** Window and door schedules list all the windows and doors and their sizes.

## 400 Series Casement Windows

**Table of Basic Casement Unit Sizes**  Scale 1/8" = 1'-0" (1:96)

[Size chart showing Unit Dimensions, Minimum Rough Opening, Unobstructed Glass, and Unobstructed Glass Transom Units Only across widths from 1'-5" (432mm) to 4'-8 1/2" (1435mm), with CTR (transom) unit identifiers CTR1510 through CTR4810. CTR units are non-venting.]

Unit widths and rough openings:
- 1'-5" (432) / 1'-5 1/2" (445) — CTR1510
- 1'-8 1/2" (521) / 1'-9" (533) — CTR1810
- 2'-0 1/8" (613) / 2'-0 5/8" (625) — CTR2010
- 2'-4 3/8" (721) / 2'-4 7/8" (733) — CTR2410
- 2'-7 1/2" (800) / 2'-8" (813) — CTR2810
- 2'-11 15/16" (913) / 3'-0 1/2" (927) — CTR3010
- 2'-9 3/4" (857) / 2'-10 1/4" (870) — CTR2910
- 3'-4 3/4" (1035) / 3'-5 1/4" (1048) — CTR3410
- 4'-0" (1219) / 4'-0 1/2" (1232) — CTR4010
- 4'-8 1/2" (1435) / 4'-9" (1448) — CTR4810

Unit model grid (by height):

| Height | Col1 | Col2 | Col3 | Col4 | Col5 | Col6 | Col7 | Col8 | Col9 | Col10 |
|---|---|---|---|---|---|---|---|---|---|---|
| 2'-0 1/8" (613) / 2'-0 5/8" (625) / 19 5/16" (491) | CR12 | CN12 | C12 | CW12 | | | | | | |
| 2'-4 3/8" (721) / 2'-4 7/8" (733) / 23 9/16" (598) | CR125 | CN125 | C125 | CW125 | CX125 | | | | | |
| 2'-11 15/16" (913) / 3'-0 1/2" (927) / 31 1/8" (791) | CR13 | CN13 | C13 | CW13 | CX13 | CXW13 | CR23 | CN23 | C23 | CW23 |
| 3'-4 13/16" (1037) / 3'-5 3/8" (1051) / 36" (914) | CR135 | CN135 | C135 | CW135† | CX135**♦ | CXW135♦ | CR235 | CN235 | C235 | CW235†♦ |
| 4'-0" (1219) / 4'-0 1/2" (1232) / 43 3/16" (1097) | CR14 | CN14 | C14 | CW14†♦ | CX14♦ | CXW14♦ | CR24 | CN24 | C24 | CW24†♦ |
| 4'-4 13/16" (1341) / 4'-5 3/8" (1356) / 48" (1219) | CR145 | CN145 | C145 | CW145†♦ | CX145♦ | CXW145♦ | CR245 | CN245 | C245 | CW245†♦ |
| 4'-11 7/8" (1521) / 5'-0 3/8" (1534) / 55 1/16" (1399) | CR15 | CN15 | C15 | CW15†♦ | CX15♦ | CXW15**♦ | CR25 | CN25 | C25 | CW25†♦ |
| 5'-4 13/16" (1646) / 5'-5 3/8" (1660) / 60" (1524) | CR155 | CN155 | C155 | CW155†♦ | CX155♦ | CXW155**♦ | CR255 | CN255 | C255 | CW255†♦ |
| 5'-11 7/8" (1826) / 6'-0 3/8" (1838) / 67 1/16" (1703) | CR16 | CN16 | C16 | CW16†♦ | CX16♦ | CXW16**♦ | CR26 | CN26 | C26 | CW26†♦ |

\* "Unobstructed Glass" measurement is for single sash only.
\*\* These units have straight arm operators, see opening specifications.
† CW series units (except CW2, CW25 and CW3 height) open to 20" clear opening width using sill hinge control bracket. Bracket can be pivoted allowing for cleaning position. CW series units are also available with a 22" clear opening width.
‡ Andersen® art glass panels are available for these units by special order only. Contact your Andersen® supplier. Panels are available for all other units on this page through normal ordering process.
♦ These units meet or exceed the following dimensions: Clear Openable Area of 5.7 sq. ft., Clear Openable Width of 20" and Clear Openable Height of 24", when appropriate hardware (straight arm or split arm) is specified.
- Casement transom units (CTR) may be rotated to be used as a casement or awning sidelight.
- Rough opening dimensions may need to be increased to allow for use of building wraps, flashings, sill panning, brackets, fasteners or other items. See page 10 for more details.
- "Unit Dimension" always refers to outside frame to frame dimension.
- Dimensions in parentheses are in millimeters.
- When ordering, be sure to specify color desired: White, Sandtone, Terratone® or Forest Green.

Left  Right  Stationary
**Venting Configuration**
Hinging shown on size table is standard. Specify left, right or stationary, as viewed from the outside. For other hinging of multiple units, contact your local supplier.

Courtesy of Iroquois Millwork Corporation

**Figure 19–3.** Typical page from a window catalog.

The rough opening dimensions are not usually given on the drawings. They should be obtained from the manufacturer. Window manufacturers do not list the same size information in their catalogs, but they do list the rough opening size. Other sizes they sometimes list are nominal size or unit dimensions (approximate overall size of the window) and glass size (size of the actual glass in the window) shown in **Figure 19–3**. If the finished doorway, the jamb, is to be built on the site by a carpenter, the rough opening size will not be available. In this case the rough opening for swing doors can be built 2 inches wider and 2 inches higher than the door. If the door is another type, you must first determine what the finished opening is to be. Manufacturers of hardware for sliding and bifold doors specify rough opening sizes for doors installed with their hardware. These manufacturer's specifications should be consulted as the opening is being framed.

Sizes of doors and windows are given with the width first and height second. To further simplify dimensioning, they are often listed as feet/inches. For example, a $2^6 \times 6^8$ door is 2 feet, 6 inches wide by 6 feet, 8 inches high.

The information in this unit is intended only to help you determine rough opening sizes. Electrical utility installers should familiarize themselves with the locations and sizes of these rough openings prior to installing the electrical wiring. More information about windows and doors can be found in Unit 25.

## Framing Openings

In stud-wall construction, it is usually necessary to cut off or completely eliminate one or more studs where windows and doors are installed. The load normally carried by these studs must be transferred to the sides of the opening. The construction over an opening that transfers this load to the sides of the opening is called the *header.* Additional studs are installed at the sides to carry this load (see **Figure 19–4**).

A set of construction drawings includes details showing the type of rough opening construction intended. The simplest type of wood construction uses two 2 × 10s or 2 × 12s with plywood spacers between to form the header (see **Figure 19–5**). A flat 2 × 4 may be nailed to the bottom of the header to reduce the height of the top of the opening, but this flat 2 × 4 is not considered as part of the size of the header.

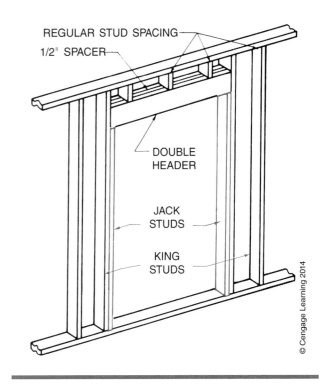

**Figure 19–4.** The jack studs support the header.

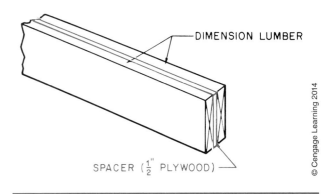

**Figure 19–5.** Solid wood header.

To conserve lumber in nonbearing walls, the header may be two 2 s 4s. The area over the header is framed with cripple studs. The area below a window also is framed with *cripple studs,* (see **Figure 19–6**). These cripple studs are installed on the normal spacing for wall studs. This may require a cripple within a few inches of a side jack stud, but is necessary to provide a nailing surface for the sheathing.

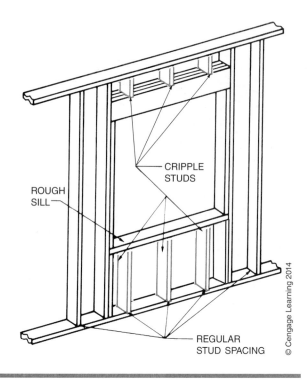

**Figure 19–6.** The spaces above and below the window opening are framed with cripple studs.

*Pocket doors* are sliding doors that slide into an opening in the wall, thereby allowing the largest possible opening without swinging into the room. Most builders install pocket doors as a pre-assembled frame (see **Figure 19–7**). The wall framing to provide the opening into which a pocket door assembly can be installed is generally the same as the framing for any other type of door. Pocket door manufacturers provide rough opening dimensions for their products.

Several systems of metal framing are available for light construction. The basic elements of light-gauge metal framing are the same as for wood framing (see **Figure 19–8**). These systems use top and bottom wall plates, studs, and floor joists. To make light-gauge framing compatible with wood, the metal members are made in common sizes for wood framing. The greatest difference is that metal framing is joined with screws instead of nails.

Headers over openings in light-gauge metal framing are usually very similar to those in wood framing. However, the system designed by the metal framing manufacturer should always be followed.

**Figure 19–7.** Pocket door frame.

> ### GREEN NOTE
> 
> *In traditional frame construction, each end of a door or window header rests on a jack stud. By using approved metal header hangers, the jack studs can be eliminated. Also, by sandwiching foam insulation between the outer layers of the header, the insulating value of the header can be greatly increased.*
> 
>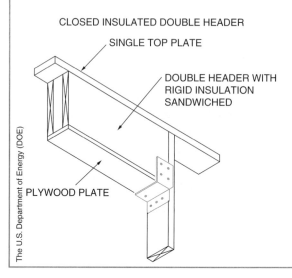

**Figure 19–8.** The parts of this steel frame are the same as those of a wood frame. Notice the window header and cripple studs.

### USING WHAT YOU LEARNED

Carpenters must calculate the lengths of all of the headers over windows and doors. Assuming the rough opening is to be 2 inches greater than the width of the door, how long must the header be over the door into the bath on the lower level of the Lake House?

This is door #002. The Schedule of Doors on Sheet 6 tells us that it is a 2'-0" × 6'-8" door. (That is a narrow doorway, but there is not enough space for a larger door.) The header rests on a 2 × 4 at each end, so the header length will be 3 inches longer than the rough opening (1½ inches at each side of the opening). Add the dimensions to find the length of the header: 2'-0" + 2" + 3" = 2'-5".

## Assignment

Refer to the Lake House drawings in your textbook packet to complete this assignment.

1. What type of header should be used over the door from the hall to bedroom #1?
2. What type of header should be used over the door from the deck to the kitchen?
3. Why should these two headers be made differently?
4. What is the length of the header over the garage overhead door? Allow for two trimmers at each side and 1¼ inches for jambs at each side.
5. What are the RO dimensions for the door from the deck into the garage?
6. Name the location and give the RO dimensions for each interior door on the Upper Level Floor Plan.
7. According to **Figure 19–3**, what are the RO dimensions for the windows in bedroom #2?
8. How long is the header over the window in bedroom #2?
9. How many cripple studs are needed beneath the windows in the south wall of the living room?
10. What is the length of the cripple studs beneath the window in bedroom #2? (Assume the bottom of the header is 6'-8½" from the top of the subfloor.)

Framing Openings in Walls

# UNIT 20
# Roof Construction Terms

## Objectives

After completing this unit, you will be able to perform the following tasks:

- Identify common roof types.
- Define the terms used to lay out and construct a roof.
- State roof pitches as a ratio of rise to run or as a fraction.

## Types of Roofs

Several types of roofs are commonly used in residential construction (see **Figure 20–1**). Variations of these roof types may be used to create certain architectural styles.

The *gable roof* is one of the most common types used on houses. The gable roof consists of two sloping sides that meet at the ridge. The triangle formed at the ends of the house between the top plates of the wall and roof is called the gable.

The *gambrel roof* is similar to the gable roof. On this roof, the sides slope very steeply from the walls to a point about halfway up the roof. Above this point, they have a more gradual slope.

The *hip roof* slopes on all four sides. The hip roof has no exposed wall above the top plates. This means that all four sides of the house are equally protected from the weather.

The *mansard roof* is similar to the hip roof, except the lower half of the roof has a very steep slope and the top half is more gradual. This roof style is used extensively in commercial construction—on stores, for example.

The *shed roof* is a simple sloped roof with no ridge. A shed roof is much like one side of a gable roof. This type of roof is used in modern architecture and for additions to existing buildings.

## Roof Construction Terms

The roof construction terms used with trussed roofs, the most common type of roof framing for light construction, are defined below and shown in **Figure 20–2**.

- **Top chord** is the top member in the truss.
- **Bottom chord** is the bottom member of the truss and acts as a ceiling joist for the space below.
- A **web** member is any of the interior bracing between the top chord and the bottom chord.
- A **gusset** is a reinforcing piece of metal or plywood fastened to the truss, where members are joined.
- **Span** is the distance between the outsides of the walls covered by a roof.
- **Rise** is the vertical distance or height of the roof.
- The **overhang** is the horizontal distance covered by the roof outside the walls.
- The **tail** is the portion of the top chord that is outside the walls.
- The **pitch** of a roof is a way of indicating how steep the roof is. The pitch of a roof is usually given as the number of inches of rise per foot of run. This is called *slope* or *unit rise*.

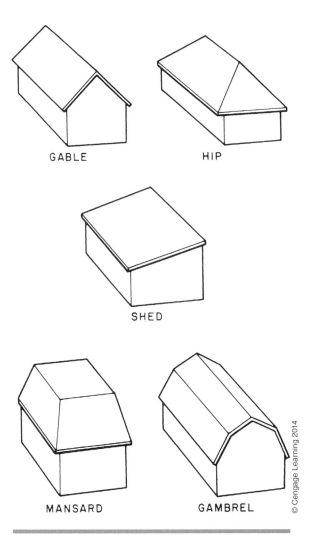

**Figure 20-1.** Common roof types.

In writing unit rise, the slope is given as the number of inches of rise for every 12 inches of run (see **Figure 20-3**). The rise per foot of run is given on drawings with the symbol . The horizontal leg of this symbol represents the run. The vertical leg represents the rise as shown in **Figure 20-4** and notice that this figure illustrates two roof pitches. These pitches are written as 10 in 12 and 5 in 12.

> ### GREEN NOTE
> *By planning the slope and overhang of the roof, the designer can minimize the amount of scrap material that is generated in its construction. The ideal dimensions for a roof section would be in 2-foot increments. Most roof framing is spaced either 16 inches on center or 2 feet on center. A roof that is multiples of 2 feet in length and width (ridge to eave) will not require extra framing if it is framed 2 feet on centers. Also, the roof decking that is usually sheets 4-foot wide material will have only 2-foot wide scraps if the width of the roof is a multiple of 2 feet. The 2-foot scraps can usually be used on the next row of decking.*
>
> *Steeper roofs have more roof area than do low-sloped roofs. The designer can make the roof width fit the 2-foot grid by adjusting the slope of the roof.*

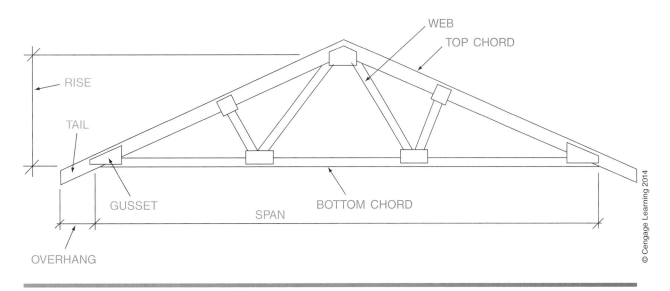

**Figure 20-2.** Parts of a roof truss.

Roof Construction Terms

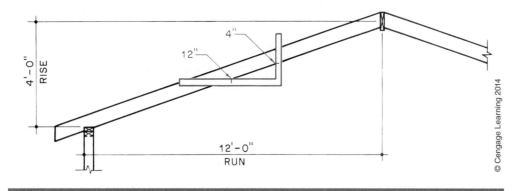

**Figure 20–3.** Rise per foot of run. This roof has a 4 in 12 slope.

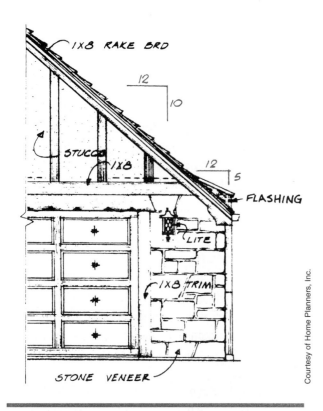

**Figure 20–4.** Roof slope is usually shown on building elevations.

Although most modern roof frames are built with **trusses**, it is valuable to know something about rafter-framed roofs, because some special situations call for **rafters** to be cut and installed on the site. The terms defined in the following list are illustrated in **Figure 20–5.**

○ The **run** is the vertical distance covered by one rafter. Run does not include any part of the rafter that extends beyond the wall. On a common two-sided roof with both sides having the same pitch, the run is one-half the span.
○ The **measuring line** is an imaginary line along which all roof dimensions are taken. The measuring line of a rafter is a line parallel to its edges and passing through the deepest part of the bird's mouth.
○ A **bird's mouth** is a notch cut in the lower edge of the rafter to fit around the top of the wall.
○ The **ridge board** is the horizontal member to which the upper ends of the rafters are connected.

When the dimensions are given for a run of other than 12 feet, the rise per foot of run can be calculated: Divide 12 by the actual run to find the proper ratio of rise. Multiply this result by the actual rise to find the rise per foot of run. For example, if the run is 14′-6″ and the rise is 5′-0″; what is the rise per foot of run? 12 ÷ 14′-6″ (14.5′) = 0.83 × 5 = 4.15. The result (4.15) is close enough to 4 that rafter calculations can be based on 4 in 12.

### GREEN NOTE

*Roof color matters. Lighter color roofs generally reflect more of the sun's energy than do darker colored roofs. Also smooth metal reflects more than does the rough surface of composition shingles. By reducing the amount of the sun's heat that is absorbed by the roof covering, the air-conditioning load of the building can be reduced.*

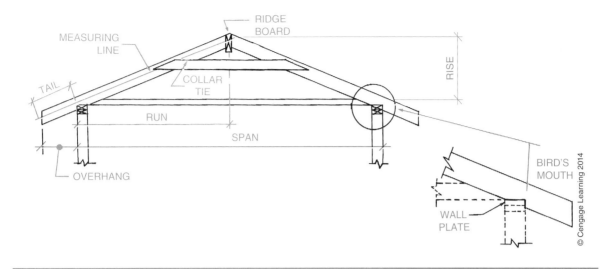

**Figure 20-5.** Common roof terms.

 **USING WHAT YOU LEARNED**

Whether the roof is famed with trusses or with site-built rafters, the same convention is used to write the slope of the roof. How would the slope of the roof shown in Figure 20–6 be written on a building elevation? The slope is shown with a symbol and numbers indicating the number of units (inches or feet) of rise for every 12 units of run. The total rise of this roof is 7'-0" in 14'-0" of run. The rise in 12 feet would be the $^{12}/_{14}$ ths of the total rise. $^{12}/_{14} \times 7 = 6$, so the slope is shown as. $\overset{12}{\diagdown}_{6}$

# Assignment

1. Give the following information for the roof shown in **Figure 20–6**:
   a. Span
   b. Run
   c. Rise
   d. Overhang
   e. Length of the rafter tails
   f. Unit rise

Refer to the Lake House drawings in your textbook packet to complete the rest of the assignment.

2. What style roof is used for most of the Lake House?
3. What is the span of the rafters over the garage?
4. What is the rise per foot of run of the rafters over the garage?
5. What is the run of the rafters over bedroom #1?
6. What is the rise per foot of run of the rafters over bedroom #1?
7. What is the overhang of the rafters over bedroom #1?

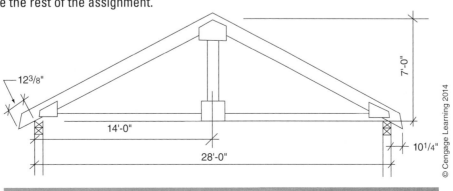

**Figure 20–6.**

# UNIT 21

# Roof Trusses

## Objectives

After completing this unit, you will be able to perform the following tasks:

○ Find information about trusses on plans, elevations, and sections.

○ Identify the appropriate truss engineering drawing for each truss in a roof frame.

○ Find key information on a truss engineering drawing.

## Truss Information on Drawings

The roof is designed to support weight, withstand weather, and give protection to the spaces below. In most roofs, the frame is made up primarily of trusses; however, some special framing situations still call for rafters. Designing trusses to withstand all the stresses that can be placed on them by wind or snow, or both, requires the skills of an engineer. Generally, the architect or designer who designs the house has some idea of what size and type of trusses will be required. The plans and elevations for the building show a lot of information about the trusses. The pitch of the roof can be seen in the building elevations (see **Figure 21–1**). Section views through the building show trusses, but the truss shown in a section might not be the truss the engineer designs. Notice that a note near the truss in **Figure 21–2** says "NYS P.E. Approved Roof Trusses @24" OC." This note indicates that the trusses will be designed by a professional engineer (P.E.) who is licensed in New York State (NYS) and that they will be spaced 24 inches from the centerline of one truss to the centerline of the next truss. You will learn in a later discussion that the actual truss specified by the engineer is not quite like the one on the building section.

## Truss Engineering Drawings

To see what the engineered trusses are, refer to the **truss drawings** from the engineer. These usually include a **delivery sheet** showing all the types of trusses used in the building (see **Figure 21–3**). This sheet usually shows the following information:

*Profile:* A small drawing showing the basic shape of the truss.
*Quantity:* Number of each type of truss needed for the building.
*Truss ID:* The engineer's letter/number designation for that truss type.
*Span:* The distance between the outside walls the truss will cover.
*Truss type:* Some trusses are referred to as *piggyback trusses*. Large trusses can be too large to transport to the construction site, so they are delivered in two parts. The top of the main truss is flat, so the piggyback truss can be applied to it after it is in place.
*Slope or pitch:* The example here shows slope or the unit rise of the roof.
*LOH:* Left overhang—the amount of the top chord that must extend beyond the building wall.
*ROH:* Right overhang—the amount of the top chord that must extend beyond the building wall.

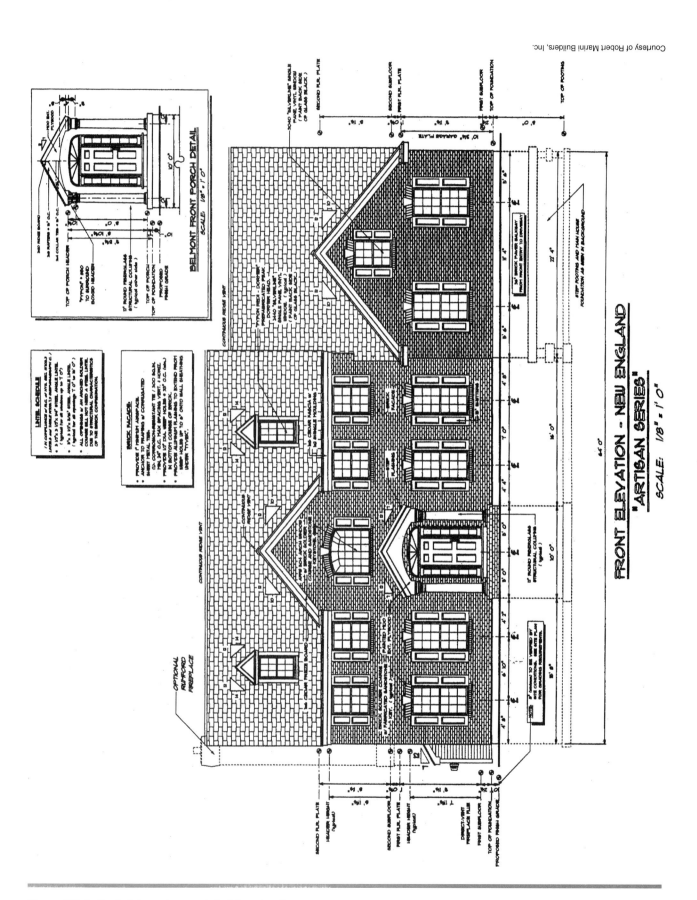

**Figure 21-1.** Roof pitch can be seen in building elevations.

Roof Trusses 143

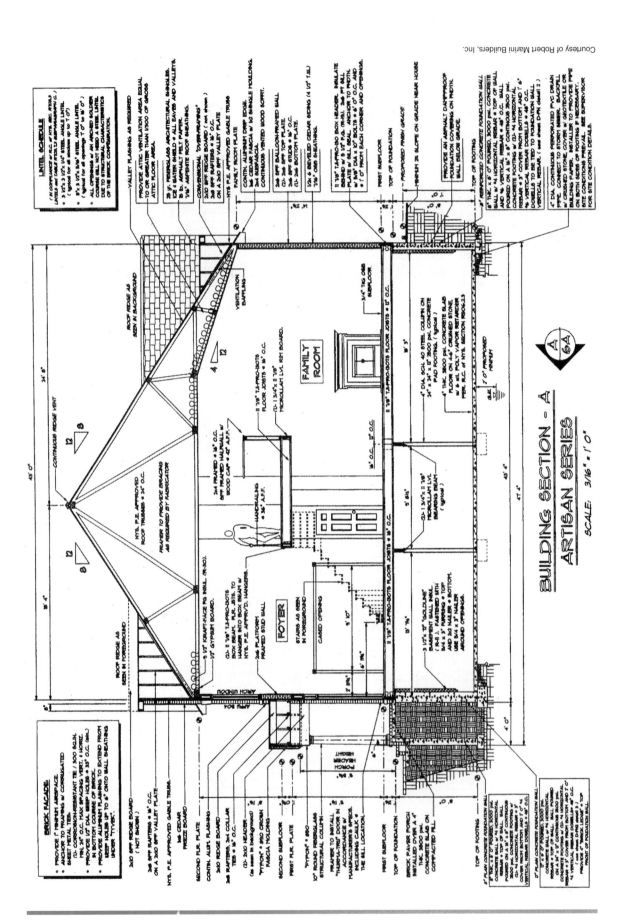

**Figure 21–2.** Building section.

| | Bellevue Builders Supply<br>500 Duanesburg Rd<br>Schenectady, NY 12306<br>Phone: (518)355-7190 Fax: (518)355-1371 | | To:<br>MARINI BLDRS. | | Delivery List | | | | |
|---|---|---|---|---|---|---|---|---|---|
| Project:<br>Model: | | Block No:<br>Lot No: | | | Job Number:<br>Page: 1<br>Date: 01-29-2004 - 3:36:11 PM<br>Project ID: A3635 | | | | |
| Contact:<br>Name:<br>Phone:<br>Fax: | Site: | Office: | Deliver To:<br>BELMONT MODEL A3635 | | Account No:<br>Designer: WTC<br>Salesperson:<br>Quote Number: | | | | |
| Tentative Delivery Date: | | | | | | | | | |

| Profile: | Qty | Truss Id: | Span: | Truss Type: | Slope | LOH | ROH | | Load By: |
|---|---|---|---|---|---|---|---|---|---|
| | 11 | T01<br>136 lbs. each | 31 - 9 - 0 | ROOF TRUSS | 8.00<br>0.00 | 0 - 0 - 0 | 0 - 10 - 8 | | |
| | 1 | T01GE<br>173 lbs. each | 31 - 9 - 0 | ROOF TRUSS<br>3 Rows Lat Brace | 8.00<br>0.00 | 0 - 0 - 0 | 0 - 10 - 8 | | |
| | 13 | T02<br>176 lbs. each | 37 - 0 - 0 | PIGGYBACK TRUSS<br>1 Row Lat Brace | 8.00<br>0.00 | 0 - 10 - 8 | 1 - 2 - 8 | | |
| | 8 | T02A<br>174 lbs. each | 37 - 0 - 0 | PIGGYBACK TRUSS<br>1 Row Lat Brace | 8.00<br>0.00 | 0 - 10 - 8 | 0 - 0 - 0 | | |
| | 1 | T02GE<br>209 lbs. each | 37 - 0 - 0 | PIGGYBACK TRUSS<br>7 Rows Lat Brace | 8.00<br>0.00 | 0 - 10 - 8 | 1 - 2 - 8 | | |
| | 1 | T02GES<br>210 lbs. each | 37 - 0 - 0 | PIGGYBACK TRUSS<br>1 Row Lat Brace | 8.00<br>0.00 | 0 - 10 - 8 | 1 - 2 - 8 | | |
| | 3 | T03<br>93 lbs. each | 22 - 0 - 0 | ROOF TRUSS | 10.00<br>0.00 | 0 - 10 - 8 | 1 - 2 - 8 | | |
| | 5 | T03A<br>90 lbs. each | 22 - 0 - 0 | ROOF TRUSS | 10.00<br>0.00 | 0 - 0 - 0 | 0 - 0 - 0 | | |
| | 1 | T03B<br>89 lbs. each | 21 - 8 - 8 | ROOF TRUSS | 10.00<br>0.00 | 0 - 0 - 0 | 0 - 0 - 0 | | |
| | 1 | T03GE<br>114 lbs. each | 22 - 0 - 0 | ROOF TRUSS | 10.00<br>0.00 | 0 - 10 - 8 | 1 - 2 - 8 | | |
| | 21 | T2PB<br>30 lbs. each | 11 - 5 - 8 | PIGGYBACK | 8.00<br>0.00 | 0 - 0 - 0 | 0 - 0 - 0 | | |
| | 2 | T2PBGE<br>34 lbs. each | 11 - 5 - 8 | PIGGYBACK | 8.00<br>0.00 | 0 - 0 - 0 | 0 - 0 - 0 | | |
| | 1 | VM01<br>63 lbs. each | 15 - 5 - 10 | ROOF TRUSS | 10.00<br>0.00 | 0 - 0 - 0 | 0 - 0 - 0 | | |

**Figure 21–3.** Truss delivery list.

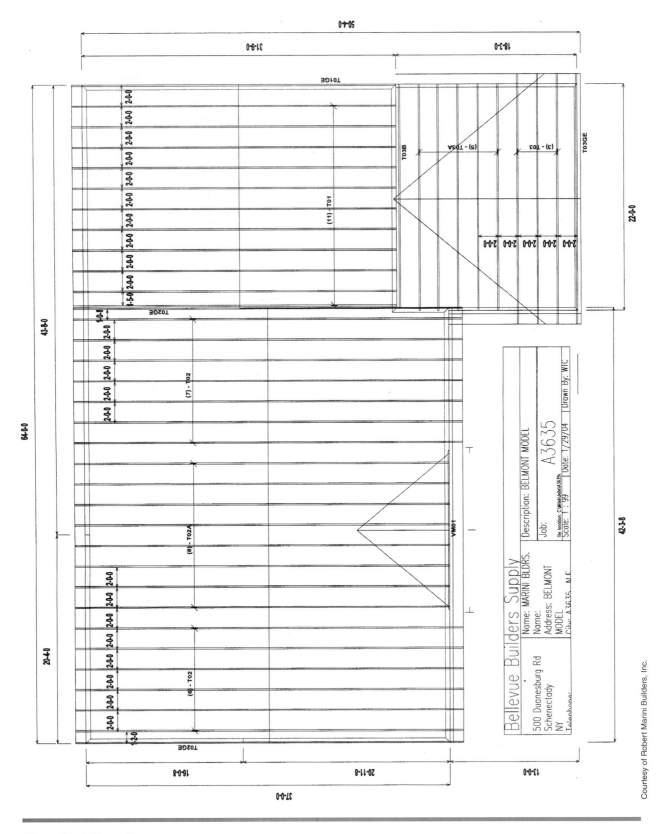

**Figure 21–4.** Truss plan.

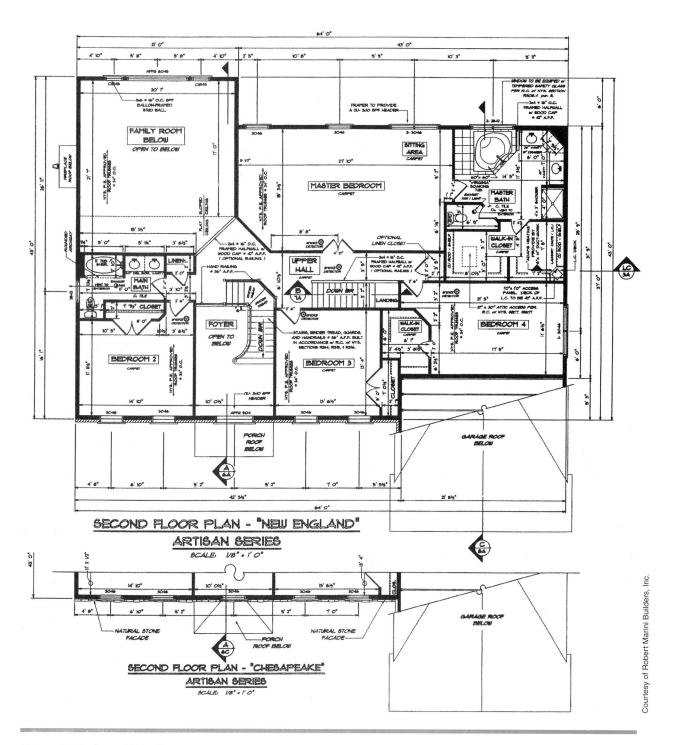

**Figure 21–5.** Second Floor Plan.

The packet of engineering drawings usually includes a truss layout plan like the one shown in **Figure 21–4**. This plan shows each truss in place on the building, with a label to indicate the truss ID. If you know from where in the building the section view in **Figure 21–2** is taken, you can find that truss on the truss layout. The section view is labeled as A/6A, so look at the second floor plan in **Figure 21–5** to see where the cut was made to take section A. The section was cut through the foyer, which is on the left side of the house. Looking again at the section, you can see the rafters for the dormer, which is above the main entrance to the house. In the truss layout shown in **Figure 21–4**, a single-line triangle indicates where the dormer will be. A notation above that triangle indicates that the eight trusses in the dormer

Roof Trusses | 147

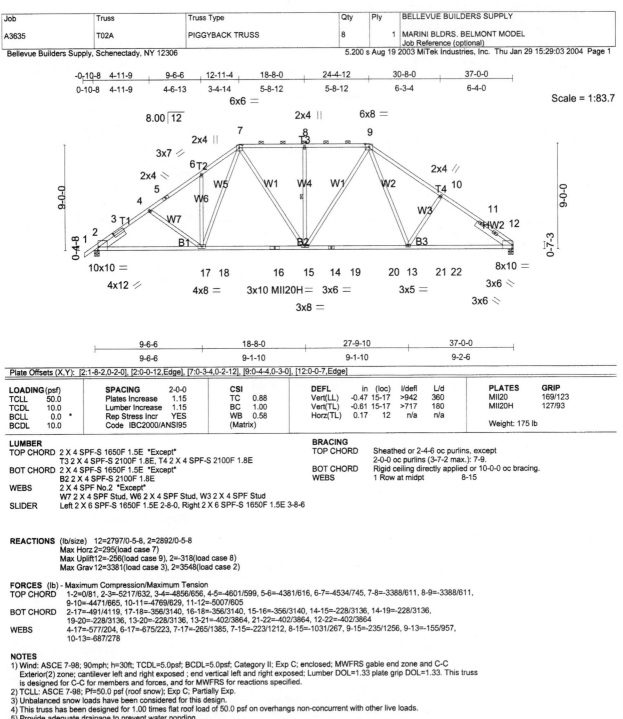

**Figure 21-6.** T02A engineering detail.

area are T02A. This is the fourth truss down from the top on the delivery sheet shown in **Figure 21–3**.

The truss engineering drawings include a detail sheet for each truss type. Truss T02A is shown in **Figure 21–6**. The truss detail includes a drawing of the truss and all the engineering data required to build the truss. Carpenters do not usually build trusses, but the same packet of engineering drawings that go to the site with the trusses is also used in the truss shop to build the trusses. It is very important for the carpenter to read all the information on the detail even though it might not all be needed to set the trusses on the building. For example, in some applications, the engineer will call for the trusses to be applied in 2-ply or 3-ply. If the trusses are plied, they are fastened together, face to face, so that all the truss members are joined in two or three layers. In this case the ply is shown as "1" in the box at the top of the detail sheet.

## Gable Ends

A roof *gable* is the triangle formed at the end of the roof by the two top chords of the truss and its bottom chord. The roof framing at a gable end must include studs, so there will be a place to attach whatever sheathing and siding are to be applied. A gable-end truss is used for this. When an overhang is needed on the gable end, it can be built as part of the gable-end truss (see **Figure 21–7**).

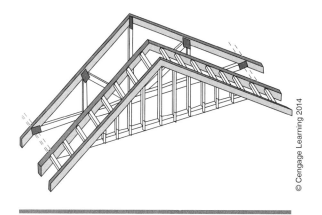

**Figure 21–7.** Gable-end truss with overhang.

### USING WHAT YOU LEARNED

Unless a building is a simple square or rectangle, there are probably several types of trusses used in its roof. That is true of the house described in this unit. It is necessary to be able to identify where each type is to be used in order to place these trusses on the building frame. For example, what truss ID is to be placed over bedroom #3 according to the drawings in this unit?

Find bedroom #3 on the floor plan in **Figure 21–5**. It is just to the left of where the garage juts out in front of the rest of the house. Now find that area on the truss plan in **Figure 21–4**. A callout on that plan tells us that there are seven trusses with ID number T02.

## Assignment

Refer to the figures in this unit as well as **Figures 21–8** and **21–9** to answer the following questions.
1. What is the slope of the garage roof?
2. What is the span of the trusses in the garage roof?
3. What are the truss IDs for the four types of trusses in the garage roof?
4. What is the difference between the two types of trusses in the garage roof?
5. Where do the T02GE trusses go?
6. Where does the T2PBGE truss go?
7. What is the on-center spacing from T02GE to the next closest truss?
8. What is the slope of the roof over bedroom #2?

Roof Trusses 149

| Job | Truss | Truss Type | Qty | Ply | BELLEVUE BUILDERS SUPPLY |
|---|---|---|---|---|---|
| A3635 | T02GE | PIGGYBACK TRUSS | 1 | 1 | MARINI BLDRS. BELMONT MODEL |
|  |  |  |  |  | Job Reference (optional) |

Bellevue Builders Supply, Schenectady, NY 12306

5.200 s Aug 19 2003 MiTek Industries, Inc. Thu Jan 29 15:29:39 2004 Page 1

Scale = 1:71.9

| LOADING (psf) | | SPACING | 2-0-0 | CSI | | DEFL | in | (loc) | l/defl | L/d | PLATES | GRIP |
|---|---|---|---|---|---|---|---|---|---|---|---|---|
| TCLL | 50.0 | Plates Increase | 1.15 | TC | 0.24 | Vert(LL) | n/a | - | n/a | 999 | MII20 | 169/123 |
| TCDL | 10.0 | Lumber Increase | 1.15 | BC | 0.09 | Vert(TL) | 0.01 | 24 | >999 | 180 | | |
| BCLL | 0.0 * | Rep Stress Incr | YES | WB | 0.39 | Horz(TL) | 0.00 | 23 | n/a | n/a | | |
| BCDL | 10.0 | Code IBC2000/ANSI95 | | (Matrix) | | | | | | | Weight: 209 lb | |

Plate Offsets (X,Y): [2:0-2-10,0-1-8], [10:0-3-0,0-0-2], [16:0-4-4,0-2-4]

**LUMBER**
TOP CHORD 2 X 4 SPF-S 1650F 1.5E
BOT CHORD 2 X 4 SPF-S 1650F 1.5E
OTHERS 2 X 4 SPF Stud *Except*
ST7 2 X 4 SPF No.2, ST7 2 X 4 SPF No.2, ST7 2 X 4 SPF No.2
ST6 2 X 4 SPF No.2, ST7 2 X 4 SPF No.2, ST7 2 X 4 SPF No.2
ST8 2 X 4 SPF No.2
WEDGE
Right: 2 X 4 SPF Stud

**BRACING**
TOP CHORD Sheathed or 6-0-0 oc purlins.
BOT CHORD Rigid ceiling directly applied or 10-0-0 oc bracing, Except:
 6-0-0 oc bracing: 29-31.
WEBS 1 Row at midpt 13-34, 12-35, 11-36, 9-37, 14-33, 15-32, 16-31

**REACTIONS** (lb/size) 2=299/37-0-0, 23=363/37-0-0, 34=280/37-0-0, 35=281/37-0-0, 36=273/37-0-0, 37=248/37-0-0, 39=279/37-0-0, 40=280/37-0-0, 41=284/37-0-0, 42=262/37-0-0, 43=338/37-0-0, 33=282/37-0-0, 32=269/37-0-0, 31=249/37-0-0, 29=284/37-0-0, 28=279/37-0-0, 27=282/37-0-0, 26=272/37-0-0, 25=305/37-0-0
Max Horz 2=279(load case 8)
Max Uplift2=-62(load case 6), 23=-35(load case 7), 34=-61(load case 7), 35=-66(load case 6), 36=-52(load case 7), 37=-18(load case 7), 39=-95(load case 8), 40=-84(load case 8), 41=-86(load case 8), 42=-83(load case 8), 43=-99(load case 8), 33=-64(load case 6), 32=-63(load case 7), 31=-7(load case 6), 29=-88(load case 9), 28=-86(load case 9), 27=-85(load case 9), 26=-85(load case 9), 25=-107(load case 9)
Max Grav 2=406(load case 2), 23=500(load case 3), 34=379(load case 2), 35=383(load case 3), 36=376(load case 3), 37=346(load case 2), 39=379(load case 2), 40=380(load case 2), 41=385(load case 2), 42=357(load case 2), 43=457(load case 2), 33=383(load case 2), 32=384(load case 2), 31=344(load case 3), 29=387(load case 3), 28=379(load case 3), 27=382(load case 3), 26=370(load case 3), 25=411(load case 3)

**FORCES** (lb) - Maximum Compression/Maximum Tension
TOP CHORD 1-2=0/77, 2-3=-259/145, 3-4=-176/132, 4-5=-147/127, 5-6=-149/111, 6-7=-76/121, 7-8=-149/125, 8-9=-150/171, 9-10=-141/170, 10-11=-52/176, 11-12=-52/176, 12-13=-52/176, 13-14=-52/176, 14-15=-52/176, 15-16=-54/175, 16-17=-158/180, 17-18=-152/115, 18-19=-58/58, 19-20=-153/48, 20-21=-152/64, 21-22=-157/68, 22-23=-228/85, 23-24=0/82
BOT CHORD 2-43=0/0, 42-43=0/0, 41-42=0/0, 40-41=0/0, 39-40=0/0, 38-39=0/0, 37-38=-0/0, 36-37=-0/0, 35-36=-0/0, 34-35=-0/0, 33-34=-0/0, 32-33=-0/0, 31-32=-0/0, 30-31=-4/0, 29-30=-54/252, 28-29=-54/252, 27-28=-54/252, 26-27=-54/252, 25-26=-54/252, 23-25=-54/252
WEBS 13-34=-339/81, 12-35=-344/86, 11-36=-335/73, 9-37=-315/33, 8-39=-342/113, 7-40=-340/104, 5-41=-343/106, 4-42=-324/102, 3-43=-394/122, 14-33=-343/84, 15-32=-343/84, 16-31=-312/23, 17-29=-350/106, 18-28=-338/106, 20-27=-342/105, 21-26=-330/102, 22-25=-372/135

**NOTES**
1) Wind: ASCE 7-98; 90mph; h=30ft; TCDL=5.0psf; BCDL=5.0psf; Category II; Exp C; enclosed; MWFRS gable end zone and C-C Exterior(2) zone; cantilever left and right exposed ; end vertical left and right exposed; Lumber DOL=1.33 plate grip DOL=1.33. This truss is designed for C-C for members and forces, and for MWFRS for reactions specified.
2) Truss designed for wind loads in the plane of the truss only. For studs exposed to wind (normal to the face), see MiTek "Standard Gable End Detail"
3) TCLL: ASCE 7-98; Pf=50.0 psf (roof snow); Exp C; Partially Exp.
4) Unbalanced snow loads have been considered for this design.
5) This truss has been designed for 1.00 times flat roof load of 50.0 psf on overhangs non-concurrent with other live loads.
6) Provide adequate drainage to prevent water ponding.
Continued on page 2

**Figure 21–8.** T02GE engineering detail.

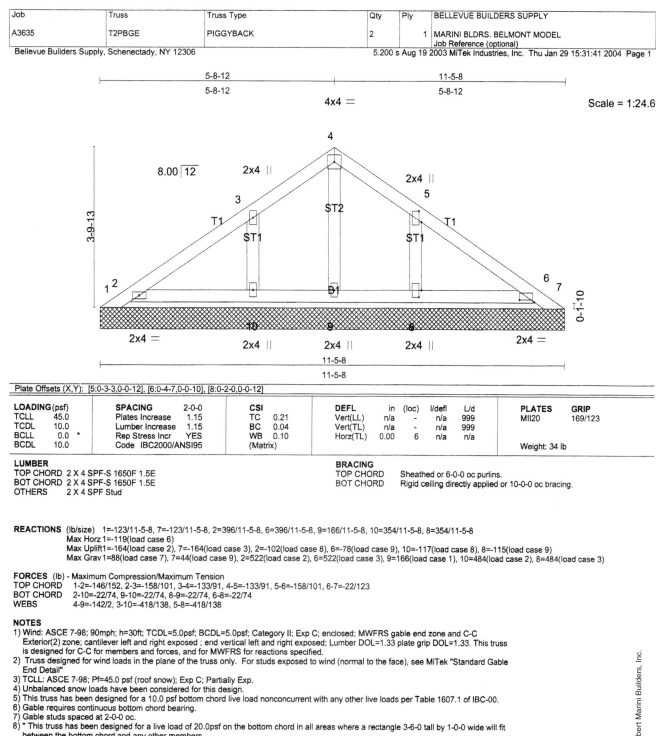

**Figure 21-9.** T2PBGE engineering detail.

# UNIT 22
# Common Rafters

## Objectives

After completing this unit, you will be able to perform the following tasks:

○ Find information about roof construction on drawings.

○ Calculate the length of common rafters.

○ Calculate the length of rafter tails when overhang is given.

## Roof Construction

In a common frame roof, the rafters, ridge board, and collar ties are the structural members. They are sized and spaced to support the weight of the roof itself plus any snow or wind that can be expected. The protection from weather is provided by sheathing (or roof decking) and roofing material (see **Figure 22–1**).

The size and spacing of the rafters vary depending on their length, pitch, and *load* (weight they must support) and are specified in Tables R802.5.1 through R802.5.8 of the *International Residential Code®*. If the rafters span a great distance or are spaced far apart, they must be deep to support their load. (The vertical dimension of rafters and joists is called *depth*). The size and spacing of the rafters are shown on a section view (see **Figure 22–2**). The ridge board is also usually shown on a section view. The ridge board should be made of stock 2 inches deeper than the rafters. The greater depth is needed because

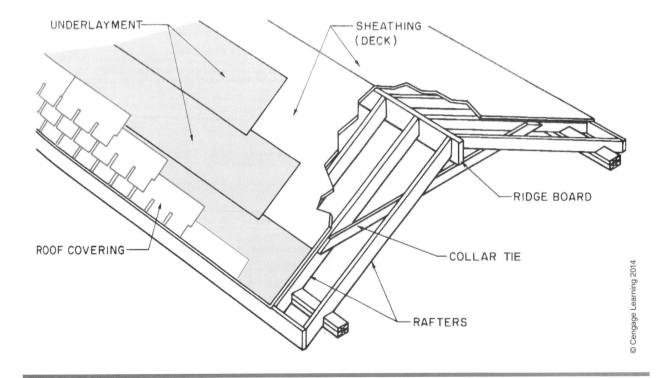

**Figure 22–1.** Elements of a roof.

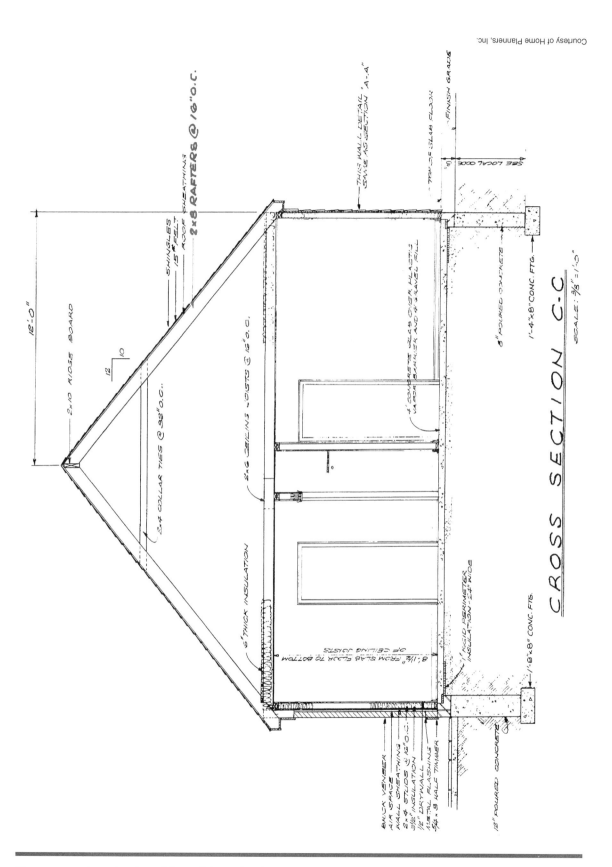

**Figure 22–2.** A section view of the roof shows the size and spacing of rafters.

Common Rafters 153

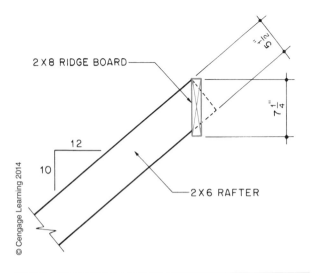

**Figure 22–3.** The ridge board must be made of wider lumber than the rafters to allow for the angled plumb cut.

### GREEN NOTE

*Raised heel rafters can be used to make more room for insulation and proper ventilation at the edge of the roof. In raised-heel-rafter construction, the ceiling joists rest on the top plate with a joist header capping their ends. A separate plate is nailed on top of the joists and header, so the bird's mouth of the rafters rests on this upper plate. In this way, the full depth of the joist is available for insulation right up to the edge of the roof.*

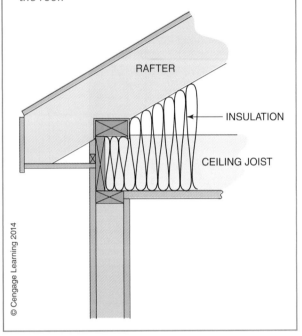

an angled cut across the rafters is longer than a square cut across the rafters (see **Figure 22–3**).

When the roof framing is complicated, a separate roof framing plan may be included (see **Figure 22–4**). Roof framing plans are used only to show the general arrangement of the framing. Therefore, unless specific dimensions are included, framing plans should not be relied upon for the lengths of the framing members.

The roof frame may also include **collar beams** or **collar ties**. Collar beams are usually made of 1-inch (nominal thickness) lumber. They are normally included on every second or third pair of rafters. This information should be shown on the roof section if collar beams are planned.

The slope of the roof may be shown on any section view that shows the rafter size. Slope is usually also shown on the building elevations.

## Roof Covering

The most common roof decking materials for houses are plywood and *oriented-strand board (OSB)*. However, in post-and-beam construction, where the extra strength is needed to span the distance between the rafters, dimensional lumber is used for the roof deck.

The material used to provide weather protection is often chosen for its architectural style (see **Figure 22–5**). Some of the most common materials are asphalt shingles, wood shingles, and terra-cotta or metal tiles. The material to be used is shown on the section view of the roof and usually on the building elevations. The roofing should be applied over a layer of asphalt-saturated building paper—sometimes called *slater's felt*.

Asphalt roofing materials, including felt and shingles, are sold by their weight per hundred square feet of coverage. One hundred square feet of roof is called a **square**. If enough shingles to cover one square weigh 235 pounds, they are called 235-lb shingles. The felt used under roofing is typically 15-lb or 30-lb weight.

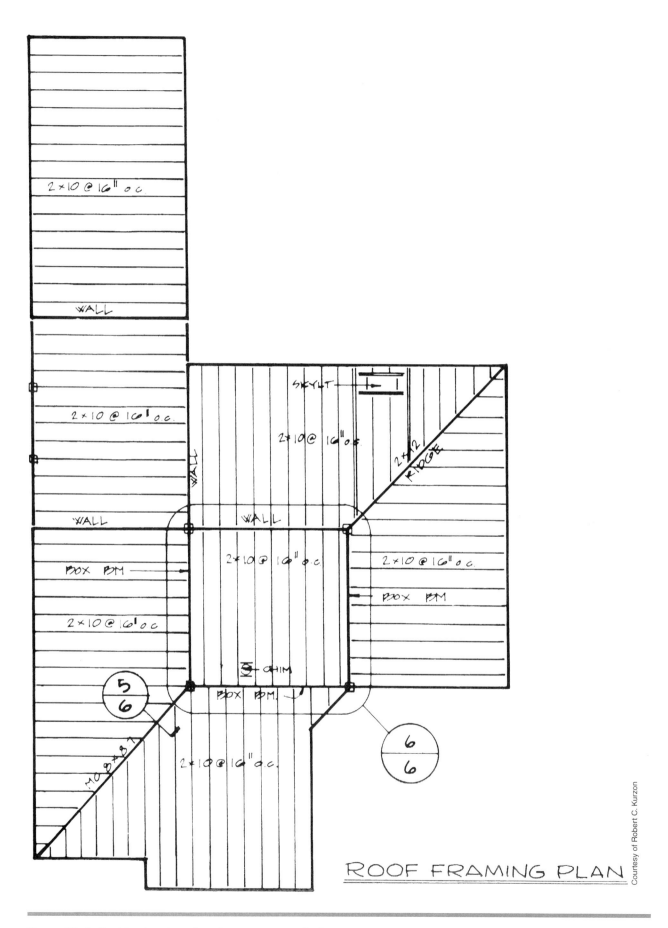

**Figure 22–4.** Roof framing plans show the arrangement of rafters.

Common Rafters

Figure 22–5. Tile roofs are popular in some parts of the country.

## Finding the Length of Common Rafters

Carpenters use a rafter table to find the length of rafters. These tables are available in handbooks and are printed on the face of a framing square (see **Figure 22–6**). To find the length of a **common rafter**, you must know the run of the rafter and its rise per foot of run.

The run can be found on the floor plan of the building. The rise per foot of run is shown on the building elevations. The length of the common rafter is then found by following these steps:

Step 1. Find the number of inches of rise per foot of run at the top of the table. These numbers are the regular graduations on the square.

Step 2. Under this number, find the length of the rafter per foot of run. A space between the numbers indicates a decimal point.

Step 3. Multiply the length of the common rafter per foot of run (the number found in Step 2) by the number of feet of run.

Step 4. Add the length of the tail, and subtract one-half the thickness of the ridge board. The result is the length of the common rafter as measured along the measuring line.

**Note:** If the overhang is given on the working drawings, it can be added to the run of the rafter instead of adding the length of the tail.

## Shed Roof Framing

Most of the Lake House has shed roof construction. That is, the common rafters cover the full span of the area they cover in a single slope. To find the length of shed rafters, treat the entire span of the rafters as run. For example, the total width of the garage is 13'-9". There is no overhang on either side of the garage, so the span of the garage rafters is 13'-9". Use 13'-9",

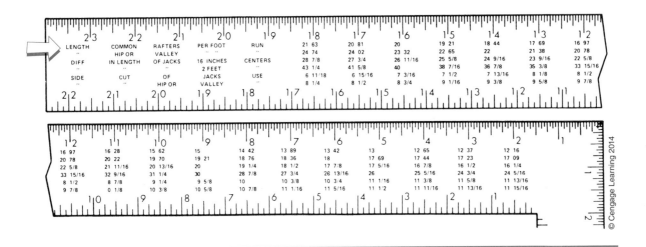

Figure 22–6. Rafter table on the face of a square: The top line is the length of common rafters per foot of run.

or 13.75 feet, as the run in calculating the length of common rafters. (See Math Review 20.)

Where shed rafters butt against a wall or other vertical surface, the drawings should include a detail to show how they are fastened. **Figure 22–7** shows three methods of fastening rafters to a vertical surface.

## Roof Openings

It is often necessary to frame openings in the roof. Where chimneys, skylights, or other features require openings through the rafters; headers and double framing members are used. This method of framing openings is similar to that used in floor framing.

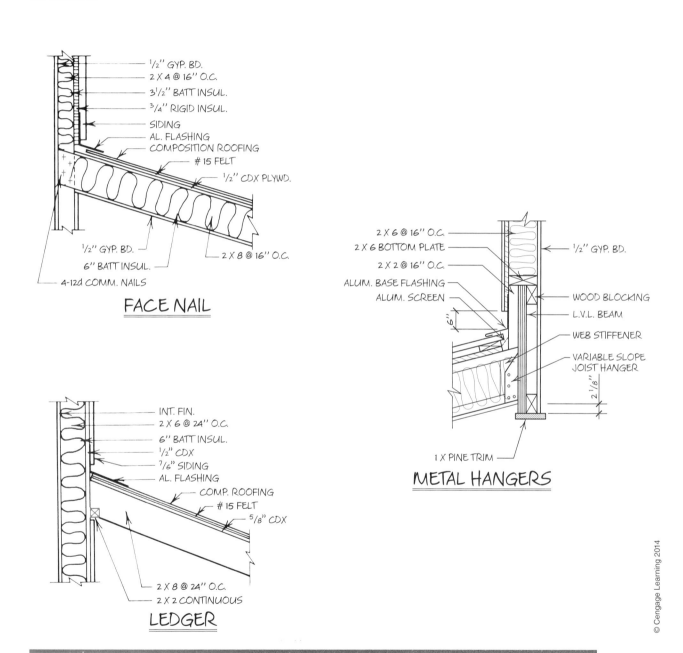

**Figure 22–7.** Details of shed roof to wall.

Common Rafters 157

### USING WHAT YOU LEARNED

Although roof trusses have become the most widely used method for framing roofs of light frame buildings, such as homes, carpenters will very often be called on to layout and use common rafters. Although the length of a rafter can be found with a construction calculator, calculators are not always available on the job site. You are encouraged to learn to use the rafter tables on a framing square. Find the length of a common rafter for the roof in **Figure 22–8**.

1. Rise per foot of run = 4″.
2. Length of common rafter per foot of run = 12.65″.
3. Run of one rafter including overhang = 16′-0″.
4. 16 × 12.65″ = 202.40″ (round off to 202½″).
5. Subtract one-half the thickness of the ridge board: 202½″ − ¾″ = 201¾″.

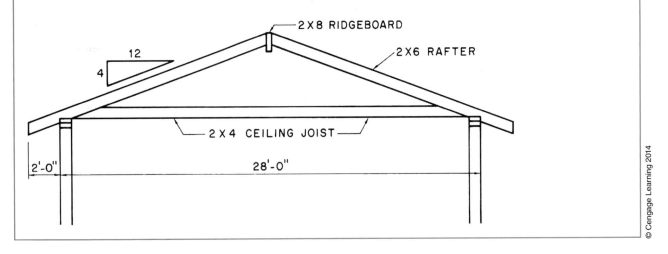

**Figure 22–8.** Find the length of a common rafter.

## Assignment

Give the following information for the rafters of the Lake House:

| Rafter Location | Thickness × Depth | Run | Rise per Foot Run | Length | O.C. Spacing |
|---|---|---|---|---|---|
| 1. Kitchen | | | | | |
| 2. Bedroom #2 and Loft | | | | | |
| 3. Living room | | | | | |
| 4. Bedroom #1 | | | | | |
| 5. Garage | | | | | |

# Hip and Valley Framing

## UNIT 23

## Hip Rafters

**Hip rafters** run from the corner of the building to the ridge at a 45° angle (see **Figure 23–1**). The length of hip rafters can be found by using a table found on most framing squares (see **Figure 23–2**). The second line of this table is used to calculate the length of hip rafters. This table is based on the unit-run-and-rise method for finding the length of common rafters, explained in Unit 22.

To calculate the length of a hip rafter, you must know the run of the common rafters in the roof and the unit rise of the roof (see **Figure 23–3**). The length of the hip rafters is then found by using the table for length of hip and valley rafters in the same way the table for the length of common rafters is used in Unit 22.

Step 1. Find the unit rise (number of inches of rise per foot of run) at the top of the table. These numbers are the regular graduations on the square.

Step 2. Under this number, find the length of the hip rafter per foot of run of the common rafters.

Step 3. Multiply the length of the hip rafter per foot of common-rafter run (the number found in Step 2) by the number of feet of run of the common rafters (one-half the width of the building).

Step 4. Subtract the ridge allowance. Because the hip rafter meets the ridge board at a 45° angle, the ridge allowance is one-half the 45° *thickness* of the ridge board (see **Figure 23–4**). The 45° *thickness* is the length of a 45° line across the thickness of the ridge board. The 45° thickness of a 1½-inch (2-inch nominal) ridge board is 2⅛ inches. Therefore, the ridge allowance for a hip rafter on a 1½-inch ridge board is 1/16 inch.

**Note:** If the hip rafter includes an overhang, add the overhang of the common rafters to the run of the common rafters.

## Objectives

After completing this unit, you will be able to perform the following tasks:

- Calculate the length of hip rafters.
- Calculate the length of valley rafters.
- Calculate the length of hip and valley jack rafters.

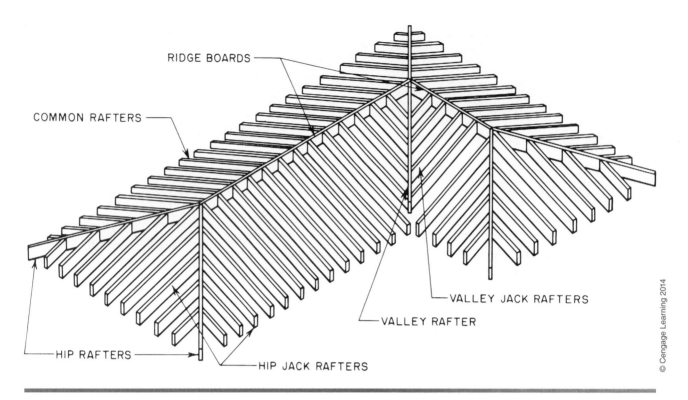

**Figure 23–1.** Parts of a hip and valley roof frame.

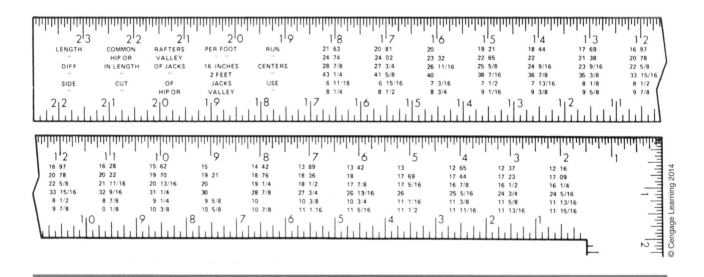

**Figure 23–2.** Rafter table on the face of a square.

**Example:** Find the length of a hip rafter for the roof shown in **Figure 23–5.**

1. Rise per foot of run = 4″.
2. Length of hip rafter per foot of common-rafter run = 17.44″.
3. Common-rafter run including overhang = 16′-0″.
4. 16 × 17.44″ = 279.04″ (round off to 279 1/16″).
5. Subtract 3/4″ ridge allowance: 279 1/16″ − 3/4″ = 278 5/16″

160 UNIT 23

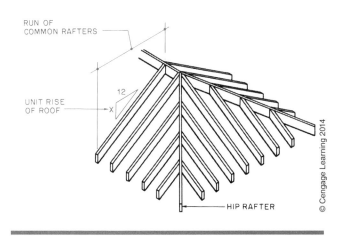

**Figure 23-3.** To use the table for the length of hip and valley rafters, use the run of the common rafters in the roof.

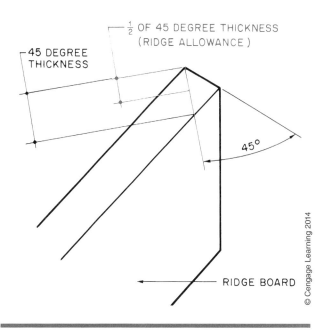

**Figure 23-4.** Use one-half of the 45° thickness of the ridge as a ridge allowance for hip rafters.

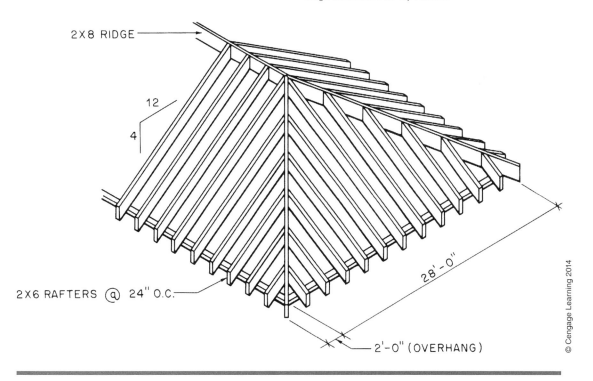

**Figure 23-5.** Find the length of a hip rafter.

The rafters that butt against the hip rafter are called hip **jack rafters**; they are cut at an angle (see **Figure 23-6**). This angled cut produces a surface that is longer than the width of the lumber from which the rafter is cut. Therefore, the hip rafters are sometimes made of wider lumber than the common rafters and jack rafters.

## Valley Rafters

The line where two pitched roofs meet is called a **valley**. The rafter that follows the valley is a **valley rafter**. It is most common for both roofs to have the same pitch. This results in the valley rafter being at a 45° angle with both ridges—the same angle as a hip rafter. Because

Hip and Valley Framing **161**

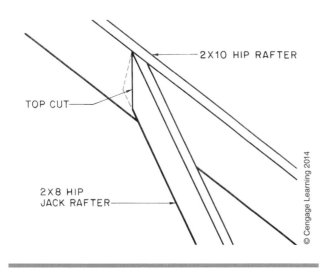

**Figure 23-6.** Because the top of the hip jack is cut on an angle, the hip rafter must be wider.

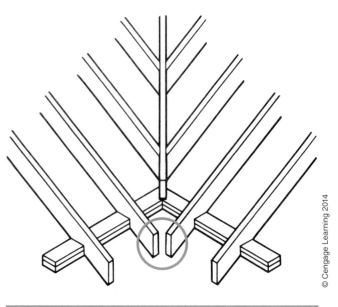

**Figure 23-7.** On some roofs the common rafter tails are close enough so no tail is needed on the valley rafters.

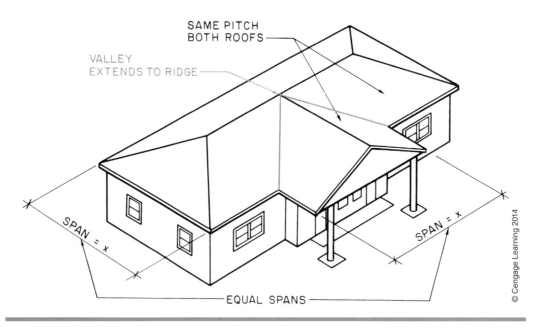

**Figure 23-8.** When both spans are the same distance and the same pitch, the valley goes all the way to the ridge.

the angles of the hip and valley rafters are the same and the pitch is the same, the same table can be used to compute their lengths.

All steps of the procedure given earlier for hip rafters can be followed to find the length of valley rafters. However, the valley rafters often have no tail even though the roof has an overhang (see **Figure 23-7**). In this case, the total length of the valley rafters is computed on the basis of the run of the common rafters excluding the overhang.

When both roofs have the same span and rise, the valley extends from the eave to the ridge (see **Figure 23-8**). When one roof has a greater span than the other, the valley does not reach the ridge (see **Figure 23-9**).

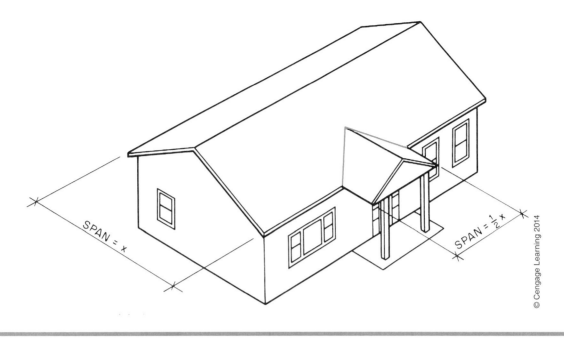

**Figure 23–9.** When the spans are not equal, the valley does not reach the ridge.

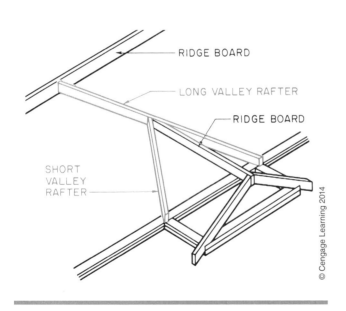

**Figure 23–10.** Framing valleys with long and short valley rafters.

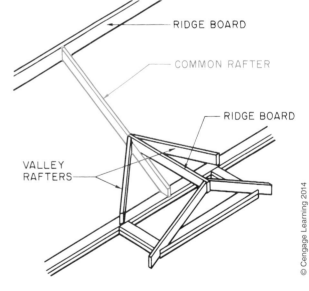

**Figure 23–11.** A valley framed against a common rafter.

In this case, the valley is framed in one of two ways. One valley rafter can extend to the ridge, and the other can butt against the first (see **Figure 23–10**). The other method is to install a common rafter on the wider roof in line with the ridge of the narrower roof. Both valley rafters can then butt against this common rafter (see **Figure 23–11**).

The length of the valley rafters is based on the run of the common rafters that have their upper (ridge) ends at the same level as the valley rafter. In other words, when the long and short valley rafters are used, the length of the long valley rafter is based on the run of the common rafters in the wider roof. The length of the short valley rafter is based on the run of the common

Hip and Valley Framing 163

rafters in the narrower roof. When both valley rafters butt against a common rafter of a higher roof, their lengths are based on the run of the common rafters in the narrower roof.

### GREEN NOTE

*The most important function of a roof is to prevent water from entering the structure. One area where leaks sometimes occur is in valleys. If the valley flashing is not wide enough for the pitch of the roof wind-blown water may be forced under the roof covering and run beneath the flashing. Most roofers also apply a self-adhering ice and water barrier to the valley area before installing the flashing. The ice and water barrier provides excellent additional protection against leaks in the valley.*

## Jack Rafters

Rafters that extend from the wall plate to a hip rafter are called *hip jack rafters,* as shown in **Figure 23–1**. Those that extend from a valley rafter to the ridge board are called *valley jack rafters,* also shown in **Figure 23–1**.

The third and fourth lines of the rafter table on most framing squares, shown in **Figure 23–2**, are used to calculate the length of jack rafters. The length of each jack rafter in a roof varies from the length of the one next to it by the same amount (see **Figure 23–12**). The amount of this variance depends on the spacing of the rafters and the pitch of the roof. The third line of the rafter table is used when the rafters are spaced 16 inches O.C.; the fourth line is used when they are spaced 24 inches O.C. As with the other lines of the rafter table, the inch numerals at the top of the square are used to indicate the unit rise of the roof. For example, if a roof has a 6 in 12 slope and the roof framing is 16 inches O.C., the difference in the length of jack rafters is $17^{7}/_{8}$ inches.

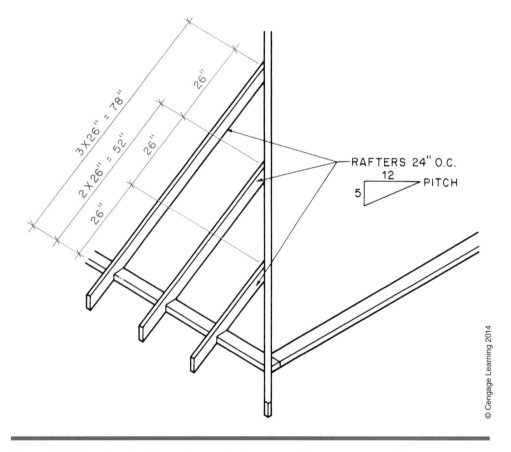

**Figure 23–12.** Each jack rafter in a string varies from the next by the same amount.

The first jack rafter should be a full space (16 inches or 24 inches) from the bottom of the hip or the top of the valley. Therefore, the first jack rafter should be the length shown on the rafter table.

The length of the tail must be added to the theoretical length from the table. The length of the tail on hip jacks is the same as the length of the tail on common rafters. Therefore, the length of the tail can be found by using the table for the length of common rafters. Simply treat the tail as a very short common rafter. If the overhang is not in even feet, divide the length of common rafter per foot of run on the table by 12. This gives you the length of the common rafter (or jack rafter tail) per inch of run.

**Example:** Find the length of the tails of the jack rafters in **Figure 23–13**.

1. Unit rise (rise per foot of run) = 4″.
2. Length of common rafter per foot of run = 12.65″.
3. Length of common rafter per inch of run = 12.65″/12 = 1.05″.
4. Run of rafter tail (overhang) = 8″.
5. Length of rafter tail = 8 × 1.05″ = 8.4″, approximately $8^{7}/_{16}$″.

Hip jack rafters are also shortened at the top to allow for the thickness of the hip rafter they butt against. This allowance is one-half the 45° thickness of the hip rafter. Valley jack rafters are shortened at the bottom to allow for the thickness of the valley rafter. This allowance is one-half the 45° thickness of the valley rafter. The valley jack rafters are also shortened at the top to allow for the thickness of the ridge board. The ridge board allowance for valley jacks is the same as the ridge board allowance for common rafters—one-half the actual thickness of the ridge board.

**Example:** Find the length of the hip jack rafters (A, B, and C) and the valley jack rafters (D, E, and F) in **Figure 23–13**.

**Hip Jack Rafters**
1. Rise per foot of run = 4″.
2. Spacing of rafters = 24″ O.C.
3. Difference in the length of jacks = $25^{5}/_{16}$″.

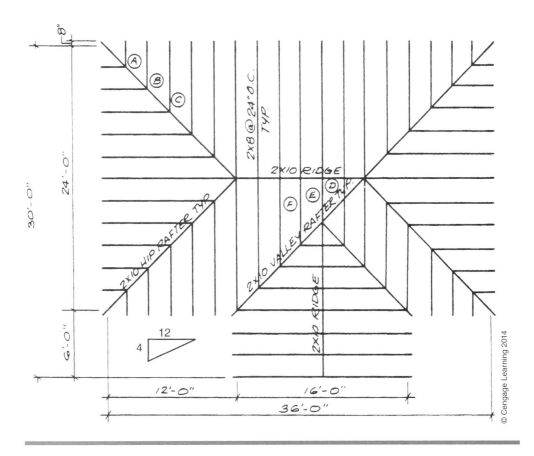

**Figure 23–13.** Find the length of the tails of jack rafters, hip jack rafters, and valley jack rafters.

4. Theoretical length of hip jack rafter A = 0 + 25⁵⁄₁₆″ = 25⁵⁄₁₆″.
5. Add tail as found in earlier example: 25⁵⁄₁₆″ + 8⁷⁄₁₆″ = 33³⁄₄″.
6. Subtract one-half the 45° thickness of the hip rafter: 33³⁄₄″-11/16″ = 32¹⁄₁₆″ (actual length of A).
7. Actual length of hip jack B = 32¹¹⁄₁₆″ (length of A) + 25⁵⁄₁₆″ (from rafter table) = 58″.
8. Actual length of hip jack C = 58″ (length of B) + 25⁵⁄₁₆″ = 83⁵⁄₁₆″.

**Valley Jack Rafters**
1. Theoretical length of valley jack rafter D = 0 + 25⁵⁄₁₆″ = 25⁵⁄₁₆″.
2. Subtract one-half the 45° thickness of the valley rafter: 25⁵⁄₁₆″-1¹⁄₁₆″ = 24¼″.
3. Subtract one-half the actual thickness of the ridge board: 24¼″-¾″ = 23½″ (actual length of D).
4. Actual length of valley jack E = 23½″ (length of D) + 25⁵⁄₁₆″ (from rafter table) = 48¹³⁄₁₆″.
5. Actual length of valley jack F = 48¹³⁄₁₆″ (length of E) + 25⁵⁄₁₆″ = 74⅛″.

## USING WHAT YOU LEARNED

Hip roofs are quite common and as is true with common rafters, it is sometimes necessary for a carpenter to be able to layout the rafters for a hip roof. In this practice exercise, we find the length of the long valley rafter in **Figure 23–14** using the rafter tables on a framing square.

A valley rafter is the same length as a hip rafter on the same roof.

1. The rise per foot of run is 6″.
2. The length of a hip (or valley) rafter per foot of run at this slope is 18″.
3. The common-rafter run of this roof is 12′-0″.
4. 12 × 18″ = 216″.
5. Subtract half the 45° thickness of the ridge board or 1¹⁄₁₆″. 216″-1¹⁄₁₆″ = 214¹⁵⁄₁₆″ or 17′-10¹⁵⁄₁₆″.

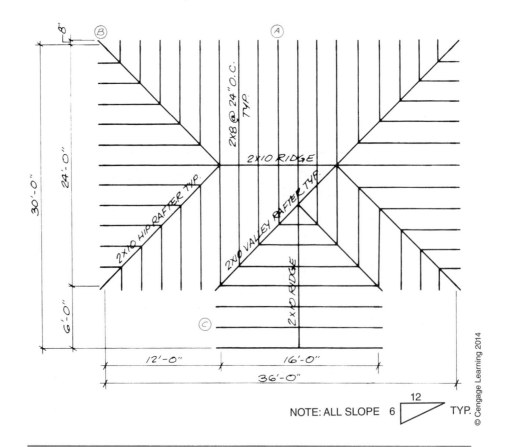

**Figure 23–14.**

## Assignment

A. Refer to **Figure 23–14** to complete questions 1–11.
1. What is the run of the common rafters at A?

*Note:* Do not include the overhang.

2. How much overhang does the roof have?
3. What is the actual length of the common rafters at A?
4. What is the actual length of the hip rafter at B?
5. What is the run of the common rafters at C?
6. What is the actual length of the common rafters at C?
7. What is the length of the short valley rafter?
8. What is the actual length of the shortest hip jack rafter?
9. What is the actual length of the second shortest hip jack rafter?
10. What is the actual length of the shortest valley jack rafter?
11. What is the actual length of the second shortest valley jack rafter?

B. Refer to the Lake House drawings (in the packet) to complete this part of the assignment.

12. What is the length of the structural steel hip rafter over the dining room?

*Note:*
- This hip rafter is a steel channel shown on Roof Framing Plan 2/6 and Details 4/6 and 5/6, and marked as MC8 $\times$ 8.7
- Remember to allow for the distance from the column centerline and the end of the rafter as dimensioned on the detail drawing.
- This roof has an unusual pitch of 2.96 in 12. This is close enough to use 3 in 12 for calculating rafter lengths.

# UNIT 24 Cornices

## Objectives

After completing this unit, you will be able to perform the following tasks:

- Describe the cornice construction shown on a set of drawings.
- List the sizes of the individual parts of the cornice shown on a set of drawings.
- Describe the provisions for attic or roof ventilation as shown on a set of drawings.

## Types of Cornices

The *cornice* is the construction at the place where the edge of the roof joins the sidewall of the building. On hip roofs, the cornice is similar on all four sides of the building. On gable and shed roofs, the cornice follows the pitch of the end (rake) rafters. The cornice on the ends of a gable or shed roof is sometimes called simply the *rake* (see **Figure 24–1**). The three main types of cornice are the *box cornice,* the *open cornice,* and the *close cornice.*

### Box Cornice

The box cornice boxes the rafter tails. This type of cornice includes a fascia and soffit (see **Figure 24–2**). The fascia covers the ends of the rafter tails. The soffit covers the underside of the rafter tails. There are three types of box cornices. These types vary in the way the soffit is applied.

**Sloping Box Cornice.** In the sloping box cornice, the soffit is nailed directly to the bottom edge of the rafter tails. This causes the soffit to have the same slope or pitch as the rafter, (see **Figure 24–3**).

**Narrow Box Cornice.** In the narrow box cornice, the rafter tails are cut level. The soffit is nailed to this level-cut surface (see **Figure 24–4**).

**Wide Box Cornice.** In a wide box cornice, the overhang is too wide for a level cut on the rafter tails to hold the full width of the soffit. In conventional wood framing, *lookouts* are installed between the rafter ends and the sidewall. The lookouts provide a nailing surface for the soffit (see **Figure 24–5**). For a metal soffit, special metal channels fastened to the sidewall and the back of the fascia hold the soffit (see **Figure 24–6**).

### Open Cornice

In an open cornice, the underside of the rafters is left exposed (see **Figure 24–7**). Blocking is installed between the rafters and above the wall plate to seal the cornice from the weather. An open cornice may or may not include a fascia.

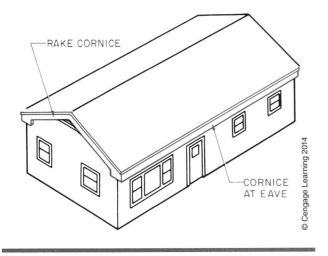

**Figure 24–1.** The cornice is the construction at the place where the roof and the sidewall meet.

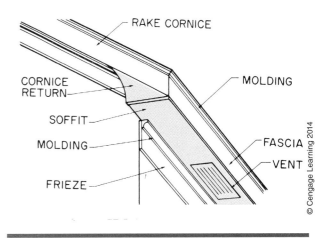

**Figure 24–2.** Parts of a box cornice.

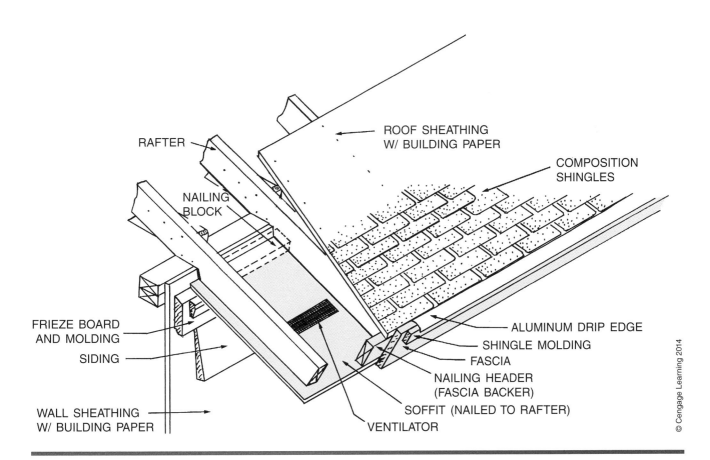

**Figure 24–3.** Sloping box cornice.

Cornices 169

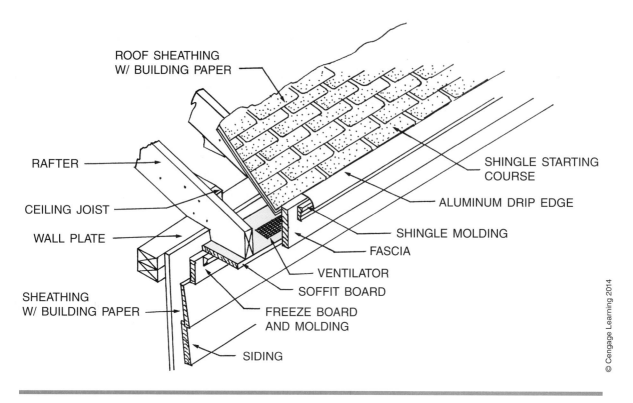

**Figure 24–4.** Narrow box cornice.

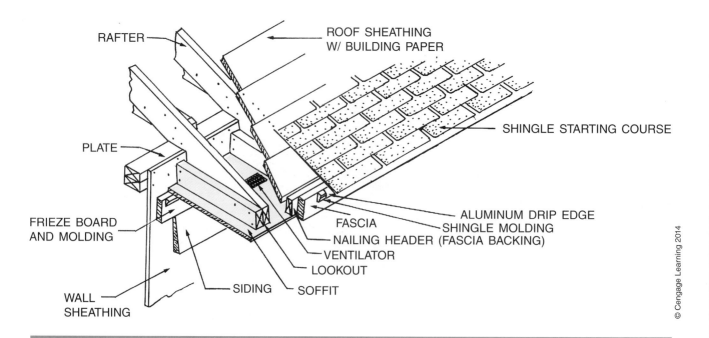

**Figure 24–5.** Wide box cornice with horizontal lookouts.

170 UNIT 24

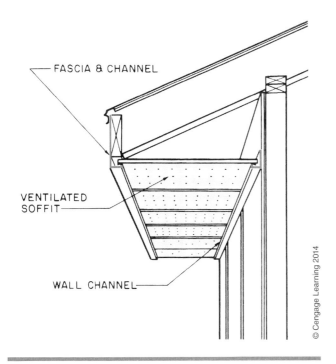

**Figure 24-6.** Metal and vinyl soffit systems include ventilated soffits and channels.

### Close Cornice

In a close cornice, the rafters do not overhang beyond the sidewall (see **Figure 24-8**). The interior may be sealed by the sheathing and siding or by a fascia. In either case, there must be some provision for ventilation.

### Cornice Returns

Any type of construction described for cornices can be used for the rake. When a sloping cornice is used, the fascia and soffit follow the line of the roof up the rake. When a level box cornice is used, a **cornice return** is necessary. This is the construction that joins the level soffit and fascia of the eave with the sloping rake (see **Figure 24-9**).

The style of cornice return is shown on the building elevations (see **Figure 24-10**). Although good architectural drafting practice requires details of all special construction, many architects do not include details of cornice returns. The carpenter is expected to know how to achieve the desired results. **Figures 24-11** and **24-12** show the construction of two popular types of cornice returns.

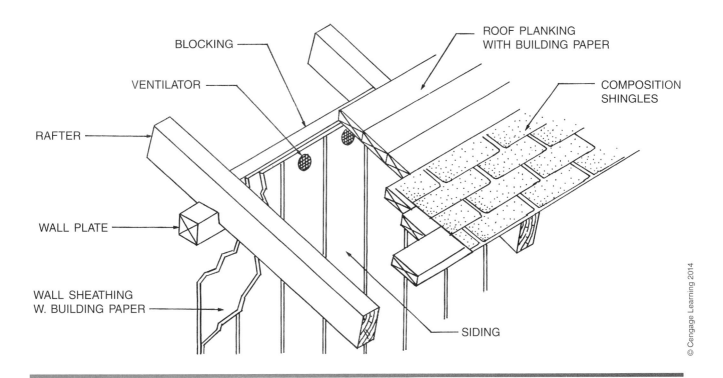

**Figure 24-7.** Open cornice.

Cornices 171

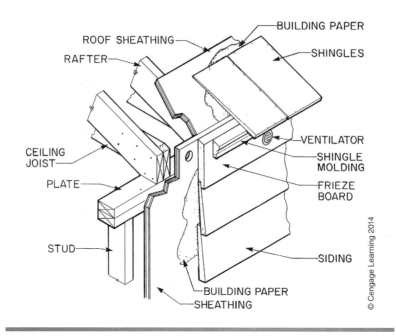

**Figure 24–8.** Close cornice.

**Figure 24–9.** Cornice return.

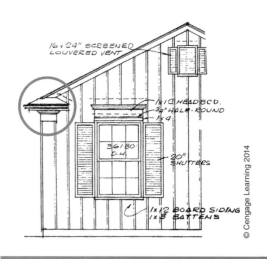

**Figure 24–10.** The cornice returns can be seen on the building elevations.

172 UNIT 24

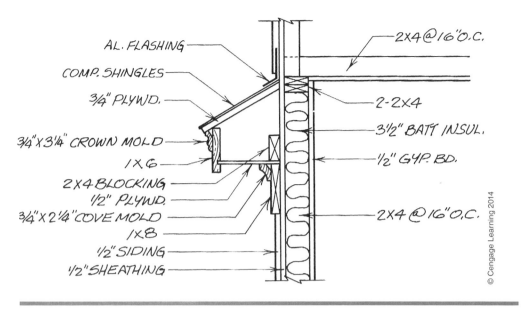

**Figure 24–11.** Section through cornice return shown in Figure 24–10.

## Ventilation

**GREEN NOTE**

*In some areas, especially where houses do not have basements, it is common to locate furnaces and water heaters in the attic. Combustion equipment, such as gas-fired furnaces and water heaters should not be located in a conditioned attic unless it is in a separate space that is sealed off from the rest of the conditioned spaces of the house. That is an attic where the insulation is placed against the roof instead of the floor. If backdrafting occurs, carbon monoxide can escape into the surrounding space. In addition to the danger posed by carbon monoxide, backdrafting can cause water vapor from burning fuel to remain in the attic and increase the chances of condensation forming in the attic.*

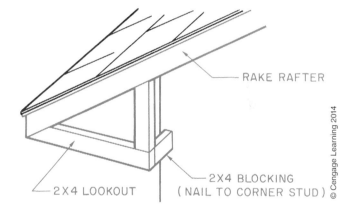

**Figure 24–12.** Framing for cornice return shown in Figure 24–9.

The cornice must allow for ventilation of the attic or roof. Attic or roof ventilation is necessary in both hot and cold weather. Without ventilation, the air in the attic becomes stagnant because it is trapped and unable to circulate.

In hot weather, this stagnant air builds up heat and makes the house warmer. Hot, stagnant air can hold a large amount of moisture. When this moisture-laden air comes in contact with the cooler roof, the moisture condenses. The condensation can reduce the effectiveness of the insulation. Condensation can also cause the wood in the attic to rot.

Attic or roof ventilation also helps prevent ice buildup in cold climates. Without ventilation, the heat from the building melts the snow that falls on the roof. As the melted snow reaches the overhang of the roof, it refreezes. Eventually an ice dam may build up. The ice dam can back up newly melted snow, causing it to seep under the shingles (see **Figure 24–13**).

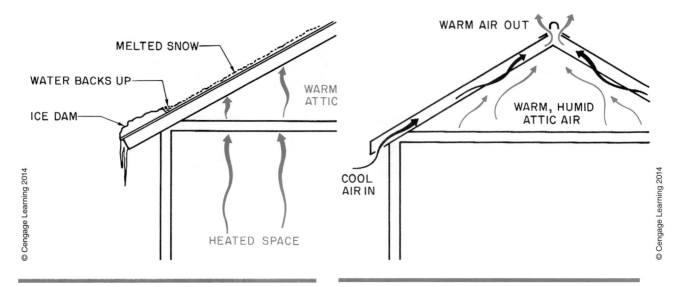

**Figure 24-13.** Ice dam.

**Figure 24-14.** Airflow through the attic.

Ventilation is created by allowing cool air to enter through the cornice and exit at the ridge or through special ventilators (see **Figure 24-14**). A section view of the sidewalls or a special roof and cornice detail shows the construction of the cornice including any ventilators. Notice that each of the soffits in **Figures 24-2, 24-4,** and **24-5** includes a ventilator.

Using TJIs—I-joists—for rafters allows for greater rafter lengths and conserves natural lumber resources, but such framing requires construction details that are not normally required with solid lumber. **Figure 24-15** shows the necessary provisions for roof framing with TJIs. Notice that many of these details include a V-cut to allow air flow.

The heated air can be allowed to exit in one of three ways. Some buildings have a ventilated, metal ridge cap. This may be shown on the building elevations or on a separate detail. When a ventilated ridge is used, an opening is left in the roof decking at the ridge. Metal roof ventilators can be installed in the roof. Ventilators can be installed in the gables. Metal roof ventilators and gable ventilators are usually shown on the building elevations only.

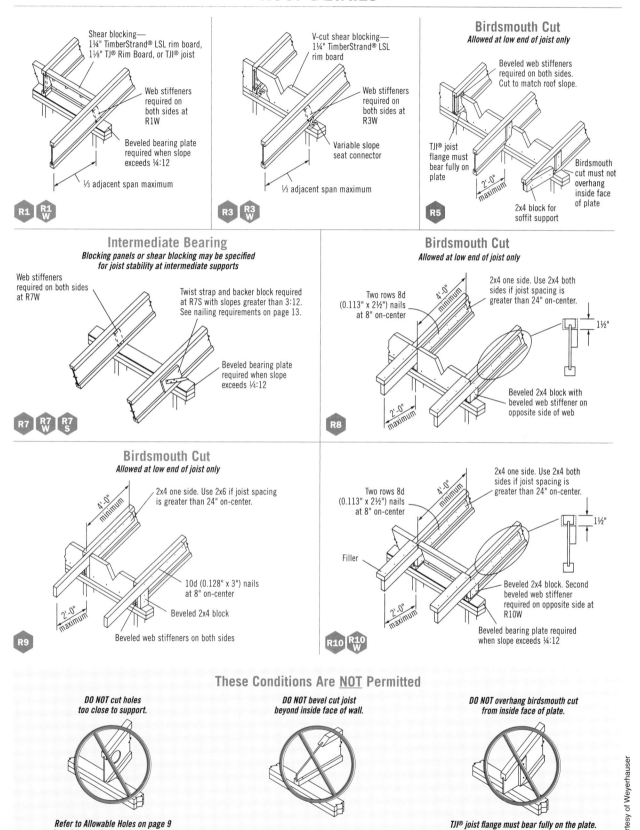

**Figure 24–15.** Roof details for framing with TJIs.

### USING WHAT YOU LEARNED

It is important to allow ventilating air to flow along the bottom of the roof from the cornice to the upper ends of the rafters. In the Lake House, what prevents birds and rodents from entering with the air?

The typical cornice construction is shown in Roof Detail 1/7. That drawing shows a continuous vent above the OSB sheathing and outside the shear blocking. This continuous vent has either screening or small enough openings so that only air can enter.

## Assignment

Refer to the Two-unit Apartment and the Lake House drawings in your textbook packet to complete the assignment.

1. Which type of cornice does the Apartment have?
2. What material is used for the Apartment cornice?
3. How wide is the Apartment soffit?
4. The Apartment fascia is made of two parts. What are they?
5. What provision does the Apartment cornice have for ventilation?
6. How does attic air exit from the Apartment?
7. Sketch the Lake House cornice and show where air enters for ventilation.
8. There are two ways that air can escape from the Lake House roof. Describe one.

# UNIT 25

# Windows and Doors

## Objectives

After completing this unit, you will be able to perform the following tasks:

- Interpret information shown on window and door details.
- Find information in window and door manufacturers' catalogs.

## Window Construction

Most windows are supplied by manufacturers as a completely assembled unit. However, the carpenters who install windows often have to refer to window details for information. Some special installations require knowledge of the construction of the window unit. Also, when a special window is required, the carpenter may build parts of it on the construction site.

## Wood Windows

The major types of windows are briefly discussed in Unit 19. All these windows include a frame and sash. The **sash** is the glass and the wood (or metal) that holds the glass. The sash is made of **rails** (horizontal parts) and **stiles** (vertical parts) (see **Figure 25–1**). The sash may also include muntins. **Muntins** are small strips that divide the glass into smaller panes. The glass is sometimes called the *lite*.

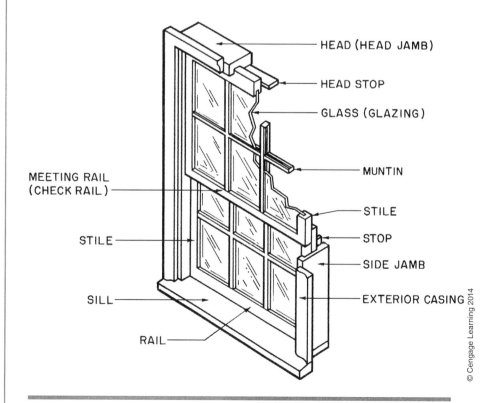

**Figure 25–1.** Parts of a window.

The window frame is made of the *side jambs,* the *head jamb,* and the sill. Stop molding is applied to the inside of the jambs to hold the sash in place. Factory-built windows also come with the exterior casing installed. The casing is the molding that goes against the wall around the frame. The interior casing and the apron, if one is included, are applied after the window is installed.

## Metal Windows

Many buildings have vinyl or metal windows. Improvements in the design of metal windows have made them competitive with wood windows in both cost and energy efficiency. The most important of these design improvements has been the development of thermal-break windows. Thermal-break windows use a combination of air spaces and materials that do not conduct heat easily to separate the exterior from the interior.

The basic parts of a metal or vinyl window are similar to those of a wood window. The sash consists of a stile, rails, and glazing. The frame is made up of side jambs, a head jamb, and a sill. However, the trim (casing) is not included as part of the window. Often the window frame itself is the only trim used on the exterior. The frame includes a nailing fin for attaching the window to the building framing.

## Window Details

All windows include the parts discussed so far. However, to show the smaller parts, which vary from one window style to another, architects and manufacturers use detail drawings. The most common type of window detail is a section (see **Figure 25–2**). All the parts can be shown in section views of the head, sill, and one side jamb. These sections also usually show the wall framing around the window.

Some of the parts that can be found on window sections are defined here. Find each of the parts on the sections and illustrations in **Figures 25–3, 25–4,** and **25–5.**

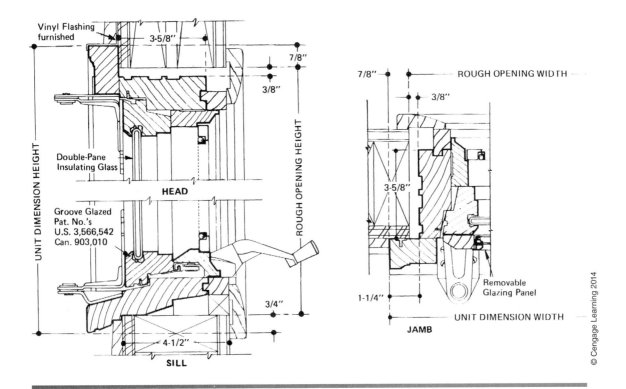

**Figure 25–2.** Typical window detail drawing.

### GREEN NOTE

*Fenestration is all of the openings in the building that are intended to allow air, daylight, people, animals, and vehicles to enter and exit the building. This is primarily windows, doors, and skylights. Even the best windows and doors tend to have poorer thermal insulating qualities than the surrounding wall. By eliminating all windows and doors, the heating and cooling load for the building could be greatly reduced, however, it would not be a pleasant home in which to live. Green home designers pay particular attention to the sizes and locations of all fenestration.*

*The thermal quality of a window is also greatly affected by its glazing. By using glass with low energy transmission properties (low-E glass) and filling the space between the layers of glass with inert gas, its heat transmission and solar properties can be improved.*

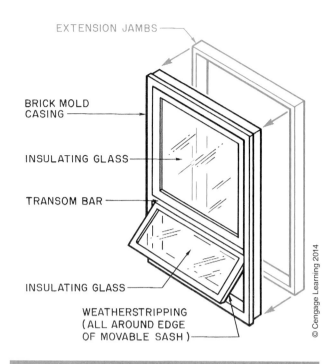

**Figure 25–3.** Window parts.

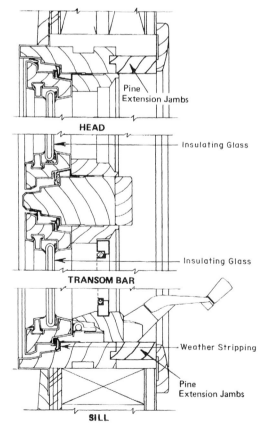

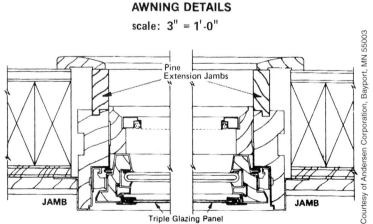

**Figure 25–4.**

Windows and Doors 179

- *Insulating glass* is a double layer of glass, creating a dead-air space. The dead air acts as an insulator. Better windows are filled with inert gas, such as argon, to provide better insulating qualities.
- *Extension jambs* are fastened to standard jambs when the window is installed in a thicker than normal wall.
- A **mullion** is a vertical section of the frame that separates side-by-side sash. If the mullion is formed by butting two windows together, it is called a *narrow mullion*. If the mullion is built around a stud or other structural support, it is called a *support mullion*.

## Door Construction

Doors include many of the same basic parts as windows (see **Figure 25–6**). A door frame consists of side jambs and a head jamb with **stop** and casing. Exterior door frames also include a sill. Many doors are made of a framework with panels (see **Figure 25–7**). The parts of **panel doors** are named similarly to the parts of a window. The vertical parts are *stiles,* and the horizontal parts are *rails.* Doors with glass or louvers are variations of panel doors. The framework is made of rails and stiles, and the glass or

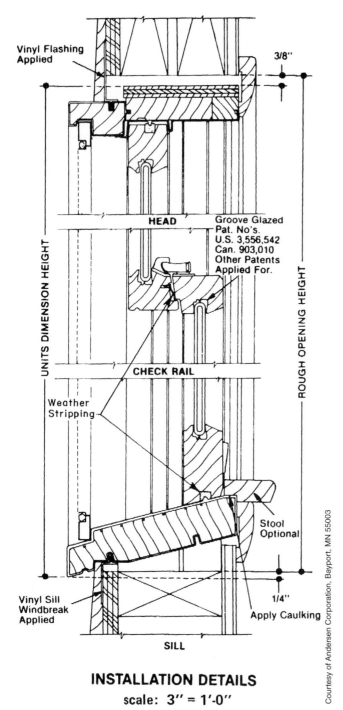

**Figure 25–5.**

- *Weather stripping* is used on windows that open and close. It forms a weather-tight seal around the sash.
- The *transom bar* is the horizontal part of a window frame that separates the upper and lower sash.
- *Meetings rails* or check rails are the rails that meet in the middle of a **double-hung window**.

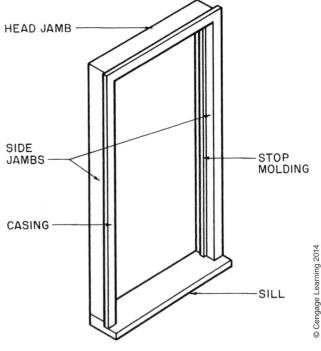

**Figure 25–6.** Parts of a door frame.

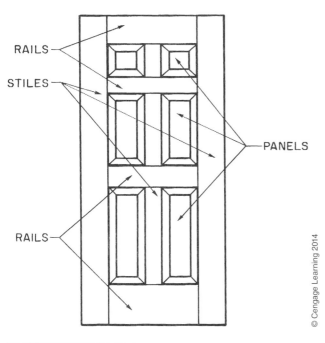

**Figure 25-7.** Construction of a panel door.

louvers replace the panels. Several manufacturers make molded doors. The most common type of molded door is made of hardboard for interior doors or steel for exterior doors which is manufactured in folds that contour the surface to look like panel doors (see **Figure 25-8**). *Hollow core doors* consist of an internal frame with flat "skin" applied to each side (see **Figure 25-9**). Some exterior doors have insulation between the two outer steel skins. These insulated doors result in considerable heating and cooling savings.

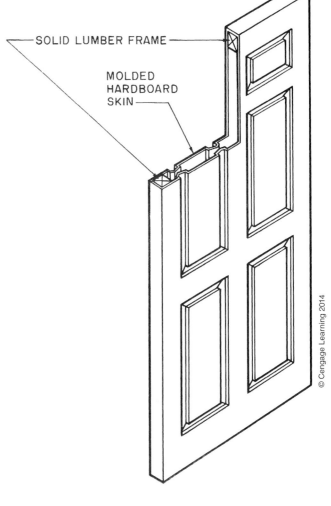

**Figure 25-8.** Molded hardboard door.

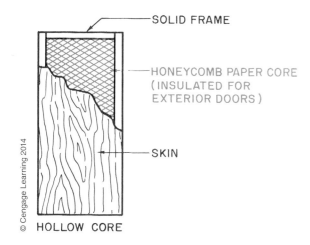
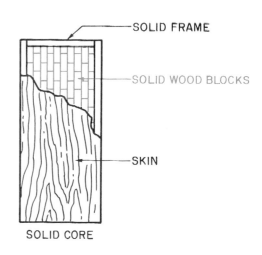

**Figure 25-9.** Flush doors.

Windows and Doors 181

## Door Details

Door details are usually less complex than window details. Where security, fire alarm, electronic lock, or special systems must be run in door and window metal frames, the corresponding installers must familiarize themselves with the actual manufacturer details and coordinate their installation with the door and window installation schedule. Carpenters rarely make doors, so all that is needed are simple details of the door frame and its trim (see **Figure 25–10**).

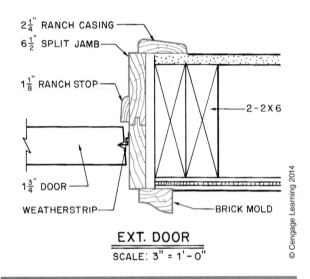

**Figure 25–10.** Typical jamb detail for an exterior door.

### GREEN NOTE

*Doors must insulate the conditioned space inside the home from the hot or cold weather outside and seal the opening against infiltration of air and water, sometimes in strong winds. If the door is glazed, that part of the door can be treated like a window and insulated with low-E glass and inert gas between the layers of glass. The rest of the door can be insulated by filling it with insulating material. To properly seal the opening requires a good weather seal on a good fitting door. If the weather seal is pinched or has gaps, air and water will enter at that point. If the door does not close evenly against the weather seal, that will cause leaks. The same principles apply to windows and skylights.*

Many doors are sold as prehung units. In these units, the frame is assembled, including the trim, and the door is hung in the frame. A section view of the jambs shows how the door is installed. For example, the door detailed in **Figure 25–10** is made with two-piece jambs. These split jambs are pulled apart; each side is then slid into the opening for installation. The stop can be either applied or integral. *Applied stop* is molding that is applied to the jambs with finish nails. *Integral stop* is milled as a part of the jamb when the jamb is manufactured.

## Reading Catalogs

It is often necessary to find specific information about windows or doors in the manufacturer's catalog or on its web site. Usually the catalog has a table of contents and web sites have a home page that lists the types of windows and doors available and illustrated. For each type of window or door, you will find some or all of the information listed below:

○ A brief description of the window type and some of the features the manufacturer wants to highlight—a little advertising.
○ Installation detail drawings.
○ Sizes available—This information usually consists of drawings of the various sizes and arrangements, with dimensions for glass size, stud or rough opening, and unit dimensions. **Figure 25–11** shows typical window size information reprinted from a manufacturer's catalog.
○ Additional information, such as optional equipment available.

Each manufacturer uses its own design to show the information. It sometimes takes a minute of study to familiarize yourself with how the manufacturer's pages are designed. Also, each manufacturer may make slightly different sizes of stock windows. If the windows used in constructing a building are from a different manufacturer than the one the architect used to make the drawings, it may be necessary to find windows that are as close as possible to the sizes shown on the drawings. Of course, if the construction specifications call for a particular manufacturer, that specification must be followed unless a change is authorized by the architect or owner.

## Table of Basic Unit Sizes  Scale 1/8" = 1'-0" (1:96)

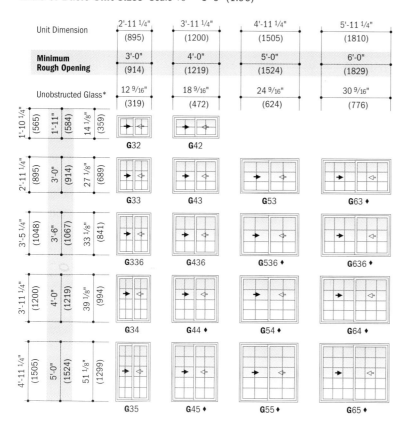

**Venting Configuration**
As viewed from the exterior. Passive sash will open after active sash has been opened.

\* "Unobstructed Glass" measurement is for single sash only.
♦ These units meet or exceed the following dimensions: Clear Openable Area of 5.7 sq. ft., Clear Openable Width of 20" and Clear Openable Height of 24".
**Rough opening dimensions may need to be increased to allow for use of building wraps, flashing, sill panning, brackets, fasteners or other items.**
"Unit Dimension" always refers to outside frame to frame dimension.
Dimensions in parentheses are in millimeters.
When ordering, be sure to specify color desired: White, Sandtone or Terratone."

## Handle Locations — Operational Force = 8 lbs.

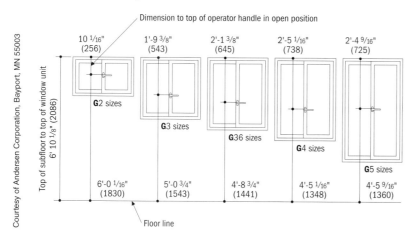

**Figure 25–11.** A typical manufacturer's catalog page showing window sizes.

### USING WHAT YOU LEARNED

Sometimes drawings specify window sizes that are very close to but slightly different from the sizes that a particular manufacturer stocks. Custom-made windows can usually be ordered in any size requested, but when the stock sizes are very close, it is much less expensive to use a stock window that is close to the size shown on the drawings. For example, what size window shown in Figure 25–11 might be used in the west wall of the dining room?

Refer to the Lake House example in your textbook packet. According to both the Upper Level Floor Plan and the West Elevation of the building, this is a "B" window on the Window Schedule. The Window Schedule on Sheet 5 tells us that a "B" window is 3'-0" × 3'-0'. On the schedule of sizes in figure 25–11, the unit dimensions are the top row and the far left column, so the closest size would be 2'-11¼" × 2'-11¼". Coincidentally, the rough opening for that window is 3'-0" × 3'-0", shown in the second row of dimensions across the top and the second column of dimensions down the left side.

## Assignment

Refer to the Lake House drawings when necessary to complete the assignment.

1. Name the lettered parts (a through f) in **Figure 25–12**.
2. What is the nominal size of the window in the south end of the Lake House dining room?
3. In the catalog sample shown in **Figure 25–11**, what is the width and height of the rough opening for a 3'-11¼" × 2'-11¼" window?
4. In the catalog sample, what are the rough opening dimensions for a window with a 24 9/16" by 39 1/8" glass size?
5. In the catalog sample in **Figure 25–11**, what is the glass size of the window in bedroom #2 of the Lake House?
6. In the catalog sample, what is the RO for the window in bedroom #2 of the Lake House?
7. Is the exterior door in the Lake House kitchen to be prehung or site hung?
8. What type and size are the doors in the Lake House playroom closet?

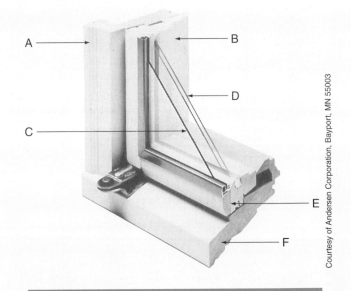

Figure 25–12.

# Unit 26: Exterior Wall Coverings

## Wood Siding

Wood is a popular siding material because it is available in a variety of patterns, it is easy to work with, and it is durable. Wood siding includes horizontal boards, vertical boards, shingles, and plywood.

Boards can be cut into a variety of shapes for use as horizontal siding (see **Figure 26–1**). These boards are nailed to the wall surface starting at the bottom and working toward the top. With wood siding, a starting strip of wood furring is nailed to the bottom of the wall. This starting strip holds the bottom edge of the first piece of siding away from the wall. Each board covers the top edge of the one below. The amount of each board left exposed to the weather is called the *exposure* of the siding (see **Figure 26–2**).

Vertical siding patterns can be created by boards or plywood (see **Figure 26–3**). These materials are applied directly to the wall with no special starting strips. However, where horizontal joints are necessary, they should be lapped with rabbet joints, or Z-flashing should be applied. The flashing can be concealed with **battens** (see **Figure 26–4**).

*Shingles* take longer to apply but make an excellent siding material. In place of a starting strip of furring, the bottom **course** (row) of shingles is doubled.

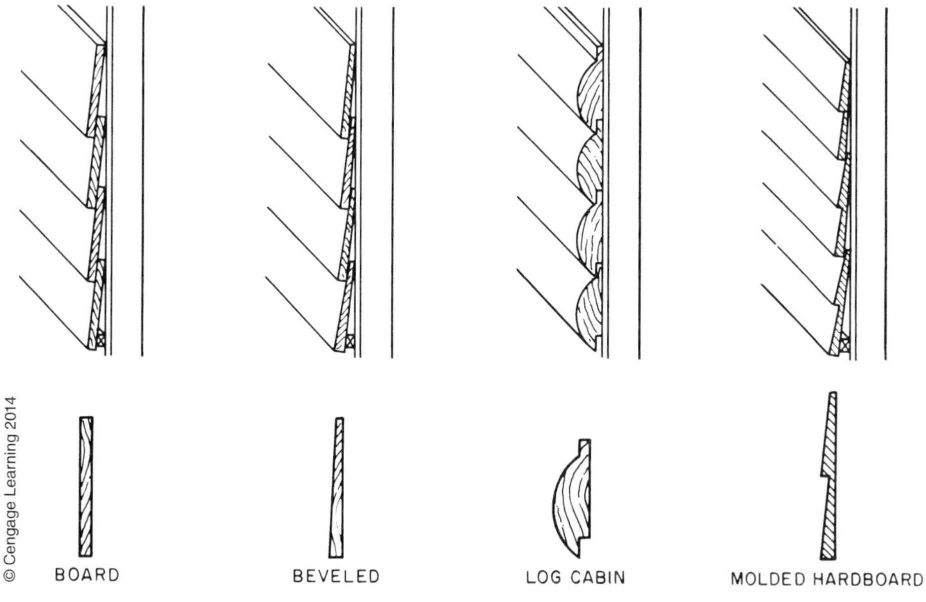

## Objectives

After completing this unit, you will be able to perform the following tasks:

○ Describe the exterior wall covering planned for all parts of a building.

○ Explain how flashing, drip caps, and other devices are used to shed water.

○ Describe the treatment to be used at corners and edges of the exterior wall covering.

BOARD    BEVELED    LOG CABIN    MOLDED HARDBOARD

**Figure 26–1.** Horizontal siding.

## Fiber Cement Siding

Fiber cement siding is made of Portland cement, ground sand, cellulose fiber, and additives mixed with water and formed into siding. Fiber cement siding will not rot, and it is termite proof.

Fiber cement siding looks like wood siding (see **Figure 26–5**). It comes in a smooth finish, a wood grain texture, and even a rough-sawn look, and it is available in widths of 6, 7½, 9½, and 12 inches. Fiber cement siding is sometimes referred to as Hardiplank®, a very popular brand of fiber cement siding. Hardiplank® comes in standard 12-foot lengths, while other siding materials come in 16-foot lengths. The 4-foot difference is because fiber cement siding is much heavier than wood siding, and anything over 12 feet would be

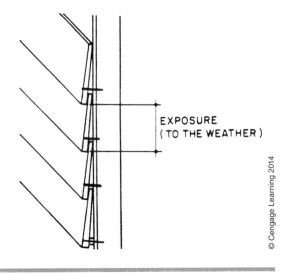

**Figure 26–2.** The exposure is the amount of the sliding exposed to the weather.

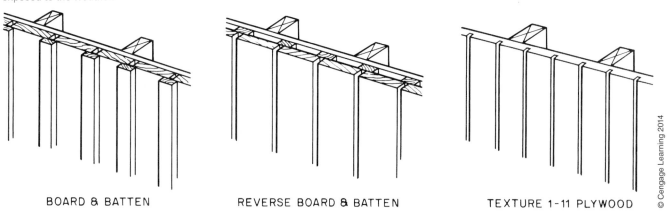

**Figure 26–3.** Three popular vertical siding patterns.

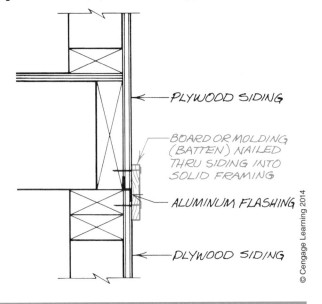

**Figure 26–4.** Horizontal hardboard siding should be covered with battens.

**Figure 26–5.** Fiber cement siding is made to look like wood siding but is more durable.

difficult for one person to handle. Although fiber cement siding looks like wood and it is installed in much the same way as wood, it is important to follow the manufacturer's installation instructions to maintain the warranty on the siding.

## Drip Caps and Flashing

It is important to prevent water from getting behind the siding, where it can cause dry rot and attract insects. Where a horizontal surface meets the siding, water is apt to collect. **Flashing** is used to prevent this water from running behind the siding. Aluminum, galvanized steel, and copper are the most common flashing materials. The flashing is nailed to the wall before the siding is applied. The lower edge of the flashing extends over the horizontal surface far enough to prevent the water from running behind the siding. Areas to be flashed are noted on building elevations. The flashing is shown on the detail drawings or building elevations (see **Figure 26–6**).

The heads of windows and doors may form a small horizontal surface. Wood **drip cap** molding can be used

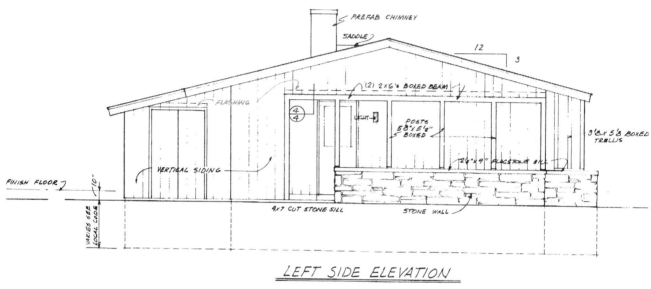

**Figure 26–6.** Flashing is shown on elevations and details.

Exterior Wall Coverings 187

### GREEN NOTE

*Together with the roofing, the exterior wall finish is the home's first line of defense against water. There are many exterior finish materials available to the homebuilder and most of them are capable of doing an excellent job of shedding water. None of them, however, will do a good job if not installed properly. Most water enters the home through the force of gravity. As it either drips off the roof or strikes the wall near the top it flows downward, always seeking the path of least resistance. Sometimes that path of least resistance is through a gap in a drip cap over a door or window, or through a poorly fitted piece of flashing. The key to success is to install all wall covering materials, house wrap, flashing, siding, masonry veneer, paint or whatever is called for with care and according to the manufacturer's instructions.*

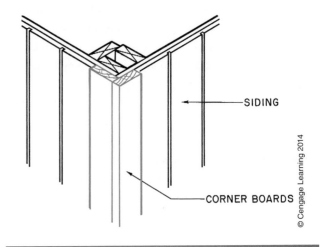

**Figure 26–8.** Corner boards can be used with most types of siding.

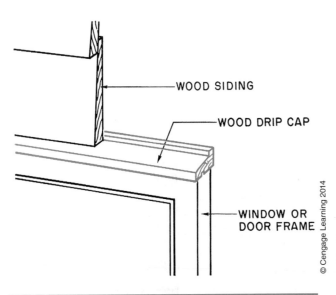

**Figure 26–7.** Drip cap.

**Figure 26–9.** Shingles are trimmed to form their own corners.

in these places to shed water (see **Figure 26–7**). Drip caps are shown on details and elevations.

## Corner and Edge Treatment

Regardless of the kind of siding used, the edges must be covered to prevent water from soaking into the end grain or running behind the siding. Around windows, vents, doors, and other wall openings, the trim around the opening usually covers the end grain of the siding. In an inside corner, a strip of wood can be used to form a corner bead for the siding. Outside corners in plywood, hardboard sheets, or vertical boards are usually built with corner boards (see **Figure 26–8**). Horizontal siding is butted against the corner boards. Shingles are trimmed to form their own outside corner (see **Figure 26–9**), or they are butted against corner boards.

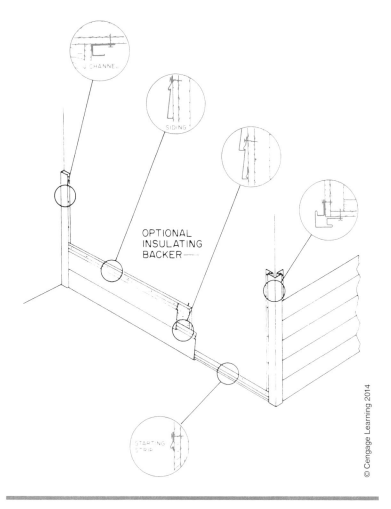

**Figure 26–10.** Accessories for metal and vinyl siding.

The corner treatment to be used can usually be seen on the building elevations. If no corner treatment is shown on the elevations, look for a special detail of a typical corner.

## Metal and Plastic Siding

Several manufacturers produce aluminum, steel, and vinyl siding and trim. The most common type is made to look like horizontal beveled wood siding. A variety of trim pieces are available for any type of application (see **Figure 26–10**).

## Stucco

*Stucco* is a plaster made with Portland cement. The wall sheathing is covered with waterproof building paper. Next the *lath* (usually wire netting) is stapled to the wall (see **Figure 26–11**). Finally, the stucco is troweled on—a rough *scratch coat* first—as shown in

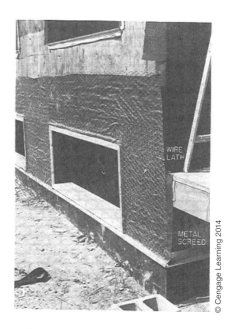

**Figure 26–11.** Wall prepared for stucco.

Exterior Wall Coverings 189

**Figure 26–12.** Scratch coat applied.

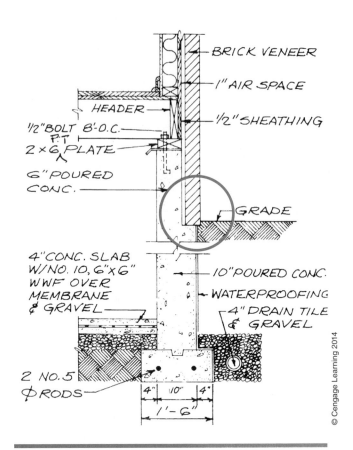

**Figure 26–13.** Typical foundation section with ledge for masonry veneer.

**Figure 26–12,** then a *brown coat* to build up to the approximate thickness, and finally a finish coat. Outside corners and edges are formed with galvanized metal beads nailed to the wall before the scratch coat is applied.

## Masonry Veneer

Masonry veneer is usually either brick or natural stone. It is called "veneer" because it is a thin layer of masonry over some kind of structural wall. The structure may be wood framing, metal framing, or concrete blocks. The masonry has to rest on a solid foundation. Normally, the building foundation is built with a 4-inch ledge to support the masonry veneer. This ledge can be seen in a typical wall section (see **Figure 26–13**).

Interesting patterns can be created by using half bricks and bricks in different positions. Some of the most common *bond patterns,* as they are called, are shown in **Figure 26–14.** If no pattern is indicated, the bricks are normally laid in running bond.

Above and below windows, above doors, and in other special areas, bricks may be laid in varying positions (see **Figure 26–15**). If bricks are to be laid in any but the stretcher position, they will be shown on the detail drawings (see **Figure 26–16**). The details for openings in masonry walls will also indicate a lintel at the top of the opening. The **lintel** is usually angle iron. It carries the weight of the masonry above the opening.

## General

The electrical installer needs to be familiar with the rough-in requirements for all outside lighting and devices and any special requirements that may be needed with the specified exterior wall finish. The required backboard or rough-in frame and box must be properly installed to prevent any finished wall damage.

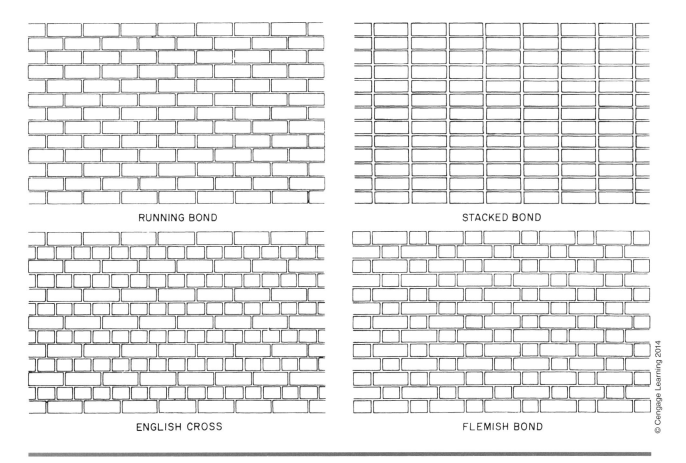

**Figure 26–14.** Frequently used bond patterns.

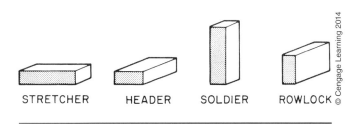

**Figure 26–15.** Brick positions.

Exterior Wall Coverings 191

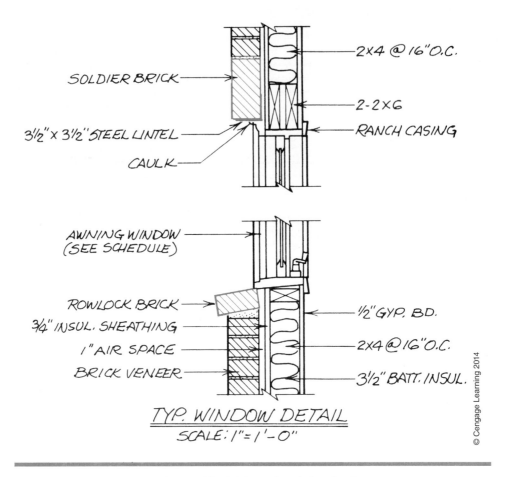

**Figure 26–16.** Notice the rowlock and soldier bricks on the window detail.

 **USING WHAT YOU LEARNED**

Before the siding is applied, the carpenters should examine the drawings to know exactly how all of the edges are to be treated. Questions to consider include:
- Where is caulk to be used?
- Where are siding edges covered with trim?
- Exactly where do the edges fall? For example, where is the bottom edge of the siding relative to the top of the foundation? This is shown on the Lake House Wall Section 3/4. The bottom of the siding aligns with the top of the foundation wall.

## Assignment

Refer to the Lake House drawings in your textbook packet to complete the assignment.
1. What material is used for the Lake House siding?
2. How are the outside corners of the Lake House siding finished?
3. Where is the bottom edge of the siding relative to the wall construction?
4. What prevents water from running under the siding at the heads of the Lake House windows?
5. Detail 1/7 of the Lake House shows aluminum screen nailed behind the top edge of the siding. What is the purpose of the opening covered by this screen?
6. Describe one use of aluminum flashing under the siding on the Lake House.

# UNIT 27 Decks

## Objectives

After completing this unit, you will be able to perform the following tasks:

○ Explain how a deck is to be supported.

○ Describe how a deck is to be anchored to the house.

○ Locate the necessary information to build handrails on decks.

Wood decks are used to extend the living area of a house to the outdoors. A deck may be a single-level platform, or it may be a complex structure with several levels and shapes. However, nearly all wood decks are made of wood planks or synthetic planks made to simulate wood laid over joists or beams (see **Figure 27–1**). The planks are laid with a small space between them, so rainwater does not collect on the deck.

The same construction methods are used for decks and porches as for other parts of the house. The parts of deck construction that require special attention or that were not covered earlier in the text are discussed here.

## Support

The deck must be supported by stable earth. The support must also extend below the frost line in cold climates. The most common method of support is by concrete columns with metal post anchors (see **Figure 27–2**). A typical metal post anchor is shown in **Figure 27–3**. Such anchors fasten posts to the concrete pier while keeping them from contacting the concrete in order to prevent decay. Indeed, all wood used in the construction of a deck should be pressure treated with a chemical to prevent such decay. Note that deck detail 3/6 for the Lake House includes a note that all wood is to be pressure treated.

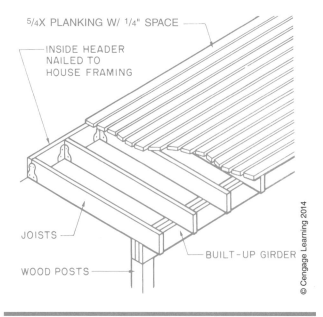

**Figure 27–1.** Typical deck construction.

There are a number of products made from combinations of wood fibers and synthetic materials (plastics) that may be used for decks and other outdoor structures. These products, called *composites,* are not affected by weather, moisture, or termites. They can generally be worked with ordinary carpentry tools and fasteners. Of course, as with all products, you should read and follow the manufacturer's instructions for installation.

Decks are usually included on the floor plans for the house. The floor plans show the overall dimensions of the decks and the locations of posts, piers, or other support (see **Figure 27–4**). If the decks are complex, they may be shown on a separate plan or detail. On the Lake House, the decks are shown on the floor framing plan.

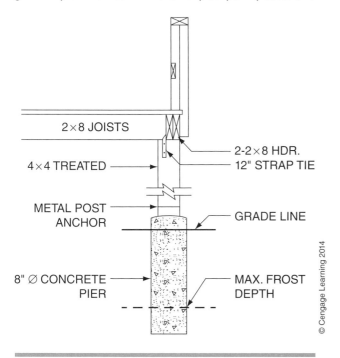

**Figure 27–2.** Foundation for a typical wood deck.

**Figure 27–3.** Metal post anchor.

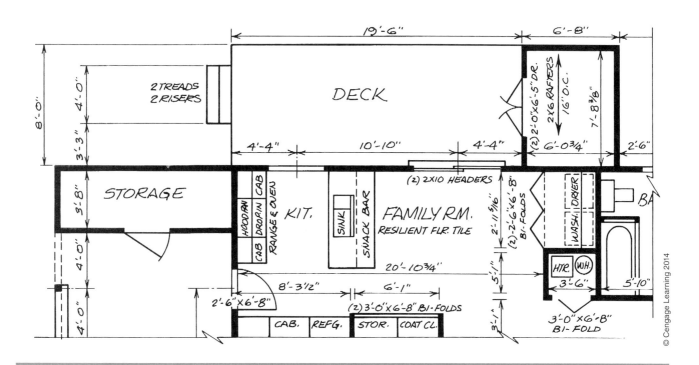

**Figure 27–4.** Deck shown on floor plan.

Typically the posts or piers support a beam or girder, which in turn supports joists. The beam may be solid wood or built-up wood. The joists can be butted against or rested on top of the beam in any of the ways discussed for floor framing.

## Anchoring the Deck

The most common ways of anchoring the deck to the building are shown in **Figure 27–5**. If the deck is at the same level as the house floor, the deck joists can be cantilevered from the floor joists. In this case, blocking is required between the joists. If the deck is secured with anchor bolts in a concrete foundation, the anchor bolts must be positioned when the foundation is poured. If the deck is to be added to an existing concrete wall, anchors made for this purpose can be inserted into drilled holes. There are many different types of anchors available, but only those rated for this use should be used to fasten the deck to the house.

The anchor bolts or through bolts hold a joist header. The deck joists either rest on top of the header or butt against it. If the joists butt against the header, they are supported by joist hangers or a ledger strip.

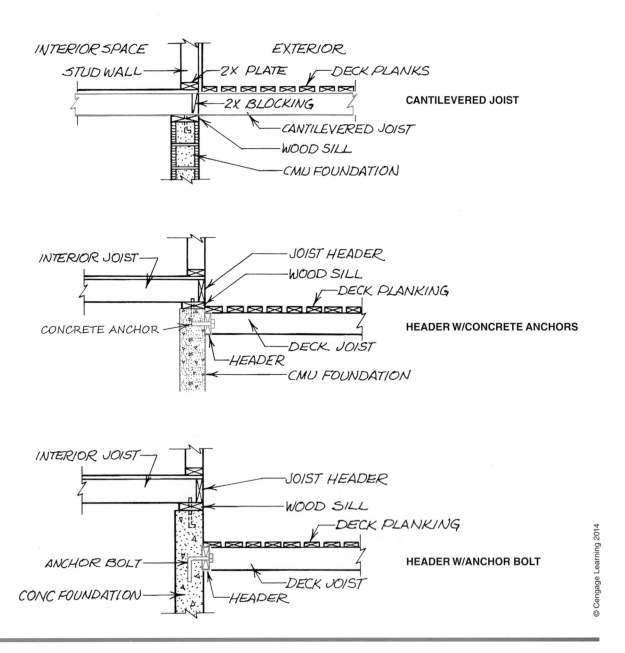

**Figure 27–5.** Three methods of anchoring the deck to the house.

### GREEN NOTE

*Decks and outdoor structures are parts of the home. In the spirit of designing and building a green home, the same principles that apply to the house should be applied to decks. Design on a 2-foot module; avoid unnecessary corners, angles, and curves; select materials that require a minimum of maintenance and do not involve unnecessary embodied energy (the energy required to manufacture and transport a product); and always observe best practices—use quality workmanship.*

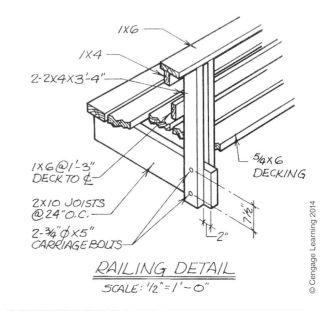

**Figure 27-6.** Railing detail.

Although all deck framing should be built of pressure-treated lumber, it is still important to prevent water from collecting on top of the header. This can be done by cementing flashing into a kerf (i.e., saw cut) in the concrete wall or with caulking.

## Railings

Most decks have a railing because they are several feet from the ground. Although metal railings are available in ready-to-install form, the architectural style of most wood decks calls for a carpenter-built wood or composite railing. The simplest type of railing is made of uprights and two or three horizontal rails. The uprights are bolted to the deck frame, and the rails are bolted, screwed, or nailed to the uprights. The style of the railing and the hardware involved are usually indicated on a detail drawing (see **Figure 27-6**).

### USING WHAT YOU LEARNED

To build a deck, it is necessary to know the sizes of all of the parts. The joists that support the deck planks must be of the correct material and cut to the correct length. On the Lake House, what are the dimensions of the joists that support the deck between the kitchen and the garage?

The basic arrangement of these joists is shown in the floor framing plan 1/6. The joists are 2 × 8s, which span from the three 2 × 8s at the west end of the house to the east end. The upper level floor plan 2/2 shows the dimension from the west edge of the deck (and the side of the garage) to the west wall of bedroom #2 to be 13'-9". Subtract 3 × 1½" for the header at the west side end of the joists to find the length of the joists. Thus, 13'-9"-4½" = 13'-4½". The joists are 2 × 8 (1½" × 7¼") by 13'-4½" long.

Decks 197

## Assignment

Refer to the Lake House drawings in your textbook packet to complete the assignment.
1. What supports the south edge of the decks located outside the Lake House living and dining rooms?
2. How far from the outside of the house foundation is the centerline of these supports?
3. How far apart are these supports?
4. How many anchor bolts are required to fasten both of these decks to the Lake House foundation?
5. What is the purpose of the aluminum flashing shown on deck detail 3/6?
6. What material is used for the railings on the Lake House south decks?
7. How many lineal feet of lower horizontal rails are there on these decks? (Do not include the cap rail.)
8. What is the total rise from the lower deck to the higher deck? Which deck is higher?
9. What supports the west edge of the deck between the Lake House kitchen and the garage?

# UNIT 28

# Finishing Site Work

As the exterior of the building is being finished or soon after it is finished, the masons, carpenters, and landscapers begin the finished landscape work. Any constructed features (called *site appurtenances*) are completed first. Then trees and shrubs are planted. Finally, lawns are planted.

## Retaining Walls

Retaining walls are used where sudden changes in elevation are required (see **Figure 28–1**). The retaining wall retains, or holds back, the earth. Where the height of the retaining wall is several feet, the earth may put considerable stress on the wall. Therefore, it is important to build the wall according to the plans of the designer. A section through the wall usually is included to show the thickness of the wall, its foundation, and any reinforcing steel

## Objectives

After completing this unit, you will be able to perform the following tasks:

○ Describe retaining walls, planters, and other constructed landscape features shown on a set of drawings.

○ Find the dimensions of paved areas.

○ Identify new plantings and other finished landscaping shown on a site plan or landscape plan.

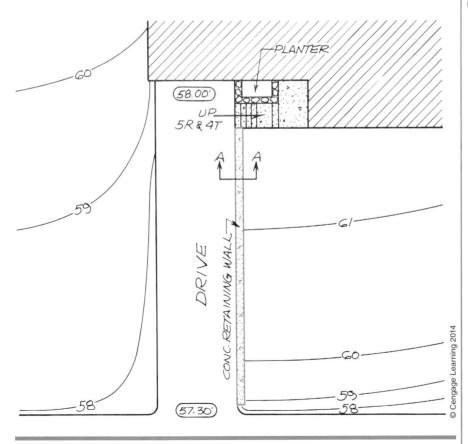

**Figure 28–1.** Retaining wall on site plan.

Finishing Site Work  199

(see **Figure 28–2**). For low retaining walls, the site plan may be the only drawing included.

A low retaining wall is sometimes built around the base of a tree when the finished grade is higher than the natural grade. This retaining wall forms a well around the tree, allowing the roots of the tree to "breathe." An example of a tree that will require a well can be seen in **Figure 28–3**, taken from the Lake House site plan. The 24-inch oak is at an elevation of approximately 333 feet, but the finished grade at this point is 336 feet. Therefore, a well 3-feet deep is required.

## Planters

Planters are sometimes included in the construction of retaining walls or attached to the building. In these cases, the information needed to build the planter is included with the information for the building or retaining wall (see **Figure 28–4**). The planter is built right along with the house or retaining wall. If a planter that is separate from other construction is included, it is usually shown with dimension on the site plan. A special section may be included with the details and sections to show how the planter is constructed (see **Figure 28–5**).

The planter should be lined with a waterproof membrane, such as polyethylene (common plastic sheeting), or coated with asphalt waterproofing. This keeps the acids and salts in the soil from seeping through the planter and staining it. The planter should also include some way for water to escape. This can be through the bottom or through weep holes. **Weep holes** are openings just above ground level. In cold climates, the planter may be lined with compressible plastic foam. This allows the earth in the planter to expand as it freezes, without cracking the planter. If the planter is to have landscape lighting, automatic watering, etc., additional waterproofing may be required where these utilities penetrate the waterproof membrane.

## Paved Areas

Paved areas on housing sites are drives, walks, and patios. Drives and walks are usually described most fully in the specifications for the project. However, the site plan includes dimensions and necessary grading information for paved areas, as shown in **Figure 28–3**. These dimensions are usually quite straightforward and easy to understand.

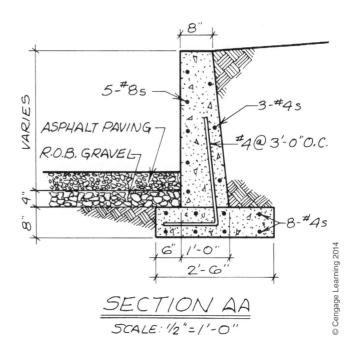

**Figure 28–2.** Typical retaining wall construction detail.

### GREEN NOTE

*Paved areas can be either impervious (water will not pass through) or pervious (water passes through to the earth below). For small areas such as walkways, the choice of pervious or impervious is not significant, but for large areas this can be an important design question. If ½ inch of rain falls on a paved area that is 100 feet by 20 feet, that is over 83 cubic feet or 617 gallons of water. If that much water drains at one some point, it can do considerable damage to the landscape in that area. If impervious pavement, such as asphalt or concrete is to be used, provisions should be made for groundwater runoff.*

*Patios* are similar to drives and walks in that they are flat areas of paving with easy-to-follow dimensions. They may differ from drives and walks by having different paving materials, such as slate, brick, and flagstone, for example. Patios may also be made of a concrete slab with different surface material.

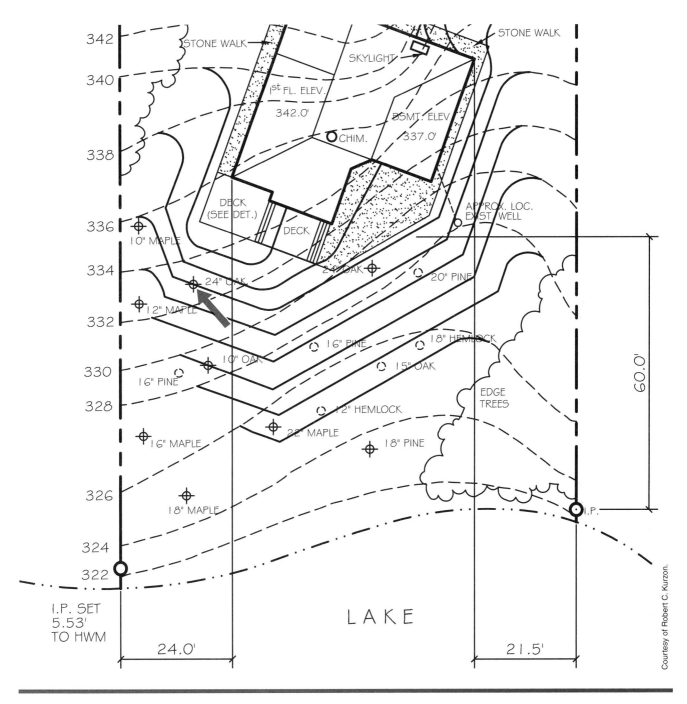

**Figure 28-3.** Typical site plan. The 24-inch oak tree near the SW corner of the house requires a well.

## Plantings

Plantings include three types: grass or lawns, shrubs, and trees. On some projects, such items are planted by the owner. When the builder/contractor does the landscaping, the trees and shrubs are planted first; then, the lawns are planted. Some or all of the trees included in the landscape design may have been left when the site was cleared. Any new trees to be planted are shown with a symbol and an identifying note, as shown in **Figure 28-3**. There is no widely accepted standard for the symbols used to represent trees and shrubs. Most architects and drafters use symbols

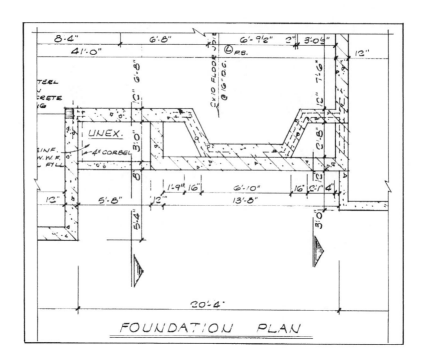

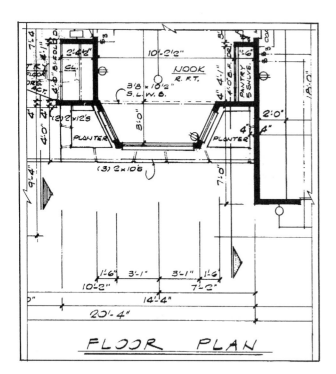
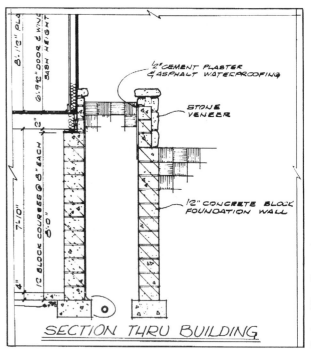

**Figure 28–4.** Because the planter is a part of the foundation, it is included on the normal drawings for the house.

that represent deciduous (leaf shedding) trees, coniferous (evergreen) trees, palms, and low shrubs (see **Figure 28–6**). The trees and shrubs may also be listed in a schedule of plantings (see **Figure 28–7**).

Grass is planted by seeding or sodding. Although a note on the site plan may indicate seeded or sodded areas, more detailed information is usually given on the schedule of plantings or in the specification.

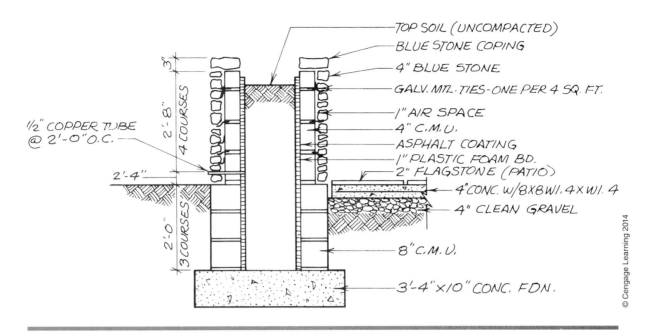

**Figure 28–5.** Section through planter and patio.

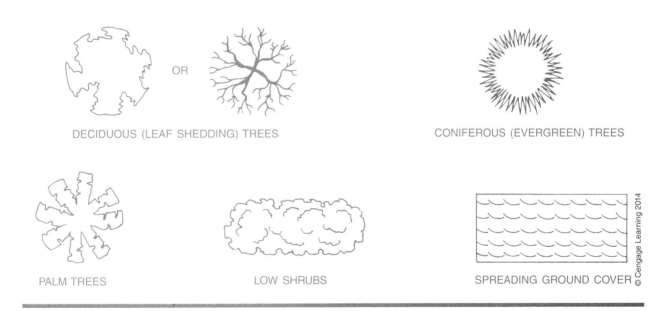

**Figure 28–6.** Typical symbols for plantings.

### GREEN NOTE

*The landscaping should be planned at the beginning, along with the rest of the home. Grading should be done to ensure that storm runoff does not damage the plantings. The area should be surveyed for invasive species. If they exist, they should be removed before any landscaping begins and certainly none should be introduced with the landscaping. Grading and planting should not harm the roots of any trees intended to remain on the site. In arid regions, design landscaping that requires little or no watering.*

Finishing Site Work

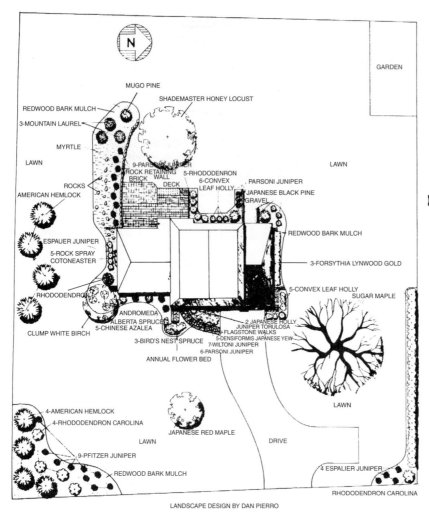

**Figure 28–7.** A landscape plan complete with plant list.

### USING WHAT YOU LEARNED

A masonry planter can be used to add plantings in a confined, easy-to-maintain area. If the planter is attached to the house, the foundation will probably be constructed at the same time as the house foundation, but the masonry finish is normally left until the house is well underway. To order the materials for the planter, the builder must know how much of each item to order. The stone coping of the planter in **Figures 28–8** and **28–9** is expensive, so it is important to be accurate in determining the quantity. How many lineal feet of coping are required for this planter?

The dimensions of the planter are shown in **Figure 28–8**. Add all of the dimensions together and subtract the 12-inch width once for each corner. (The coping does not overlap at the corners.) That is, 16'-0" + 2'-4" + 4'-0" + 2'-8" + 12'-0" + 5'0" − 6'-0" (corners) = 36'-0".

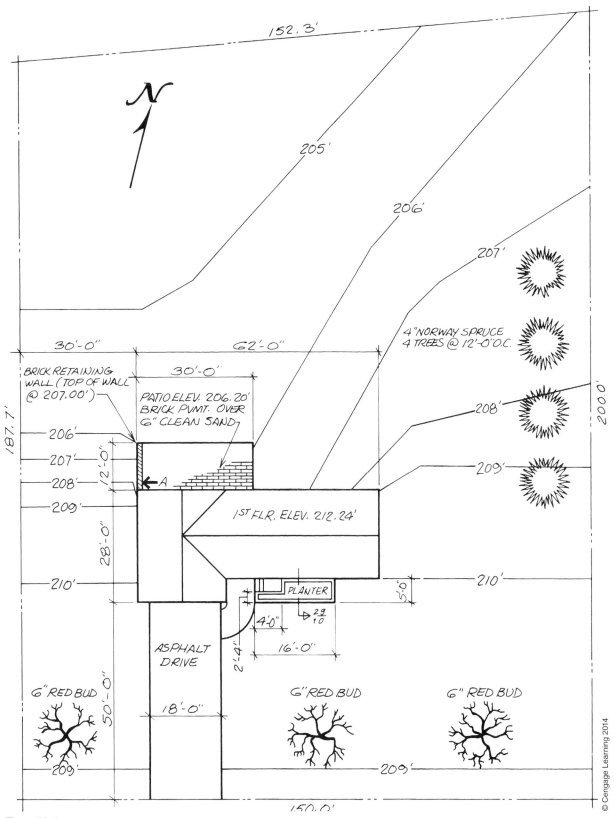

Figure 28–8.

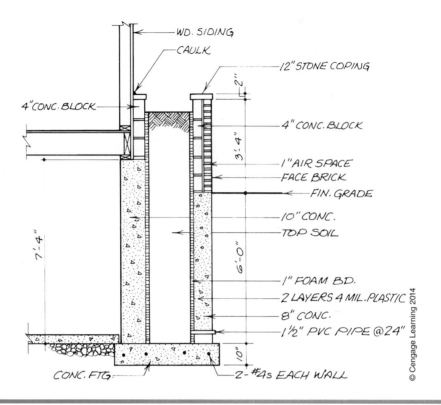

Figure 28–9.

## Assignment

Refer to **Figures 28–8** and **28–9** to complete the assignment.

1. What is the height of the retaining wall above the patio surface at A?
2. How long is the retaining wall?
3. Of what material is the retaining wall constructed?
4. What is the width and what is the length of the patio?
5. What materials are used in the construction of the patio?
6. Describe the weep holes in the planter.
7. How is the planter treated to prevent acids and salts from staining its surface?
8. How many deciduous trees are to be planted?
9. What is the area of the driveway?
10. Assuming that the driveway is 4 inches thick, how many cubic yards of asphalt does it require? (See Math Review 22.)

# UNIT 29

# Fireplaces

## Basic Construction and Theory of Operation of Wood-Burning Fireplaces

A fireplace can be divided into four major parts or zones: foundation, firebox, throat area, and chimney (see **Figure 29–1**). Each of these zones has a definite function. To understand the construction details, it is necessary to know how these zones work.

## Objectives

After completing this unit, you will be able to perform the following tasks:

- Describe the foundation, firebox, throat, and chimney of a fireplace using information from a set of construction drawings.

- Explain the finish of the exposed parts of the fireplace, using information from a set of construction drawings.

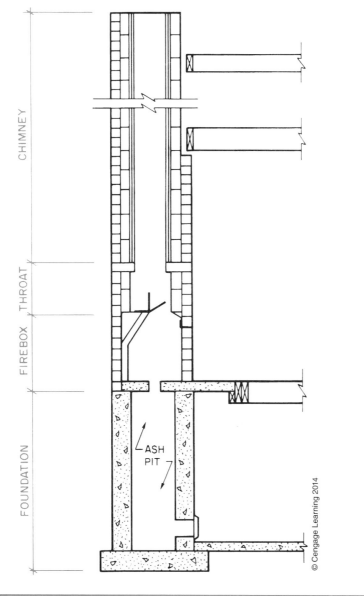

**Figure 29–1.** Four zones of a fireplace.

Fireplaces 207

## Foundation

The fireplace foundation serves the same purpose as the foundation of the house—it supports the upper parts and spreads the weight over an area of stable earth. The foundation consists of a footing and walls capable of supporting the necessary weight (see **Figure 29–2**). The fireplace foundation sometimes houses an ash pit. The *ash pit* is a reservoir to hold ashes that are dropped through an ash dump (a small door) in the floor of the fireplace (see **Figure 29–3**). When the foundation includes an ash pit, a *cleanout door* is installed near the bottom of the ash pit (see **Figure 29–4**).

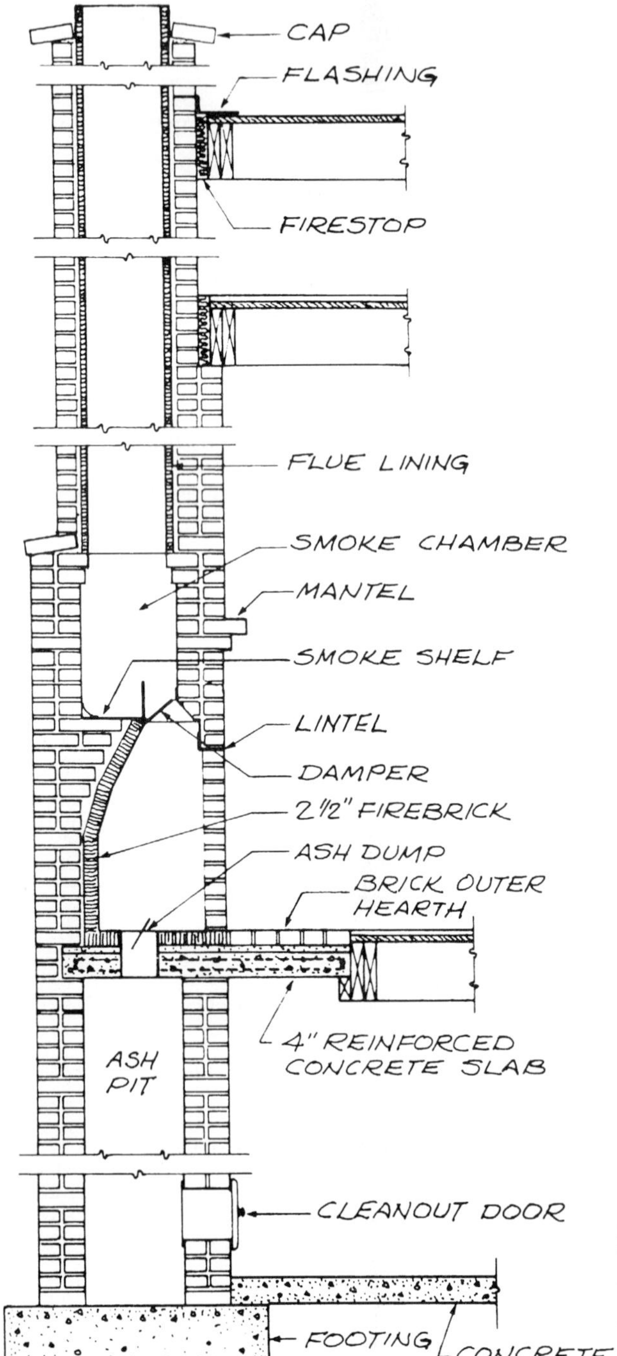

## Firebox

The firebox is the area where the fire is built. In all masonry fireplaces, the firebox is constructed of two layers of masonry, as shown in **Figure 29–2**. Each layer is called a wythe. The floor of the firebox consists of firebricks laid over a concrete base. The concrete base may extend beyond the face of the firebox

**Figure 29–2.** Section view of a masonry fireplace.

**Figure 29–3.** Ash dump.

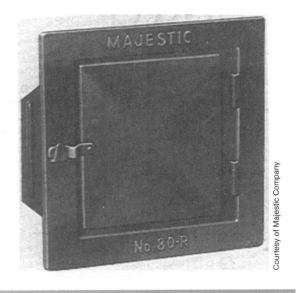

**Figure 29–4.** Cleanout door.

to support the hearth. The **hearth**, which may be tile, stone, brick, or slate, forms a noncombustible floor area in front of the fireplace. The outer walls of the firebox are most often common brick. The inner walls are of firebrick to withstand the heat of the fire. The back wall of the firebox slopes in to direct the smoke and gas into the throat area. The masonry over the fireplace opening is supported by a steel lintel. The lintel is long enough so that 4 inches of it can rest on the masonry at each end.

## Throat

The throat of the fireplace is the area where the firebox narrows into the chimney. Modern wood-burning fireplaces are built with a metal damper in the throat (see **Figure 29–5**). The **damper** is a door that can be closed to prevent heat from escaping up the chimney when the fireplace is not in use. The damper is placed on top of the firebox with 1-inch clearance on all four sides. This clearance allows the metal damper to expand as it gets hot.

The flat area behind the damper (above the sloped back of the firebox) is called the *smoke shelf*. The smoke shelf is especially important for the proper operation of the fireplace. The cold air coming down the chimney hits the smoke shelf and turns back up with the rising hot gas and smoke from the firebox. This helps carry the smoke and gas up the chimney (see **Figure 29–6**). If the smoke shelf is not built properly and kept clean, the falling cold air can force the smoke and gas back into the firebox.

## Chimney

The chimney carries the smoke and hot gas from the throat to above the house. The top of the chimney must be high enough above the roof, trees, and other nearby obstructions to ensure that the air flows evenly across its top. According to the International Residential Code®, the chimney must extend at least 2 feet through the roof and 2 feet above anything within 10 feet. However, the dimensions on the drawings should always be followed.

To ensure fire safety and a smooth inner surface, masonry chimneys are lined with a clay **flue**. This flue is installed in 2-foot sections as the chimney is built. A 1-inch air space is allowed between the flue lining and the chimney masonry (see **Figure 29–7**).

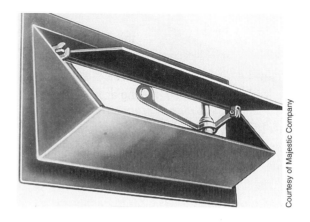

**Figure 29–5.** Damper.

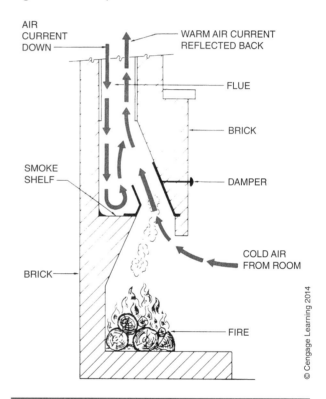

**Figure 29–6.** The smoke shelf turns the incoming air back up the chimney.

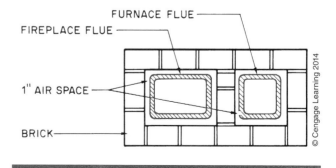

**Figure 29–7.** Plan view of a two-flue chimney.

Fireplaces 209

### GREEN NOTE

*Wood-burning fireplaces are not very efficient in heating a structure. They are primarily decorative. Their efficiency can be improved somewhat by the use of a prefabricated metal fireplace liner. There is a design of fireplaces that uses massive amounts of masonry and a flue that winds the smoke in an indirect route up the chimney. The fire and hot flue gases heat the masonry which slowly radiates its heat to the surrounding spaces. This type of fireplace can be up to 90 percent efficient, but it requires large amounts of masonry materials and is expensive to build.*

*Gas-burning fireplaces are more efficient, but still not as efficient as a good furnace or boiler. Some gas-burning fireplaces have an electronic ignition, eliminating the need for a pilot flame that is always burning, consuming fuel, and emitting greenhouse gases.*

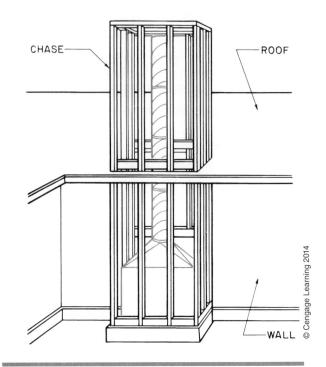

**Figure 29–8.** The framed enclosure for a metal chimney is called a chase.

In recent years insulated, metal chimneys have become quite popular. These chimneys are lightweight; they do not require massive foundation for their support. It also takes less time to install them. Because the outer wall of a metal chimney remains cool, it can be enclosed in wood (see **Figure 29–8**). The chimney sections are slipped together and fastened with sheet metal screws; then, the chimney is framed with wood and covered with the specified siding and trim. A chimney enclosure of this type is called a *chase*.

## Prefabricated Metal Fireplaces

Constructing masonry fireplaces is time consuming. Their great weight requires massive foundations. Engineered, metal fireboxes have been developed that can be installed in very little time and require only modest foundations (see **Figure 29–9**). The prefabricated units are available from several manufacturers and in a variety of styles. However, they are all similar in that they have double walls and a complete throat with the damper in place. Most also have a firebrick floor.

The double wall improves the heating ability of the metal fireplace. Cool air enters the space between the walls through openings near the bottom. The air absorbs heat from the fire, and because warm air naturally rises,

**Figure 29–9.** Prefabricated metal fireplace.

it exits through openings near the top of the unit (see **Figure 29–10**). The outside surfaces of the prefabricated unit are cooled by the circulating air; the unit can be enclosed in wood if recommended by the manufacturer. For a more traditional appearance, the exposed face of the fireplace can be covered with masonry veneer.

## Gas-Burning Fireplaces

Gas-burning fireplaces are a feature of many new homes and are becoming more common than their wood-burning ancestors. A gas fireplace is an appliance that is installed as a unit. The appliance is installed in a rectangular metal box called a *chase*, which is vented to the outside, and then connected to the gas supply. Any style of architectural trim can be applied to the front of the chase after the unit is installed.

Gas fireplaces are efficient, but some heat is lost through the metal surfaces. It is very important that the manufacturer's specifications and instructions be followed to ensure safe venting and to keep combustible materials a safe distance from hot surfaces. Gas-burning fireplaces can be vented straight through the building wall (see **Figure 29–11**), or vertically through a chase (see **Figure 29–12**). Clearances and framing details are provided in the manufacturer's instructions.

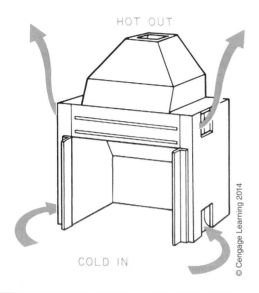

**Figure 29–10.** In a heat-circulating fireplace, cold air enters the double wall at the bottom, and warm air exits at the top.

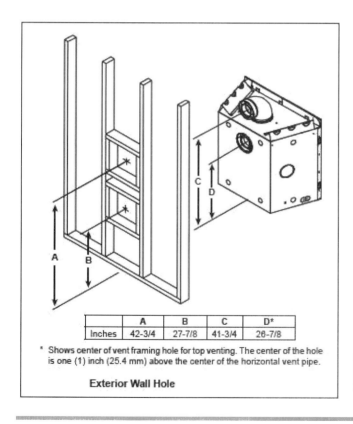

|  | A | B | C | D* |
|---|---|---|---|---|
| Inches | 42-3/4 | 27-7/8 | 41-3/4 | 26-7/8 |

*Shows center of vent framing hole for top venting. The center of the hole is one (1) inch (25.4 mm) above the center of the horizontal vent pipe.

**Exterior Wall Hole**

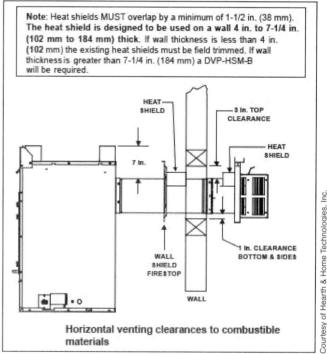

Horizontal venting clearances to combustible materials

**Figure 29–11.** Venting through the wall.

Fireplaces 211

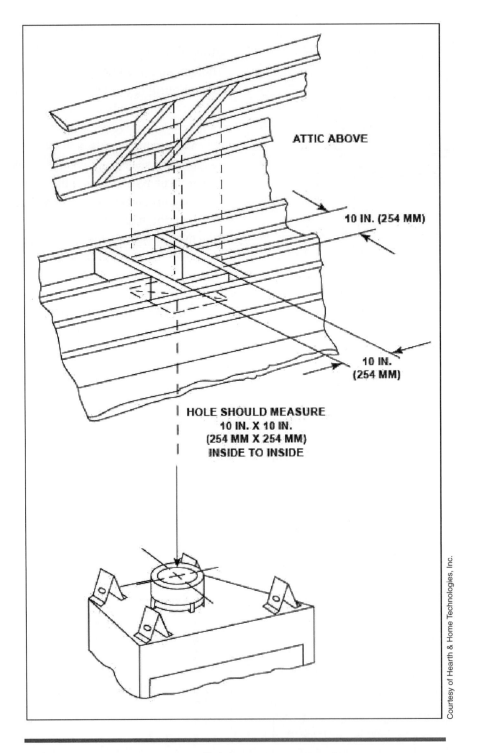

**Figure 29–12.** Venting through a vertical chase.

## Fireplace Drawings

Where a gas fireplace is to be installed, most of the information about the fireplace is found in the manufacturer's installation instructions. Gas appliances are fairly lightweight, so there are no special foundation requirements. Wood-burning fireplaces, however, are constructed on site from masonry materials, so the building plans and elevations must supply all the construction information.

The foundation is normally included on the foundation plan for the building. The dimensions and notes show the location of the fireplace foundation, its size, and the size and type of material to be used. The floor plan of the house shows where the fireplace is located and its overall dimensions. The building elevations show the chimney.

More detailed information about the fireplace is shown on the fireplace details, which usually include a cross section (see **Figure 29–13**). The following information is often included on a section view of the fireplace:

○ Dimensions of the firebox
○ Materials used inside the firebox (firebrick)
○ Materials used for the outside of the firebox and chimney
○ Ash dump, if any is included
○ Lintel over the opening
○ Dimensions of the hearth
○ Mantel, if any is included
○ Dimension from the smoke shelf to the flue
○ Size of the flue
○ Materials for the chimney

An elevation of the fireplace shows the exterior finish of the fireplace (see **Figure 29–14**). Only those features that could not be adequately described on the section view are called out on the elevation. However, this view shows the exterior finish—the mantel and the trim—better than the section view does.

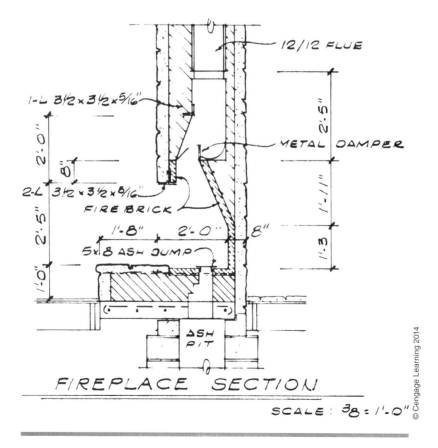

**Figure 29–13.** A section view shows the construction of the firebox and throat area.

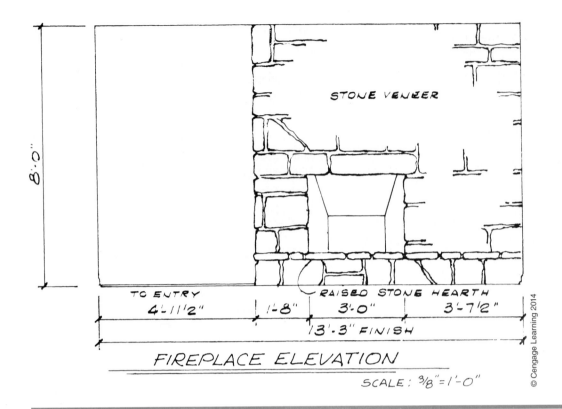

**Figure 29–14.** An elevation view is a good guide to the finished appearance of the fireplace.

 **USING WHAT YOU LEARNED**

Natural stone products are usually ordered well in advance of the time they are needed on the site and cut to finished size by the supplier or a company that specializes in working with these materials. This means that the precise size of each cut piece must be specified when the piece is ordered. What is the width and length of the granite piece on top of the Lake House fireplace? Allow 4 inches for the width of the soldier-course bricks, including mortar. There are three fireplace details on sheet 7. Section 4/7 shows that the granite is to fit inside a soldier course of bricks (bricks laid in a vertical position.) The overall size of the fireplace, without the hearth, is shown in Fireplace Plan 5/7 as 3'-4" by 7'-0". Subtract 4 inches from each side (8 inches from the width and 8 inches from the length) for the soldier course bricks and the size of the granite is 2'-8" by 6-4".

## Assignment

Refer to the Lake House drawings in your textbook packet to complete the assignment.

1. What type of fireplace does the Lake House have?
2. How wide is the opening of the firebox?
3. How high is the opening of the firebox?
4. What is the opening next to the fireplace?
5. Determine the overall width and length of the fireplace, including the hearth.
6. Of what material is the hearth constructed?
7. What is used for a lintel over the firebox opening? (Include dimensions.)
8. Briefly describe the foundation of the fireplace.
9. How far above the highest point on the roof is the top of the chimney?
10. What is the total height from the playroom floor to the top of the chimney?
11. What is the overall height of the brickwork involved in the fireplace construction?
12. The top of the fireplace is covered with granite on ¾-inch plywood. How much clearance is there between that plywood and the chimney?

# UNIT 30

# Stairs

## Stair Parts

In order to discuss the layout and construction of stairs, you need to know the parts of stairs described below and shown in **Figure 30–1**:

- *Stringers* are the main support members. The assembly made up of the stringers and vertical supports is called a **stair carriage**.
- **Treads** are supported by the stringers. The treads are the surfaces one steps on.
- *Risers* are the vertical boards between the treads.
- A *landing* is a platform in the middle of the stairs. Landings are used in stairs that change directions or in very long flights of stairs.
- The *run* of the stairs is the horizontal distance covered by the stairs.
- The *rise* of the stairs is the total vertical dimension of the stairs.
- The **nosing** is the portion of the tread that projects beyond the riser.

## Objectives

After completing this unit, you will be able to perform the following tasks:

- Identify the parts of stairs.
- Calculate tread size and riser size.

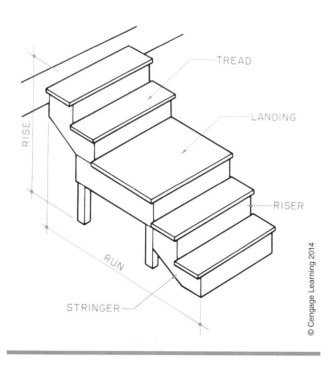

**Figure 30–1.** Basic stair parts.

Stairs 215

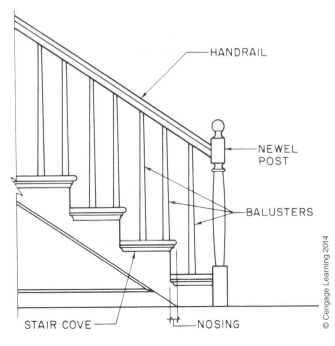

**Figure 30–2.** Stair trim and balustrade.

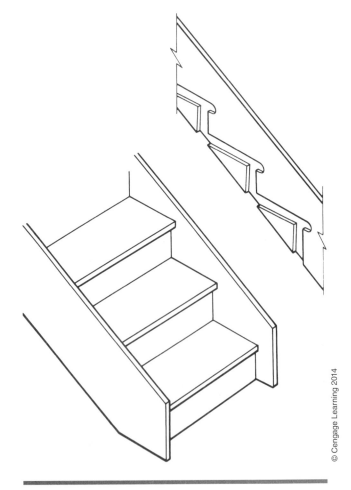

**Figure 30–3.** Open stringers and open risers.

**Figure 30–4.** Housed stringers and closed risers.

- The underside of the nosing may be trimmed with molding called *stair cove*.
- A *handrail* is usually required on any stairs that are not completely enclosed.
- **Balusters** are the vertical pieces that support the handrail at each step.
- The *newel post* is a heavier vertical support used at the bottom of the stair.
- The balusters, newel post, and handrail together are called a *balustrade*.

The trim parts and balustrade are shown in **Figure 30–2**.

## Types of Stairs

Stairs can be built with open stringers or housed stringers. *Open stringers* are cut in a sawtooth pattern to form a surface for fastening the treads (see **Figure 30–3**). *Housed stringers* are routed out, so the treads fit between them (see **Figure 30–4**). Stairs are also called open or closed depending on whether the space between the treads is enclosed with risers. Both the kind of stringers and the risers are shown on a section

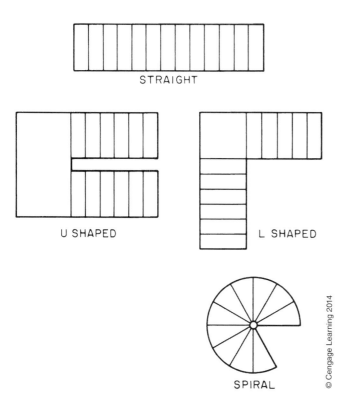

**Figure 30–5.** Stair layouts.

through the stairs. If the stringers are housed, the top of the far stringer shows on the section view. When stair lighting is specified to be installed in the stair riser, the electrical installer must coordinate with the carpenters to avoid a possible conflict with the center or mid-span stringer.

Stairs are also named according to their layout, as seen in a plan view (see **Figure 30–5**). Straight stairs are the simplest design. They may or may not include a landing. L-shaped and U-shaped stairs are used where space does not permit a straight run. L-shaped and U-shaped stairs have a landing at the change in directions. The carriages for L-shaped and U-shaped stairs include vertical supports under the landing (see **Figure 30–6**). Spiral stairs require the least floor space. They are made up of winder treads, usually supported by a center column.

Temporary stairs used only during construction are usually of cruder construction. They are often made of 2 × 10 stringers with wooden cleats to support the treads.

Service stairs in unoccupied spaces, such as a cellar, are often made of 2 × 12 open stringers with 2 × 10 treads. Stairs for exterior decks are also sometimes built with 2 × 12 open stringers (see **Figure 30–7**). However, the carpenter should remember that these stairs are important to the overall appearance of the deck.

## Calculating Risers and Treads

Nearly all stair details include a notation showing the number of treads with their width and the number of risers with their height, as shown in **Figure 30–6**. Occasionally only the total run, total rise, number of treads, and number of risers are given on the drawings, as shown in **Figure 30–7**. Notice that there is one more riser than there are treads. The builder must calculate the size of the treads and risers. The steepness of the stairs depends on the relationship between tread width and riser height. If the treads are wide and the risers low, the stairs are gradual. If the treads are narrow and the risers high, the stairs are steep. To climb stairs with wide treads and high risers requires an uncomfortably long stride. To climb stairs with narrow treads and low risers requires an uncomfortably short stride. Building codes specify the maximum height of risers and depth of treads. Section 3.11.5.3 of the *International Building Code*® specifies that risers cannot be more than 7¾ inches high and treads must be at least 10 inches deep. Although lower risers and deeper treads are allowed by the code, very low risers combined with very deep treads can make for awkward stairs. Stairs are designed by architects to comply with these rules and to be comfortable to climb and descend.

To build the stairs in **Figure 30–7** within these rules, the risers must be 7½ inches high and the treads must be 10 inches wide. The height of the risers can be found by dividing the total rise by the number of risers (3'-1½" or 37½ ÷ 5 = 7½"). (See Math Review 10.) The width of the treads is found by dividing the total run by the number of treads (3'-4" or 40" ÷ 4 = 10").

To calculate the size of the treads and risers in stairs with landings, treat each part as a separate stair. However, the treads and risers should be the same size in each part.

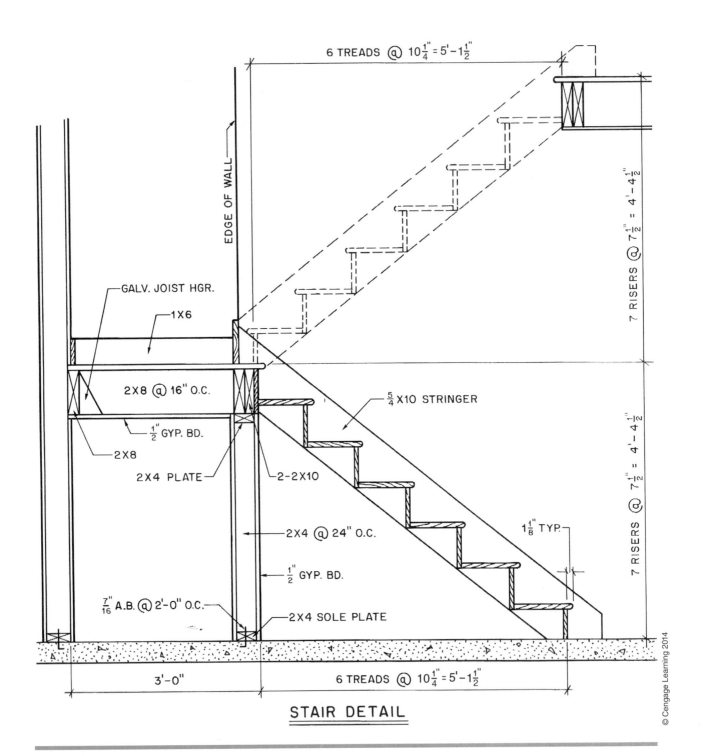

**Figure 30–6.** Stair detail with 2 × 4 studs to support the landing.

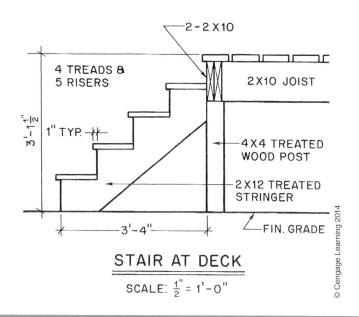

**Figure 30-7.** Typical detail for deck stairs.

### USING WHAT YOU LEARNED

Although interior staircases are often built off-site by a millwork company, stairs attached to exterior decks are usually built on-site by a carpenter. This means that the carpenter must be able to determine the sizes of all the stair parts by studying the construction drawings. How many risers and the desired height are required for the stairs going from the ground to the lower south deck of the Lake House? These stairs are described on Floor Framing Plan 1/6. A callout indicates that there are four risers and each riser is 8 inches high.

## Assignment

Refer to the Lake House drawings (in the packet) to complete the assignment.

Questions 1 through 6 refer to the stairs between the south decks of the Lake House.

1. What is the total run of the stairs between the decks?
2. What is the width of each tread?
3. What is the length of each tread?
4. What is the rise of each step in these stairs?
5. How many stringers are used under these stairs?
6. What size material is used for the stringers?

Questions 7 through 13 refer to the stairs from the kitchen to the bedroom level in the Lake House.

7. How many risers are there, and how high is each riser?
8. How many treads are there, and how wide (front to back not including nosing) is each tread?
9. What is the total rise of the stairs?
10. Is this stair built with open or housed stringers?
11. The railing at this stair extends the length of the kitchen hall. If vertical railing supports are spaced at 16" O.C., how many vertical supports are used?
12. How long is each vertical railing support?
13. What material is used for the horizontal rails?

# UNIT 31

# Insulation and Room Finishing

## Objectives

After completing this unit, you will be able to perform the following tasks:

- Identify the insulation to be used in walls, floors, and ceilings.
- Identify the wall, ceiling, and floor covering material to be used.
- List all the kinds of interior molding to be used.

## Insulation

Thermal insulation is any material that is used to resist the flow of heat. In very warm climates, thermal insulation is used to resist the flow of heat from the outside to the inside. In cold climates, thermal insulation is used to resist the flow of heat from the inside to the outside. Insulating material is rated according to its ability to resist the flow of heat. The measure of this resistance is the R value of the material. The higher the **R value**, the better the material insulates. Typical R values for sidewall or attic insulation range from R-3 to R-38 (see **Figure 31–1**).

### GREEN NOTE

*The R-value of insulation can range from R-3.14 per inch for some batt insulation to as high as R-7.20 per inch for polyisocyanurate foam, but there are other factors to consider such as embodied energy and ease of installation. Most important is the quality of the installation, especially for batt insulation, which is still the most common type of insulation in frame homes. Compressing batt insulation only 5 percent can reduce its R-value by up to 50 percent; and gaps are, of course, uninsulated openings. Spray insulations are much less susceptible to compression.*

Thermal insulation is generally available in four forms. *Foamed-in-place* materials are synthetic compounds that are sprayed onto a surface and that then produce an insulating foam by a chemical reaction (see **Figure 31–2**). *Rigid boards* are plastic foams that have been produced in board form at a factory. Another common type of rigid insulation is made of fiberglass that is manufactured in rigid form instead of as flexible blankets (see **Figure 31–3**). Rigid boards are frequently used for foundation insulation, under concrete slabs, and for sheathing. *Blanket insulation* is in the form of flexible rolls or *batts* usually made of fiberglass wool (see **Figure 31–4**). Loose *pouring insulation* can be any of a variety of materials that can be poured into place and has good insulating property. A common pouring insulation is loose fiberglass (see **Figure 31–5**).

The insulation is shown in the building sections. If the insulation is to be installed between studs, joists, or rafters, it will be sized accordingly. For example, if

| Type | Typical Thicknesses and R Values | Comments |
| --- | --- | --- |
| Fiberglass blankets & batts | 3/4", R-3<br>2 1/2", R-8<br>3 1/2", R-11<br>3 1/2", R-13<br>3 1/2", R-15<br>6", R-19<br>9", R-30<br>12", R-38 | Flexible blanket-like material. Available with or without vapor barrier on one side |
| Fiberglass blowing wool | R value depends on depth of coverage, but is slightly less than fiberglass blankets. | Other types of blowing or pouring insulation include cellulose and mineral wool. |
| Rigid fiberglass board | 1", R-4.4 | Material is similar to fiberglass blankets, but with rigid binder to create rigid boards. Usually faced with aluminum foil. |
| Closed-cell urethane foamed board | 1/2", R-3.12<br>3/4", R-4.70<br>1", R-6.25<br>1 1/2", R-7.81<br>2", R-12.5 | Plastic that has been cured in a foamed state to introduce bubbles of air. This creates a rigid board. |
| Foamed-in-place urethane | R value depends on thickness depth, but is approximately the same as rigid urethane boards. | Other plastic materials may also be foamed in place. |
| Polyisocyanurate foam | 1", R-7.20<br>2", R-14.40 | Usually available as rigid foam boards. |

**Figure 31–1.** Common types of thermal insulation.

**Figure 31–2.** Foamed-in-place insulation.

batts are to be used in a 2 × 4 wall, the insulation will be indicated as 3½ inches thick—the width of a 2 × 4. Where insulation is used in ventilated spaces, there should be room for the necessary air circulation (see **Figure 31–6**).

**Figure 31–3.** Rigid insulation.

Where the insulation includes a vapor barrier, such as kraft paper, foil, or polyethylene, the vapor barrier is installed on the heated side of the wall. This prevents the moisture in the warm air from passing

Insulation and Room Finishing 221

Courtesy of Owens Corning Fiberglas Corporation

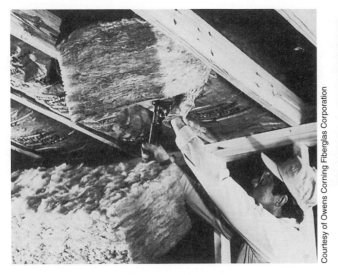

**Figure 31–4.** Blanket insulation.

Courtesy of Owens Corning Fiberglas Corporation

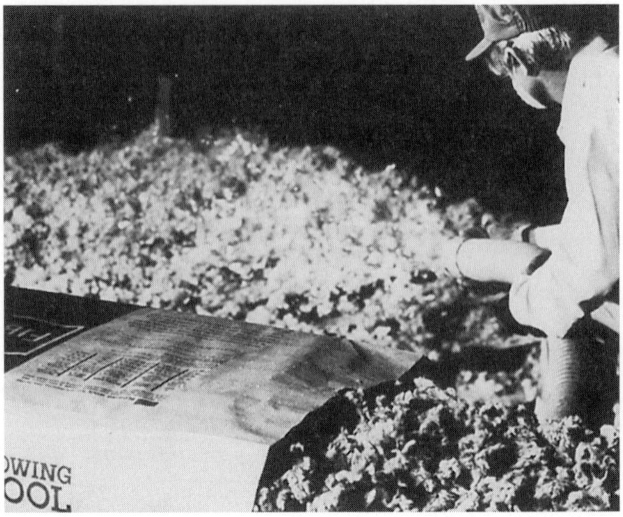

**Figure 31–5.** Blowing wool.

Courtesy of Home Planners, Inc.

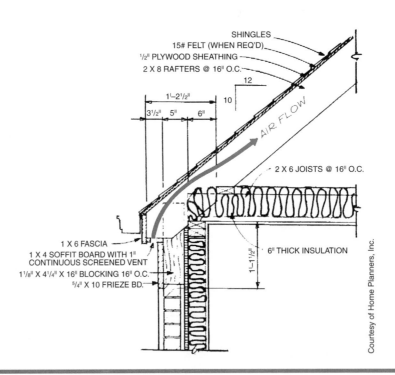

**Figure 31–6.** The insulation must not block the airflow.

through the wall and condensing on the cold side of the wall. Such condensation can reduce the R value of the insulation and cause painted surfaces to blister and peel.

## Wall and Ceiling Covering

By far the most widely used wall surface material is **gypsum wallboard.** The most common thicknesses are $^3/_8$ inch, ½ inch, and $^5/_8$ inch. If gypsum board (sometimes called *Sheetrock™*) is to be used over framing that is spaced more than 16 inches O.C. or over masonry, the designer may call for furring. **Furring** consists of narrow strips of wood, usually spaced at 16 or 12 inches O.C. to which the wall covering is fastened (see **Figure 31–7**). The furring, if any is to be used, and the thickness of the gypsum wallboard are indicated on the wall sections.

*Ceramic tile* also is frequently used on bathroom walls. Ceramic tile may be installed over a base of water-resistant gypsum board, plaster, or plywood.

Gypsum wallboard is the most common ceiling treatment in new home construction. Suspended ceilings are common in commercial construction. Suspended ceilings consist of panels or ceiling tiles supported in a lightweight metal framework. The metal framework is suspended several inches below the ceiling framing on steel wires (see **Figure 31–8**).

Building sections or wall sections usually include a typical wall and ceiling. This is representative of what is planned for most of the walls and ceilings in the house. However, there may be some exceptions, such as the bathrooms or kitchen, where water-resistant wallboard is required. Somewhere within the contract documents, you should find a complete list of all room finishes. This may be on one of the drawings, or it may be written into the specifications. **Figure 31–9** shows a room finish schedule that might be included on the drawings. It is common for things like finish color to be left for the owner to choose.

## Finished Floors

A list of possible floor materials would be very long. However, most of the materials fall into one of the following categories: wood, carpet, ceramic, masonry, and resilient materials such as vinyl tile. The finished floor covering is easily found in a schedule of room finishes, but the underlayment for each category is different. **Underlayment** is any material

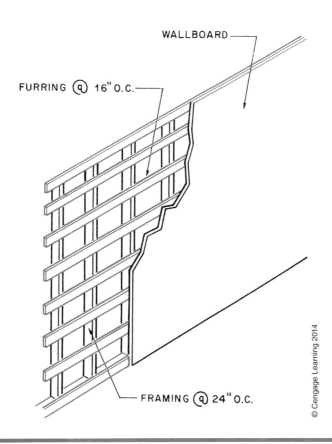

**Figure 31–7.** Furring is used to provide a nailing surface when framing is spaced too wide or over masonry and concrete.

Insulation and Room Finishing 223

**Figure 31–8.** Suspended ceiling.

that is used to prepare the **subfloor** to receive the finished floor.

Architectural drafters differ in how they indicate what underlayment is to be used and how the finished floor is to be installed. Some do not include any underlayment on the drawings but rely on the builder's knowledge of good construction practices. This is a dangerous practice. If you find this situation, the architect should be asked for clarification. Sometimes when the drawings do not describe the underlayment, the specifications do. Some drafters indicate the underlayment on the floor plans. Other drafters include the details and section of each area with different types of finished floors, so the underlayment can be shown there.

## Interior Molding

Molding is used to decorate surfaces, protect the edges and corners of surfaces, and conceal joints or seams between surfaces. Molding may be made of wood, plastic, or metal and is available in many shapes and styles. The shapes of commonly used wood molding have been standardized. Each shape is identified by a number, **Figure 31–10**.

Most interior molding is shown on detail drawings. The following are some of the most common uses for interior molding:

- Window casing
- Window **stool**
- Door casing
- Base (bottom of wall)
- **Cove** (top of wall)
- Chair rail (middle of wall)
- Trim around the fireplace mantel
- Trim around built-in cabinets

| ROOM | FLOOR | WALLS | CEILING |
|---|---|---|---|
| KITCHEN | QUARRY TILE | GYP. BD. W/WALL PAPER | 12"X 12" TILE |
| DINING ROOM | OAK PARQUET | GYP. BD. | GYP. BD. |
| LIVING ROOM | CARPET/PART. BD. | GYP. BD. | GYP. BD. |
| FAMILY ROOM | CARPET/PART. BD. | HD. BD. PANEL/GYP. BD. | GYP. BD. |
| BEDROOM #1 | CARPET/PART. BD. | GYP. BD. | GYP. BD. |
| BEDROOM #2 | CARPET/PART. BD. | GYP. BD. | GYP. BD. |
| BEDROOM #3 | CARPET/PART. BD. | GYP. BD. | GYP. BD. |
| BATH #1 | CERAMIC TILE | CERAMIC TILE/GYP. BD. | GYP. BD. |
| BATH #2 | CERAMIC TILE | CERAMIC TILE/GYP. BD. | GYP. BD. |
| CLOSETS | CARPET/PART BD. | GYP. BD. | GYP. BD. |
| FOYER | SLATE | GYP. BD. W/WALL PAPER | GYP. BD. |

**Figure 31–9.** Room finish schedule.

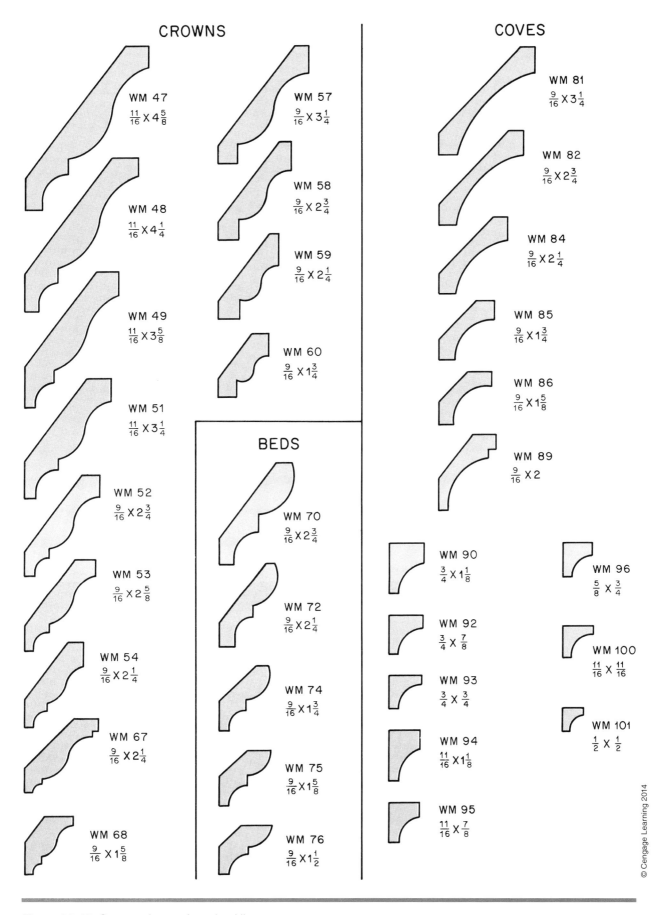

**Figure 31-10.** Common shapes of wood molding.

## QUARTER ROUNDS

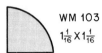
WM 103
$1\frac{1}{16} \times 1\frac{1}{16}$

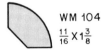
WM 104
$\frac{11}{16} \times 1\frac{3}{8}$

WM 105
$\frac{3}{4} \times \frac{3}{4}$

WM 106
$\frac{11}{16} \times \frac{11}{16}$

WM 107
$\frac{5}{8} \times \frac{5}{8}$

WM 108
$\frac{1}{2} \times \frac{1}{2}$

WM 109
$\frac{3}{8} \times \frac{3}{8}$

WM 110
$\frac{1}{4} \times \frac{1}{4}$

## HALF ROUNDS

WM 120
$\frac{1}{2} \times 1$

WM 122
$\frac{3}{8} \times \frac{11}{16}$

WM 123
$\frac{5}{16} \times \frac{5}{8}$

WM 124
$\frac{1}{4} \times \frac{1}{2}$

## FLAT ASTRAGALS

WM 133
$\frac{11}{16} \times 1\frac{3}{4}$

WM 134
$\frac{11}{16} \times 1\frac{3}{8}$

WM 135
$\frac{7}{16} \times \frac{3}{4}$

## BASE SHOES

WM 126  $\frac{1}{2} \times \frac{3}{4}$

WM 129  $\frac{7}{16} \times \frac{11}{16}$

WM 127  $\frac{7}{16} \times \frac{3}{4}$

WM 131  $\frac{1}{2} \times \frac{3}{4}$

## SHELF EDGE/SCREEN MOLD

WM 137  $\frac{3}{8} \times \frac{3}{4}$

WM 141  $\frac{1}{4} \times \frac{5}{8}$

WM 138  $\frac{5}{16} \times \frac{5}{8}$

WM 142  $\frac{1}{4} \times \frac{3}{4}$

WM 140  $\frac{1}{4} \times \frac{3}{4}$

WM 144  $\frac{1}{4} \times \frac{3}{4}$

## GLASS BEADS

WM 147  $\frac{1}{2} \times \frac{9}{16}$

WM 148  $\frac{3}{8} \times \frac{3}{8}$

## BASE CAPS

WM 163  $\frac{11}{16} \times 1\frac{3}{8}$

WM 167  $\frac{11}{16} \times 1\frac{1}{8}$

WM 164  $\frac{11}{16} \times 1\frac{1}{8}$

WM 172  $\frac{5}{8} \times \frac{3}{4}$

WM 166  $\frac{11}{16} \times 1\frac{1}{4}$

## BRICK MOLD

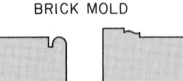

WM 175  $1\frac{1}{16} \times 2$

WM 180  $1\frac{1}{4} \times 2$

WM 176  $1\frac{1}{16} \times 1\frac{3}{4}$

## DRIP CAPS

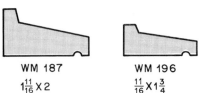

WM 187  $1\frac{11}{16} \times 2$

WM 196  $\frac{11}{16} \times 1\frac{3}{4}$

WM 188  $1\frac{1}{16} \times 1\frac{5}{8}$

WM 197  $\frac{11}{16} \times 1\frac{5}{8}$

## CORNER GUARDS

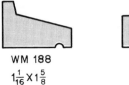
WM 199  $1 \times 1$

WM 200  $\frac{3}{4} \times \frac{3}{4}$

WM 201  $1\frac{5}{16} \times 1\frac{5}{16}$

WM 204  $1\frac{5}{16} \times 1\frac{5}{16}$

WM 202  $1\frac{1}{8} \times 1\frac{1}{8}$

WM 205  $1\frac{1}{8} \times 1\frac{1}{8}$

WM 203  $\frac{3}{4} \times \frac{3}{4}$

WM 206  $\frac{3}{4} \times \frac{3}{4}$

## BATTENS

WM 224  $\frac{9}{16} \times 2\frac{1}{4}$

WM 229  $\frac{11}{16} \times 1\frac{5}{8}$

## ROUNDS

WM 232  $1\frac{5}{8}$
WM 233  $1\frac{5}{16}$
WM 234  $1\frac{1}{16}$

## SQUARES

WM 236  $1\frac{5}{8} \times 1\frac{5}{8}$
WM 237  $1\frac{5}{16} \times 1\frac{5}{16}$
WM 238  $1\frac{1}{16} \times 1\frac{1}{16}$
WM 239  $\frac{3}{4} \times \frac{3}{4}$

**Figure 31–10.** (Continued)

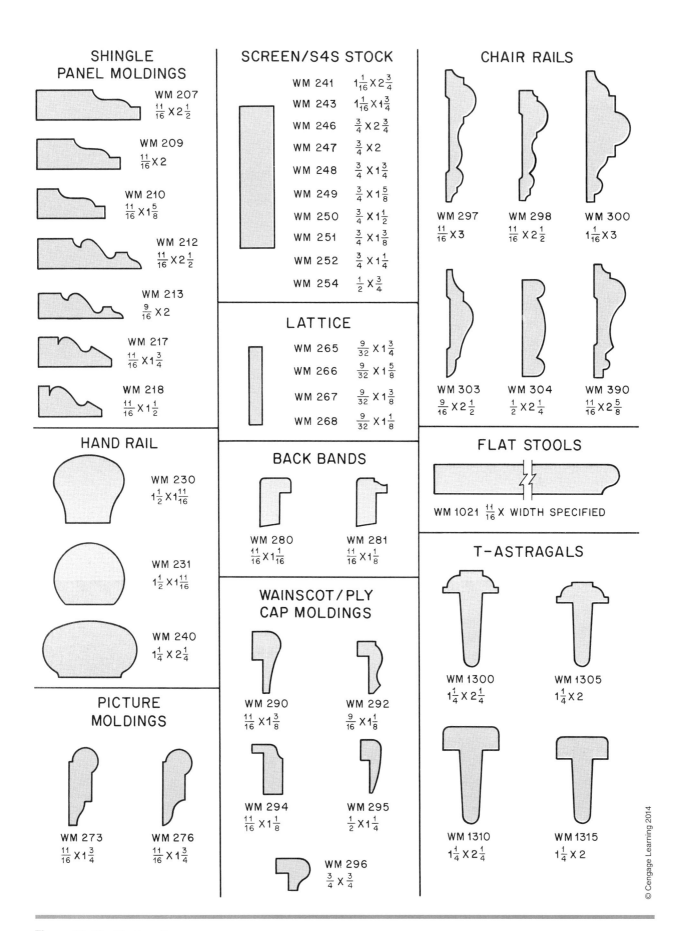

**Figure 31–10.** *(Continued)*

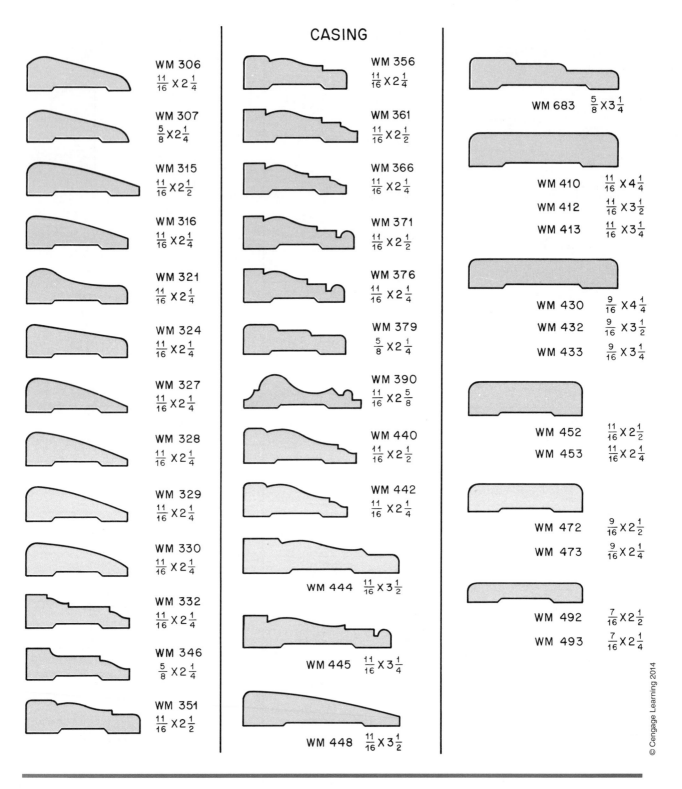

**Figure 31–10.** *(Continued)*

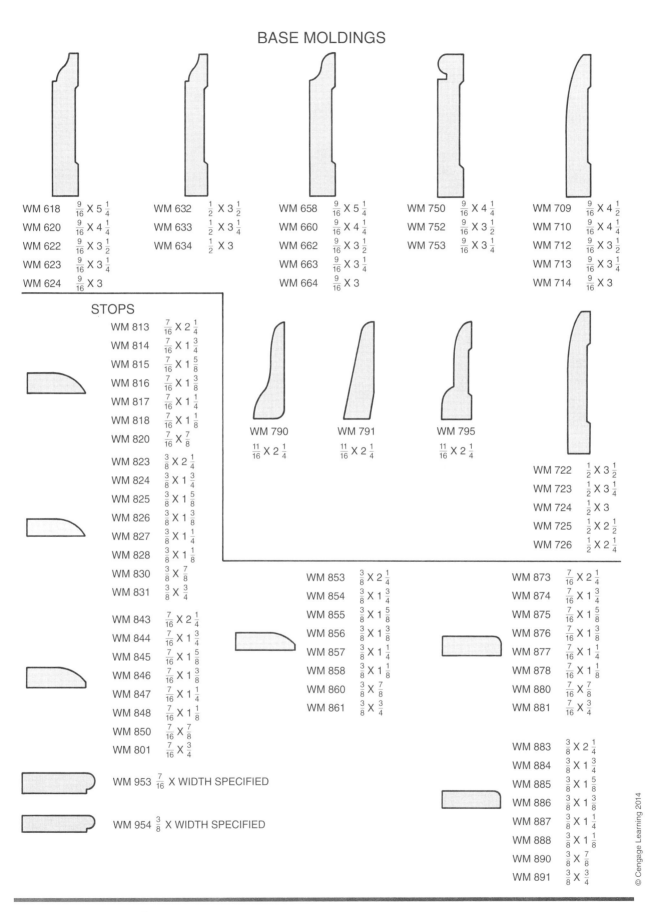

**Figure 31–10.** *(Continued)*

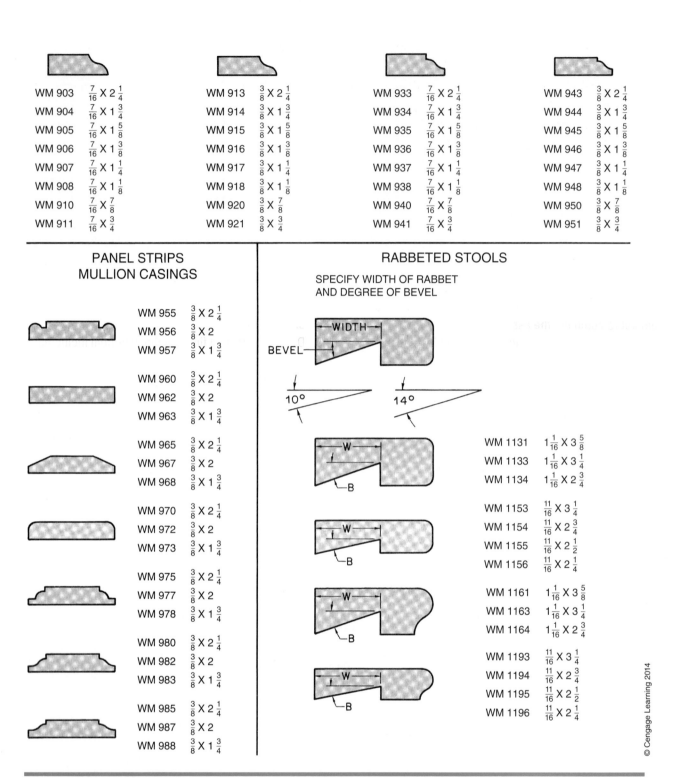

**Figure 31-10.** *(Continued)*

**USING WHAT YOU LEARNED**

It is important to search the construction drawings to determine what kind and size of material is used for interior trim throughout the house. In the Lake House, what kind and size of material is used as trim at the bottoms of the walls in the lower level, where the floors are concrete? Wall Section 3/4 shows 1" × 4" cellular PVC base trim at this location. Cellular PVC is not affected by the moisture that may be in the concrete.

## Assignment

Refer to the Lake House drawings in your textbook packet to complete the assignment.

1. What size or rating and what kind of insulation are to be used in each of the following locations?
   a. Framed exterior walls
   b. Roof
   c. Under heat sink
   d. Masonry walls of playroom
2. What type of molding is to be used as casing around interior doors?
3. What type of molding is used at the bottom of interior walls and partitions?
4. What kind of trim is used to cover the lower edges of exposed LVL beams?
5. Describe the wall finish in the playroom including:
   a. What the wall finish material is fastened to
   b. The kind of material used for wall finish
6. What material is used for ceiling finish in the playroom?
7. What material is used for subflooring on typical framed floors?
8. What is the finished floor material at the heat sink?
9. What covers the interior faces of LVL beams?
10. What is the finished wall material in the bedrooms?

# UNIT 32 Cabinets

## Objectives

After completing this unit, you will be able to perform the following tasks:

○ List the sizes and types of cabinets shown on a set of drawings.

○ Identify cabinet types and dimensions in manufacturer literature.

## Showing Cabinets on Drawings

The layout of the cabinets can be determined by reading the floor plan and cabinet elevations together. The floor plan normally includes reference marks indicating each cabinet elevation (see **Figure 32–1**). The floor plan shows the location of major appliances and may include some overall layout dimensions.

When the project has separate electrical drawings, the electrical installer must coordinate these designated outlet locations between the electrical and architectural drawings. There may be electrical outlets shown on the architectural drawings and not shown on the electrical drawings. These variances must be resolved early in the rough-in stage of construction.

More complete cabinet information is usually shown on the cabinet elevations (see **Figure 32–2**). Cabinet elevations are drawn for each direction from which cabinets can be viewed. These elevations show the types of cabinets, their sizes, and their arrangement. Cabinet types and sizes are recognized by a combination of drawing representations and commonly used letter/number designations. Some drawings rely heavily on standard dimensioning and

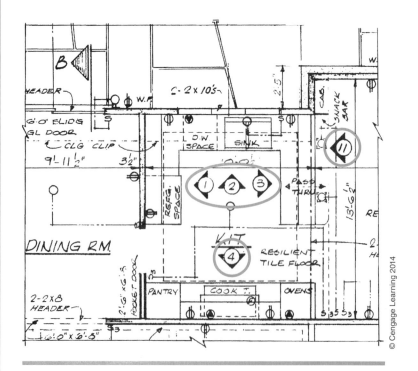

**Figure 32–1.** Kitchen floor plan with key to cabinet elevations.

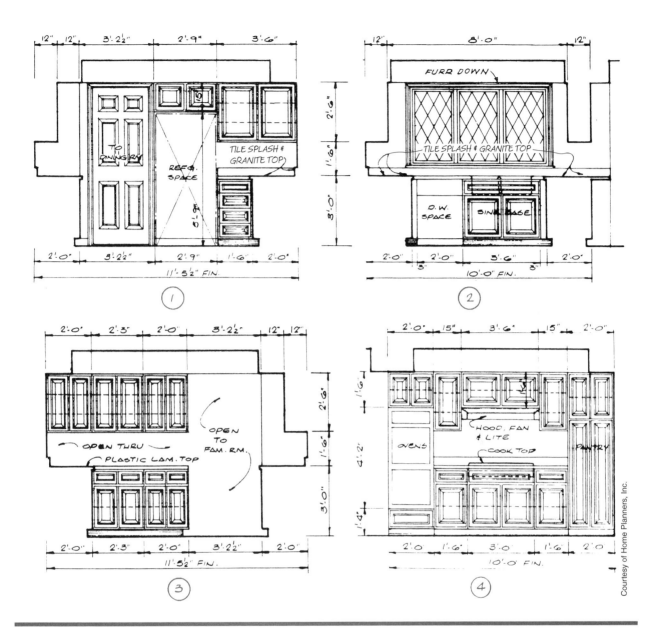

**Figure 32–2.** Elevations for Figure 32–1.

pictorial representation. This is true of the case shown in **Figure 32–2**. Other drawings include a letter/number designation for each cabinet, as shown in **Figure 32–3**. The letter part of the designation represents the type of cabinet. Architects and drafters vary in how they use letter designations, but they are usually easy to understand after a moment's study. Some typical letter designations are shown in **Figure 32–4**.

The numbers in the cabinet designation represent the dimensions of the cabinet. Base cabinets are a standard height (usually 34½ inches, resulting in 36 inches when a 1½-inch countertop is added) and a standard front-to-back depth (usually 24 inches) Therefore, base cabinet designations only include two digits to represent width in inches. Wall cabinets are a standard front-to-back depth (usually 12 inches), but the width and height vary. Therefore, wall cabinet designations include four digits—two for width and two for height. The first and second digits usually indicate width, and the third and fourth digits usually indicate height. **Figure 32–5** explains a typical cabinet designation.

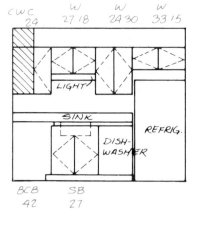

KITCHEN ELEVATIONS

**Figure 32–3.** Letter/number designations for cabinets.

| LETTER | CABINET TYPE |
|---|---|
| W | WALL CABINET |
| WC | CORNER WALL CABINET |
| B | BASE CABINET |
| D OR DB | DRAWER BASE CABINET |
| BC | BASE CORNER CABINET |
| RC | REVOLVING CORNER CABINET |
| SF | SINK FRONT (ALSO FOR COOKTOP) |
| SB | SINK BASE (ALSO FOR COOKTOP) |
| U | UTILITY OR BROOM CLOSET |
| OV | OVEN CABINET |

**Figure 32–4.** Key to typical cabinet designations.

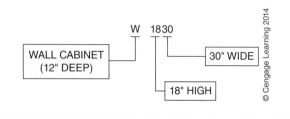

**Figure 32–5.** Typical cabinet designations.

In addition to cabinets, most cabinet manufacturers provide a variety of accessories:

- Shelves are finished to match the cabinets, so they can be used in open areas.
- Valances are prefinished decorative pieces to use between wall cabinets (over a sink, for example).
- Filler pieces are prefinished boards used to enclose narrow spaces between cabinets.
- A variety of molding may be used for trim.

**USING WHAT YOU LEARNED**

Workers from many building trades must be able to understand cabinet size designations. Electricians, for example, need to know where the edge of a particular cabinet will be in order to install an outlet. Plumbers need to know cabinet types and sizes in order to locate piping for sinks, dishwashers, and icemakers. Of course, the carpenters who install the cabinets need to know the right location for installing each cabinet.

Describe the cabinet located directly above the dishwasher. The notation is W2418. The W indicates that it is a standard wall cabinet. The first two numerals, 24, indicate that it is 24 inches wide. The 18 indicates that it is 18 inches high.

## Assignment

Refer to the Lake House drawings in your textbook packet to complete the assignment.

Make a list of all of the Lake House kitchen cabinets, including type of cabinet, height, width, and depth.

# UNIT 33

# Lake House Specifications

## Objectives

After completing this unit, you will be able to perform the following tasks:

○ Find specific information in construction specifications.

○ Apply information from specifications to specific construction drawings.

Construction drawings provide much information about how a building is to be constructed—the positions and sizes of the parts and materials to be used. Some information is more practically recorded in written form and the construction specifications, often referred to as the **specs** provide that information. For example, it is not practical to show the grade of lumber to be used for roof framing or to indicate who is responsible for clean-up after construction is finished. For single-family homes, all of the necessary technical information is often conveyed on the drawings. In this case, there are likely to be extra notes on the drawings to convey the information that would otherwise be conveyed in separate specifications. The drawings for the Hidden Valley townhouse in Section 3 are a good example of notes on drawings that provide technical specifications. In this unit, we consider specifications that are written separately from the drawings.

## CSI Format

Large and complex buildings can require a lot of information to be conveyed in the specs. Specs for a large commercial construction project can be hundreds of pages. To make it easier to find information, specification writers follow the Construction Specifications Institute's MasterFormat (**CSI format**) for organizing their specs. MasterFormat standardizes titles and section numbers for organizing data about construction requirements, products, and activities. It can include up to 48 divisions with numbered sections in each division (see **Figure 33-1**).

While this format makes organization of large projects easier to follow, it would be very cumbersome for a small project, like a single-family home. A more practical and often used system is to partially follow the CSI format. The major divisions follow the CSI format, but the numbering of subsections is modified or dropped altogether.

The format of the specs for the Lake House is an adaptation of CSI format. The major divisions follow CSI numbering, with unused divisions omitted. The subsections have been rearranged to better fit the needs of the project. Each of the technical divisions begins with a brief description of the work covered by that division. Also included are a description of the materials needed for that work and subsections for each type of work in the applicable division. To find a particular part of the construction project in the specs, use the following steps:

1. Look up the appropriate division on the Table of Contents.
2. Read the "Work Included" at the beginning of the division to be sure you have the right division.
3. Find the description of the necessary materials right after the "Work Included."
4. Read the applicable subsection covering the work you want to do.

| Division | Title | Division | Title |
|---|---|---|---|
| 1 | General Requirements | 27 | Communications |
| 2 | Existing Conditions | 28 | Electronic Safety and Security |
| 3 | Concrete | Divisions 29 and 30 are reserved for future expansion. | |
| 4 | Masonry | 31 | Earthwork |
| 5 | Metals | 32 | Exterior Improvements |
| 6 | Wood, Plastics, and Composites | 33 | Utilities |
| 7 | Thermal and Moisture Protection | 34 | Transportation |
| 8 | Openings | 35 | Waterway and Marine Construction |
| 9 | Finishes | | |
| 10 | Specialties | Divisions 36 through 39 are reserved for future expansion. | |
| 11 | Equipment | 40 | Process Integration |
| 12 | Furnishings | 41 | Material Processing and Handling Equipment |
| 13 | Special Construction | | |
| 14 | Conveying equipment | 42 | Process Heating, Cooling, and Drying Equipment |
| Divisions 15 through 20 are reserved for future expansion. | | 43 | Process Gas and Liquid Handling, Purification and Storage Equipment |
| 21 | Fire Suppression | | |
| 22 | Plumbing | 44 | Pollution Control Equipment |
| 23 | Heating, Ventilating, and Air Conditioning | 45 | Industry Specific Manufacturing Equipment |
| | | 46 | Water and Wastewater Equipment |
| Division 24 is reserved for future expansion. | | | |
| 25 | Integrated Automation | Division 47 is reserved for future expansion. | |
| 26 | Electrical | 48 | Electrical Power Generation |

**Figure 33–1.** MasterFormat for numbering construction standards.

## SPECIFICATIONS FOR LAKE HOUSE

### CONTENTS

| | | | |
|---|---|---|---|
| GENERAL CONDITIONS | | THERMAL & MOISTURE PROTECTION | 07000 |
| GENERAL REQUIREMENTS | 01000 | DOORS & WINDOWS | 08000 |
| SITE WORK | 02000 | FINISHES | 09000 |
| CONCRETE | 03000 | FIREPLACE | 10000 |
| MASONRY | 04000 | EQUIPMENT | 11000 |
| STRUCTURAL STEEL | 05000 | MECHANICAL | 15000 |
| WOOD & PLASTIC | 06000 | ELECTRICAL | 16000 |

(Divisions 12000, 13000, and 14000 are not used.)

Lake House Specifications 237

## GENERAL CONDITIONS

The General Conditions of the Contract for Construction, AIA Document A 107, whether or not bound herein, are hereby incorporated into and made a part of this contract and these specifications.

## 01000 GENERAL REQUIREMENTS

A. ARCHITECT'S SUPERVISION
The architect will have continual supervisory responsibility for this job.

B. TEMPORARY CONVENIENCES /
The general contractor shall provide suitable temporary conveniences for the use of all workers on this job. Facilities shall be within a weathertight, painted enclosure complying with legal requirements. The general contractor shall maintain all temporary toilet facilities in a sanitary condition.

C. PUMPING
The general contractor shall keep the excavation and the basement free from water at all times and shall provide, maintain, and operate at his own expense such pumping equipment as shall be necessary.

D. PROTECTION
The general contractor shall protect all existing driveways, parking areas, side walks, curbs, and existing paved areas on, or adjacent to the owner's property.

E. GRADE LINES, LEVELS, AND SURVEYS
The owner shall establish the lot lines.
The general contractor shall:
1. Establish and maintain bench marks.
2. Verify all grade lines, levels, and dimensions as shown on the drawings, and report any errors or inconsistencies before commencing work.
3. Layout the building accurately under the supervision of the architect.

F. FINAL CLEANING
In addition to the general room cleaning, the general contractor shall do the following special cleaning upon completion of the work:
1. Wash and polish all glass and cabinets.
2. Clean and polish all granite.
3. Clean and polish all hardware.
4. Remove all marks, stains, fingerprints, and other soil or dirt from walls, woodwork, and floors.

G. GUARANTEES
The general contractor shall guarantee all work performed under the contract against faulty materials or workmanship. The guarantee shall be in writing with duplicate copies delivered to the architect. In case of work performed by subcontractors where guarantees are required, the general contractor shall secure written guarantees from these subcontractors. Copies of these guarantees shall be delivered to te architect upon completion of the work. Guarantees shall be signed by both the subcontractor and the general contractor.

H. FOREMAN
The general contractor shall have a responsible foreman at the building site from the start to the completion of construction. The foreman shall be on duty during all working hours.

I. FIRE INSURANCE
The owner shall effect and maintain builder's risk completed value on this job.

## 02000 SITE WORK

WORK INCLUDED
This work shall include, but shall not be limited by the following:
A. Clearing the site.
B. Excavating, backfilling, grading, and related items.
C. Removal of excess earth.
D. Protection of existing trees to remain on the site.

All excavation and backfilling required for heating, plumbing and electrical work will be done by the respective contractors and are not included under site work.
It is the contractor's responsibility to field inspect existing conditions to determine the scope of the work.

## 02100 CLEARING

A. Clean the area within the limits of the building of all trees, shrubs, or other obstructions as necessary.
B. Within the limits of grading work as shown on the drawings remove such trees, shrubs, or other obstructions as are indicated on the drawings to be removed, without injury to trunks, interfering branches, and roots of trees to remain. Do cutting and trimming only as directed. Box and protect all trees and shrubs in the construction area to remain; maintain boxing until finished grading is completed.
C. Remove all debris from the site; do not use it for fill.

## 02200 EXCAVATION

A. Carefully remove all sod and soil throughout the area of the building and where finish grade levels are changed. Pile on site where directed. This soil is to be used later for finish grading.
B. Do all excavation required for footings, piers, walls, trenches, areas, pits, and foundations. Remove all materials encountered in obtaining indicated lines and grades required.
Beds for all foundations and footings must have solid, level, and undisturbed bed bottoms. No backfill will be allowed and all footings shall rest on unexcavated earth.
C. The contractor shall notify the architect when the excavation is complete so that he may inspect all soil before the concrete is placed.
D. Excavate to elevations and dimensions indicated, leaving sufficient space to permit erection concrete forms, walls, waterproofing, masonry, and inspection of foundations. Protect the bottom of the excavation from frost.

## 02260 BACKFILL

A. All outside walls shall be backfilled to within 6 inches of the finished grade with clean fill. Backfill shall be thoroughly compacted.
B. Unless otherwise directed by the architect, no backfill shall be placed before the first floor framing is in place. No backfill shall be placed until all walls have developed such strength to resist thrust due to filling operations.

## 02270 GRADING

A. Do all excavating, filling, and rough grading to bring the entire area outside of the building to levels shown on the drawings.
B. Where existing trees are to remain, if the new grade is lower than the natural grade under the trees, a sloping mound shall be left under the base of the tree extending out as far as the branches; if the grade is higher, a well shall be constructed around the base of the tree to provide the roots with air and moisture.
C. After rough grading has been completed and approved, spread topsoil evenly to the previously stripped area. Prepare the topsoil to receive grass seed by removing stone, debris, and unsuitable materials. Hand rake to remove water pockets and irregularities. Seeding will be done by the owner.
D. Furnish and place run of bank gravel as approved under all floor slabs.

## 03000 CONCRETE

WORK INCLUDED
Provide all materials, labor, equipment, and services necessary to furnish, deliver, and install all work of this section, as shown on the drawings, as specified herein, and/or as required by job conditions including, but not limited to the following:
A. Concrete for all footings and piers.
B. Concrete for all foundation walls.
C. Concrete for all slabs on ground.
D. Concrete for slab at heat sink.
E. Furnishing and installation of all required anchors.
F. Supplying, fabrication, and placement of all reinforcing bars and mesh and wire reinforcement for concrete where shown, called for, or required with proper supporting devices.
G. Erection of all forms required for concrete work and removal upon completion of the work.
H. The finishing of all concrete work as hereinafter specified.
I. Porous fill below slabs on ground.

## 03010 MATERIAL

A. Fine Aggregate
Fine aggregates for concrete shall consist of natural sand having clean, hard, sharp, uncoated grains free from injurious amounts of dust, lumps, soft or flaky particles, shale, alkali, organic matter, loam, or other deleterious substances.
B. Coarse Aggregate (stone)
Coarse aggregates shall consist of crushed stone or gravel having clean, hard, strong, durable, uncoated particles, free from injurious amounts of soft, friable, thin, elongated or laminated pieces, alkali, organic, or other deleterious matter.
C. Water
All water used in connection with concrete work shall be clean and free from deleterious materials or shall be water used for drinking daily.
D. Portland Cement
Portland cement shall be an approved domestic brand complying with Standard Specifications for Portland Cement, ASTM Designation C-150, Type 1. Only one brand of cement shall be used throughout the course of the work.
E. All concrete is to be machine mixed in an approved mixer with a water metering device. Concrete is to reach a compressive strength of 2500 psi after 28 days.
F. Reinforcement
All reinforcing, unless otherwise shown or specified, shall conform to ASTM A-615, Grade 60. Wire mesh reinforcing shall have a minimum ultimate tensile strength of 70,000 psi, and shall conform to ASTM Specifications A-185, latest edition.

## 03320 INSPECTION & PLACING

A. All reinforcing shall be free of rust, scale, oil, or other coatings that tend to reduce the bond to concrete. All reinforcing is to be tied with 18 gauge wire at intersections and shall be held securely in position during the pouring of concrete.
B. The architect will inspect all footing beds, forms, and reinforcing, just prior to placing concrete for footings, foundation, and slabs.
C. All concrete shall be placed upon clean surfaces, and properly compacted fill, free from standing water. The concrete shall be compacted and worked into corners and around reinforcing.
D. All concrete to be true and level as indicated on drawings to within ± ¼ inch in 10 feet.

## 03330 FINISHING

A. Slabs in occupied spaces shall be troweled smooth and free of trowel marks.
B. Slabs in unoccupied spaces will have wood float finish.

## 04000 MASONRY

WORK INCLUDED
This work shall include but not be limited by the following:
A. Brickwork
B. Concrete block work
C. Mortar for brick and block work

## 04010 MATERIALS

A. Delivery and storage
All material shall be delivered, stored, and handled so as to prevent the inclusion of foreign materials and the damage of the materials by water or breakage.
Packaged materials shall be delivered and stored in the original packages until they are ready for use.

Lake House Specifications

B. Materials showing evidence of water or other damage shall be rejected.
C. Brick shall be chosen by the owner from approved samples.
D. Concrete block shall be load bearing, hollow, concrete masonry units and shall conform to the standard specifications of ASTM C-145-71.
E. Mortar used for laying brick and concrete block shall consist of (1) one part masonry cement to (3) parts sand. The mortar ingredients shall comply with the following requirements:
  1. Masonry cement: ASTM C-91 T Type 2.
  2. Aggregates: ASTM C-144.
  3. Water: Clean, fresh, free from acid, alkali, sewage, or organic material.

## 04220 INSTALLATION

A. All work shall be laid true to dimensions, plumb, square, and in bond and properly anchored.
B. Joints shall be finished as follows:
C. All brick shall be laid on full mortar bed with a shoved joint. All joints shall be completely filled with mortar. All horizontal and vertical joints shall be raked 3/8 inch deep.
D. All mortar joints for concrete block masonry shall have full mortar coverage on vertical and horizontal face shells.
E. Vertical joints shall be shoved tight. Full mortar bedding shall have ruled joints.
F. Concealed work shall have joints cut flush.
G. Fill voids in top course with masonry and set anchor bolts as shown on construction drawings.
H. Protection: Cover the walls each night and when the work is discontinued due to weather.

## 04240 CLEANING AND POINTING

A. Point up all the voids and open joints with mortar. Remove all of the excess mortar and dirty spots from the entire surface.
B. Upon completion all brickwork shall be thoroughly cleaned with clean water and stiff fiber brushes and then rinsed with clean water. The use of acids or wire brushes is not permitted.

## 05000 STRUCTURAL STEEL

WORK INCLUDED
This work shall include but not be limited by the following:
A. Structural tube columns.
B. Welded flanges and plates.
C. Structural steel rafters and beams.
D. Stanchions at railings: This contractor shall supply fabricated stanchions to be installed by others.
E. Grouting base plates.

## 05010 MATERIALS

A. All structural steel to conform to ASTM A-36.
B. Welding electrodes to conform to American Welding Society A5.1, E70 series.

## 05100 FABRICATION & ERECTION

A. Drilling or punching of holes in columns, beams, and rafters shall not be permitted unless approved.
B. Welds shall be by qualified operators and shall achieve complete penetration without voids, cracks, or porosity.
C. Concealed structural steel shall have one coat of approved rust-resistant primer.

## 06000 WOOD & PLASTICS

WORK INCLUDED
All lumber, plywood, rough hardware, trim, paneling, and finish carpentry joinery and millwork required or implied by drawings and/or specifications. Cabinets and countertops are not included in this section.

## 06010 MATERIAL

A. Grade or trademark is required on each piece of lumber; only official marks of association under whose rules it is graded will be accepted.
B. Plywood shall conform to U.S. Product Standards PS-66 and shall be branded or stamped with type and grade.
C. Moisture content shall not exceed 19% for framing lumber, 12% for plywood, and 8% for finish millwork.
D. Work that is to be finished or painted shall be free from defects or blemishes on surfaces exposed to view that will show after the finish coat of paint or stain is applied. Defective materials not up to specifications for quality and grade for its intended use, or otherwise not in proper condition, shall be rejected.
E. Rough lumber shall be dressed (4) four sides, air dried, well seasoned, sound, and free from splits, cracks, shakes, and wanes, loose or unsound knots, and decay and excessive warp. Species and grades shall be those listed:

Douglas fir or hem-fir for rough carpentry
  – Each piece marked as to grade and free from defect.
  – No. 1 light framing with not more than 25% No. 2 framing allowed for all lumber 2×6 or larger.
  – No. 2 construction grade for studs.

Treated lumber: Southern yellow pine, ACQ-D treated per AWPA standards. Finish lumber and millwork: clear white pine or ponderosa pine.

F. All nails, spikes, screws, bolts, joist hangers, and timber connectors as indicated, noted, or detailed on drawings, and as required to produce a safe, substantial, and workmanlike job in all respects.

## 06100 ROUGH CARPENTRY

A. Install all rough wood framing, nailers, edge members, curbs, blocking, grounds, rough sills, backing, furring, and the like as indicated, detailed, noted, or required to properly support, back up, and complete the work of this section and of any or all trades under these contracts.

- Securely attach and anchor to adjacent construction as detailed or as approved if detail is not provided.
- Shim to line if so required to provide a uniform base for any other work.

B. Provide double studs adjacent to and headers of size indicated over all openings.

C. TJI joists and rafters to be installed according to manufacturer's instructions.

## 06200 FINISH CARPENTRY

A. Provide all rabbets, splines, ploughs, and other cuts as detailed or required for neat, solid, fitting and joining.

B. Finish millwork where indicated to have a clear finish shall be dressed and sanded, free from machine and tool marks, abrasions, raised grain and other defects on surfaces exposed to view. Construction and workmanship of millwork items shall conform to or exceed the requirements of AWI and good shop practices.

C. Joints shall be tight and sop formed as to conceal shrinkage.

D. Interior millwork, running finish, and trim shall be in as long lengths as practicable, shall be spliced only where necessary, and only when approved by the Architect. All such splices shall be beveled and jointed where solid fastenings can be made.

## 07000 THERMAL & MOISTURE PROTECTION

WORK INCLUDED

A. Dampproofing basement walls
B. Vapor barriers
C. Thermal insulation
D. Roofing
E. Flashing
F. Caulking and sealants

## 07010 MATERIALS

A. Dampproofing on basement walls to be Sonneborne Building Products, Semi-mastic Hydrocide 600 or approved equal.

B. Vapor barriers under concrete slabs to be 4 mill thick polyethylene.

C. Vapor barriers on insulated walls and roofs to be 4 mill thick polyethylene.

D. Insulation exposed to earth shall be Dow Styrofoam SM, R-5.4 per inch.

E. All rigid board insulation not exposed to earth shall be Owens Corning, High-R Sheathing, R-7.2 per inch.

F. Batt insulation shall be Owens Corning Eco Touch Fiberglas unfaced, or approved equal.
   - 3½" thickness, R-11
   - 6" thickness, R-19
   - 9" thickness, R-30

G. Metal flashing, 28 gauge Aluminum

H. Caulking to be acrylic polymer conforming to F.F. TT-S-00230

## 07150 DAMPPROOFING

A. Apply two coats of asphalt Dampproofing over all concrete or masonry wall surfaces to receive earth backfill.

B. Apply polyethylene vapor barrier over gravel fill at all concrete slabs.

C. Vapor barrier sheets to be lapped 6" minimum.

D. Apply polyethylene vapor barrier to the heated side of all insulated frame walls and roofs.

## 07200 THERMAL INSULATION

A. Pack all voids and cavities in exterior walls and roof. Avoid compressing batt insulation.

B. Cut and fit insulation and vapor barriers as necessary for snug fit.

C. Allow minimum air space of 1/2" between insulation and roof sheathing.

D. Rigid insulation is to be nailed and glued with Dow Mastic number 11, according to the manufacturer's instructions.

## 07300 ROOFING

A. Roofing shall be Image II, Standing Seam, 24 gauge steel, color by owner from approved samples.

B. Roofing to installed according to manufacturer's instructions.

## 07600 FLASHING

A. Apply factory painted metal drip edge at all roof edges.

B. Chimney to be flashed at roof with Majestic number 9-6-12 galvanized flashing.

C. Flash all pipes at roof with neoprene flashing of the proper size.

D. Flash at all intersections of roofs and vertical surfaces an as otherwise shown on construction drawings.

## 07900 CAULKING

Caulk all windows, doors, and other openings.

## 08000 DOORS & WINDOWS

WORK INCLUDED

All doors, door frames, windows, skylights, trim for each, and all hardware not included elsewhere.

## 08010 PRODUCTS

A. Hollow metal doors to be manufactured by Pease, or approved equal.

B. Wood doors by Iroquois Millwork or approved equal. Birch plywood skin with phenolic impregnated Kraft core.

C. All door frames to be of clear pine in standard patterns as shown on the construction drawings. Side lites to be Iroquois, Weather Guard SL with 5/8" insulating glass.

D. Windows to be Andersen, Series 200. Windows to include aluminum screens by the same manufacturer.

E. Skylight to be Skylight Concepts number CMDADE1850.

F. Door trim to be standard WM patterns milled from clear pine.
G. Contractor shall allow $1,500 for locksets, latches, bifold hardware, hinges, weather stripping, medicine cabinets, closet rods, and shower curtain rods.

### 08100 DOORS

A. Frames to be plumb and square with accurately fitted joints. Set exposed nails with a nail set.
B. Accurately align doors with frames and adjust hardware as necessary for smooth operation.
C. Install molding as shown on construction drawings with accurately mitered corners.

### 08500 WINDOWS & SKYLIGHTS

A. Install all windows true and plumb, and according to manufacturer's recommendations to produce a weathertight installation.
B. Install all hardware and accessories and check all moving sections for smooth operation.

### 08900 HARDWARE

A. Install all door and window hardware according to the manufacturer's recommendations and check for smooth operation.
B. Install a closet rod in each closet. Closet rods to be secured through wall finish to blocking installed with rough carpentry.
C. Install shower curtain rods over tub and shower stall. Shower curtain rods to be secured through wall finish to blocking installed with rough carpentry.
D. Install a medicine cabinet (by owner) over each lavatory.

### 09000 FINISHES

WORK INCLUDED
A. Gypsum wallboard
B. Ceramic tile in baths and toilet rooms
C. Quarry tile floors
D. Painting and varnishing

### 09250 WALLBOARD

MATERIAL
A. All wallboard material to be the product of one manufacturer; U.S. Gypsum, Flintkote, or approved equal. Drywall in baths, toilet rooms, and tub room to be moisture resistant.

INSTALLATION
A. Gypsum wallboard shall be installed with joints centered over framing or furring.
B. Fasten gypsum wallboard with power-driven drywall screw located not over 12" O.C. at all edges and in the field.
C. Outside corners are to be protected with metal corner bead.
D. Finish all joints with a minimum of three coats of joint compound and standard gypsum board reinforcing tape in accordance with the manufacturer's printed instructions.
E. Dimples at screw heads shall receive three coats of compound.

### 09300 TILE

MATERIAL
A. Wall tile to be American Olean Tile Company standard grade bright glazed in a color selected. Bathroom accessories to be same manufacturer and color.
B. Floor tile in bath, toilet, and tub rooms to be American Olean unglazed 1'×1" ceramic mosaic tile in color selected.
C. Quarry tile to be installed in this specification will be 6"×6" shale-and-clay tile provided by owner.
D. Marble thresholds at all doors adjacent to mosaic tile floors shall be Vermont Marble 7/8"×3 1/2".

INSTALLATION
A. Layout ceramic tile on walls and floors so that no tiles less than half-size occur.
B. Cut and fit tile around toilets, tubs, and other abutting devices.
C. Install all floor tile by thin-set method in accordance with TCA recommendations.
D. Install wall tile in mastic cement conforming to the recommendations of the tile manufacturer.
E. Grout all tile work to completely fill joints.
F. Clean all tile surfaces to present a workmanlike job.
G. Install 12 bathroom accessories as follows:
- 2 soap dishes
- 3 toilet paper holders
- 7 towel bars

### 09900 PAINTING

MATERIAL
A. Exterior stain: One coat Minwax exterior stain or approved equal, in color by owner.
B. Interior walls – flat: Two coats Martin Senour, Bright Life latex flat or approved equal, in colors by owner.
C. Interior walls – semi gloss: Two coats Martin Senour, Bright Life latex semi gloss or approved equal, in colors by owner.
D. Interior painted woodwork: Two coats Martin Senour, Bright Life latex semi gloss or approved equal, in colors by owner.
E. Metal: Two coats DeRusto rust resistant enamel or approved equal, colors by owner.
F. Polyurethane: Two coats United Gilsonite Laboratories, ZAR gloss.
G. Primers: All primer to be that recommended by manufacturer of top coat.

APPLICATION
A. Repair all minor defects by patching, puttying, or filling as normally performed by painting contractors.
B. Prime uncoated wood surfaces with tinted primer and touch up previously painted surfaces.
C. Sand all surfaces smooth before each coat of paint to produce a smooth and uniform job at completion.
D. Protect all adjacent surfaces and other work incorporated into project against damage or defacement.

### GREEN NOTE

*Volatile organic compounds (VOCs) are emitted as gases from certain solids or liquids. VOCs include a variety of chemicals, some of which may have short- and long-term adverse health effects. Many of the products used in the construction and maintenance of homes for decades emit VOCs: paint and varnish, carpets, drapes, and cleaning agents, to name a few. It has only been in recent years that we have been aware of these VOCs. The level of VOCs in some products is no regulated by law. Most of the products named above are now available with either very low or zero VOCs and it is common for construction specifications to specify the acceptable levels of VOCs in products to be used in construction.*

E. Coat all surfaces according to the following painting schedule and as indicated on the drawings. All colors are to be selected by the owner.

| Exterior Walls and Trim | Stain |
| --- | --- |
| Metal Railings | Rust-Resistant Enamel |
| Playroom | Walls Flat/Ceiling Flat |
| All Baths | Walls Semigloss/Ceilings Semigloss |
| Halls and closets | Walls Flat/Ceiling Flat |
| Living room | Walls Flat/Ceiling Flat |
| Dining room | Walls Flat/Ceiling Flat |
| Kitchen | Walls Semigloss/Ceilings Semigloss |
| Bedrooms | Walls Flat/Ceiling Flat |
| Loft | All Except Floor And Ladder: Flat |
| Loft floor and ladder | Polyurethane |
| Interior doors | Polyurethane |
| Interior stairs | Polyurethane |
| Unscheduled interior trim | Semigloss |

## 10000 FIREPLACE

**WORK INCLUDED**
Provide and install fireplace, chimney, and accessories as indicated on the construction drawings. Related masonry and granite cap are not included in this section.

## 10310 EQUIPMENT

A. Majestic Company fireplace number SR36A.
B. Majestic Company chimney with 8" flue.
C. All chimney accessories required to conform with printed instructions of Majestic Company.

## 10320 INSTALLATION

Fireplace and chimney are to be installed according to printed instructions of Majestic Company and as indicated on the construction drawings. All equipment is to be installed level and plumb and finished to provide a workmanlike appearance.

## 11000 EQUIPMENT

**WORK INCLUDED**
A. Provide and install kitchen cabinets. This section does not include kitchen sink or related plumbing.
B. Provide and install vanity cabinets. This section does not include lavatories or related plumbing.
C. Provide and install granite countertops, vanity tops, fireplace cap. Provide necessary cutouts for kitchen sink, by plumbing contractor; cooktop, by owner; oven, by owner; and chimney, by masonry contractor.

## 11910 MATERIAL

A. Cabinets and vanities shall be Merillat, Classic style.
B. Granite countertops and fireplace cap to be supplied by Granite Mountain Stone Design in color selected by owner from approved samples. Granite to be 3 cm thickness.

## 11920 INSTALLATION

A. Cabinets shall be installed level and true with no less than 4 screws per base unit and present a workmanlike appearance.
B. Granite shall be adhered to cabinet frames with a full bead of silicone.
C. All granite shall be installed with as few joints as possible. Joints shall be tight and filled with resinous material according to the manufacturer's instructions. Exposed edges shall be polished to match the finish of the top.
D. Provide necessary cutouts for installation of kitchen sink, cooktop, and grill.
E. Provide cutout for fireplace chimney with 1/8" clearance.

## 15000 MECHANICAL

This division includes heating, ventilating, air conditioning, and plumbing. It is omitted here because these topics have not been covered earlier in the textbook.

## 16000 ELECTRICAL

This division covers all aspects of electrical work. It is omitted here because these topics have not been covered earlier in the textbook.

### USING WHAT YOU LEARNED

The specifications for a construction project are part of the contract for the work to be performed, so it is very important to carefully read and fully understand what is written in the specs. The key to accurately applying the specs is to know where to find the information they contain. Look at the Table of Contents if there is one and find the division you are working on. Where in the lake house specifications (see **Figure 33–2**) would you find instructions for how to treat remaining trees that are above the natural grade when the site is being graded?

In Division 02000, Section 02270 is grading. Item B of that section says, "Where existing trees are to remain, if the new grade is lower than the natural grade under the trees, a sloping mound shall be left under the base of the tree extending out as far as the branches."

## Assignment

Refer to the Lake House specifications in **Figure 33–2** to complete the assignment.

1. What division covers vapor barriers under slabs?
2. What is the material and thickness of the vapor barrier under the heat sink?
3. What material and rating is the thermal insulation under the heat sink?
4. What is the strength requirement for the concrete in the heat sink?
5. How is the concrete slab in the heat sink to be finished?
6. What make and model number is the skylight?
7. What division of the specs includes installing shower curtain rods?
8. What brand and type of paint is to be used on the kitchen walls?
9. Who is to choose the color of the paint for the kitchen walls?
10. What brand and type of paint is to be used on the trim around the skylight?

# Test

**A. Refer to the Lake House drawings in your textbook packet to determine each of the following dimensions.**

1. Northwest corner of the site to the high-water mark at the west boundary
2. Septic tank to the nearest property line
3. Width of the walk on the west side of the house
4. Width of the south end of the drive (in front of the garage)
5. Depth of the earth fill over the septic drainfield
6. Elevation at the butt of the most southerly maple tree to remain
7. Outside of the garage foundation (width × length)
8. Outside of the garage footing (width × length)
9. North to south spacing of the square steel columns
10. Length of the northwest square steel column
11. Length of the haunch under the slab between the north steel columns
12. East-west dimension inside the lower level bathroom at the widest end
13. Closet in bedroom #1 (width × length)
14. Inside of bedroom #1 (width × length)
15. Deck between the kitchen and the garage (width × length)
16. Length of the joists in the loft
17. Height of the foundation wall at the overhead garage door
18. Elevation at the bottom of the concrete piers for the kitchen deck
19. Difference in elevations at the bottom of the south garage footing and the bottom of the nearest house footing
20. Elevation at the bottom of the deepest excavation
21. Width of the cabinet over the refrigerator
22. Width of the cabinet closest to the living room stairs
23. Total thickness of a typical exterior frame wall (allow ½ inch for siding)
24. Lineal feet of soldier-course bricks in the fireplace
25. Lineal feet of 2 × 10 lumber in the eave of the corrugated roof
26. Finish floor to the top surface of the freestanding closet/cabinet unit in the kitchen (include surface material)
27. Length of the studs in the wall separating the playroom from the crawl space under the kitchen
28. Length of the steel dowels that anchor the foundation walls to the footings
29. Length of the C8 × 11½ structural steel beam
30. Outside surface of the foundation under the dining room to the centerline of the deck footings
31. Door in the south end of the garage (width × height × thickness)
32. Length of the 2 × 4 studs under the living room bench
33. Width of the treads (front to back excluding nose) in the stairs from the kitchen to the bedroom level
34. Rough opening for the door from the hall into bedroom #1 (width × height)
35. Thickness of the concrete at the haunches

**B. Describe the material at each of the following locations in the Lake House. Include such considerations as the kind of material, nominal size of masonry units, and nominal size of lumber.**
1. Tile field pipe
2. Reinforcement in the concrete slabs
3. Rungs in the ladder to loft
4. Hip rafter in the northeast corner of house
5. Roof insulation
6. Roof deck
7. Exterior wall studs
8. Rafter headers at skylight
9. Purlins under the corrugated roof
10. Finished surface on the west wall of playroom
11. Reinforcement in the footings
12. Anchor bolts in the wood sill
13. Wood sill
14. Vapor barrier under the concrete slab
15. Subfloor in the bedrooms
16. Floor joists in the kitchen
17. Floor joists in the bedrooms
18. Stair stringers between the decks
19. Railing around the south decks
20. Top course of the foundation at heat sink
21. Expansion joint at the edge of heat sink
22. Rafters above the loft
23. Posts supporting the LVL beams
24. Girder under the west wall of bedroom #1
25. Base plates on the posts supporting LVL beams
26. Girder under the kitchen deck joists
27. Finished surface of the living room bench
28. Housed stringer on the north side of stair to bedrooms
29. Stair treads
30. Door casings (interior)
31. Door casings (exterior)
32. Window casings (interior)
33. Glazing (light) in the window on south side of bedroom #1
34. Planks on the kitchen deck
35. Foundation at the west side of the kitchen deck

**C. Answer each of these questions about the Lake House.**
1. How many cubic yards of concrete are required for the concrete slab at elevation 337.00 feet?
2. What is the length of the concrete piers between the kitchen and the garage?
3. How many lineal feet of 1× pine trim are needed to cover the bottom of the LVL beam over the kitchen?
4. How long are the rafters in the corrugated roof? (Refer to the illustrated rafter table.)
5. How long is the hip rafter at the northeast corner of the roof? (Refer to the illustrated rafter table.)

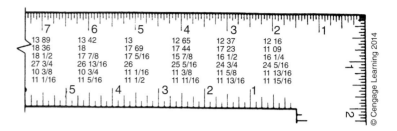

**D. Answer the following questions.**

1. Who should design roof trusses?
   a. architect
   b. carpenter
   c. engineer
   d. any of the above

2. Which drawing will include information about the slope of a roof?
   a. building elevation
   b. truss delivery sheet
   c. truss detail drawing
   d. all of the above

3. What is a piggyback truss?
   a. a truss that is designed to be used as the top half of a large truss
   b. a truss design that uses extra web members
   c. a truss that is to be placed against the side of another truss
   d. a particular brand of roof truss

4. Which sheet would be the easiest one on which to find the required number of trusses of a particular type?
   a. floor plan
   b. truss delivery sheet
   c. truss detail sheet
   d. building elevation

5. What is the most reliable place to find the spacing between trusses?
   a. truss layout
   b. floor plan
   c. truss detail sheet
   d. truss delivery sheet

Part II Test 247

# PART 3

# MULTIFAMILY CONSTRUCTION

Part III provides an opportunity for you to extend your ability to read construction drawings. Once you have mastered the contents of Parts I and II, you should be able to read and thoroughly understand most residential construction drawings. Part III helps you apply the skills developed earlier to other types of construction as well as the work of the mechanical and electrical trades.

The Town House drawings in your textbook packet were selected for reference in Part III because they represent quality construction in a geographic region where many construction practices are different from those in the rest of North America—and because they are more complex than the drawings referenced thus far, providing an opportunity to sharpen your skills. The Town House has more floors and construction details. These features can be seen in the Town House drawings, which can make them more difficult to interpret. As you work through the units of Part III and the corresponding drawings, take the time to understand all the information about a particular building feature and study all the drawings related to it.

A complete set of drawings for the Town House is far too large to include in the packet that accompanies this textbook. Therefore, selected sheets and portions of sheets are used here. To use the available space most efficiently, some of the drawings have been reorganized and combined—with parts of two sheets printed on one. The original sheet and drawing numbers have been retained; therefore, all references printed on the drawings are applicable.

# UNIT 34

# Orienting the Drawings

## Objectives

After completing this unit, you will be able to perform the following tasks:

○ Locate a particular building or plan within a large development.

○ Explain the relationships between drawings for construction projects where several plans are to be adjoined in one building.

○ Visualize a building design by reading the drawings.

## Identifying Buildings and Plans

Multifamily dwellings are often built in large developments—with many similar buildings in a single development. Some developments are completed in phases. One phase is completely constructed and begins earning income for the developer before the next stage is started. **Figure 34–1** shows the site plan for Hidden Valley, the development that is the design basis for the Town House drawings in your textbook packet. Hidden Valley is to be developed in four phases. Each phase is outlined with a heavy broken line on the site plan.

The first phase of Hidden Valley includes 11 buildings. Each building is labeled as to building type and the parts that make up the building (see **Figure 34–2**). Building Type II is made up of four separate units, each with its own floor plan. Each unit is built like a separate building joined to the next (see **Figure 34–3**). Each plan is identified by a letter or letter and numeral. (The term *plan* is used to refer to a particular arrangement of rooms or design. This should not be confused with the use of *plan* to refer to a type of drawing.)

The plans in building Type II are A, B1, B2, and A R. Each plan type is described on separate drawings. Drawings 1 through 5 are for Plan A. Drawings 6 through 10 are for Plan B.

One technique that is used to create similar, yet different, plans is to reverse them. A reversed plan is created by building the plan as though seen in a mirror. The reversed plan has the same features and the same dimensions, but their arrangement is reversed (see **Figure 34–4**). In building Type II of the

### GREEN NOTE

*One way of reducing the impact of housing on the environment is to increase the housing density in a development. Moving people closer together, leaves more space for the unbuilt environment. Many zoning ordinances require housing developments to include a certain percentage of green space. Housing density can be further increased by the use of multifamily buildings. Also, because these dwelling units share walls and have fewer exterior walls, they tend to be more energy efficient.*

**Figure 34–1.** Site plan.

Hidden Valley development, the south end is a reversal of Plan A. This is designated on the site plan and on **Figure 34–2** as A R. The letter *R* indicates a reverse plan.

## Organization of the Drawings

The drawings for a large project must be systematically organized, so information can be found easily. Architects follow similar pattern in organizing their drawings. The general order of drawing sheets is similar to that of specifications: site plans are first; foundation plans and floor plans, next; building elevations, third; then, structural and architectural details; mechanical, next to last; and electrical, last. The cover sheet usually includes an index or table of contents for the drawing set (see **Figure 34–5**).

The index for the Hidden Valley drawing does not indicate mechanical or electrical drawings. In some parts of the United States, it is common practice not to prepare separate drawings for mechanical or electrical work on town houses. The essential information for these trades is included on the drawings for the other trades.

## Visualizing the Plan

The first step in becoming familiar with any plan should be to mentally walk through the plan. This technique is described in Unit 18 for the Lake House. Refer to the Hidden Valley drawings as you mentally walk through Plan A of the Town House.

Orienting the Drawings 251

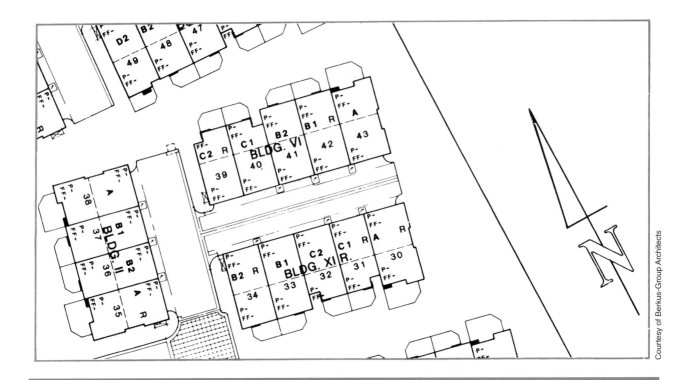

**Figure 34–2.** This is a section of the site plan at the size it was drawn.

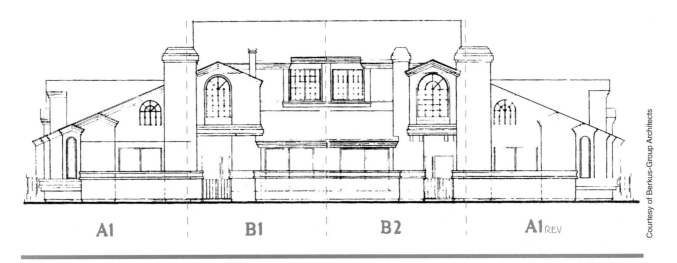

**Figure 34–3.** Building Type II includes four housing units.

Enter through the overhead garage door on the east side of the lowest level. The garage is an open area with stairs leading up to the main floor. Next to the stairs is an area of dropped ceiling beneath the stairs on the main floor. A note on Drawing 2 refers to Section B.

This note introduces a new consideration for anyone reading construction drawings. Although the architect reviews the drawings carefully, there is always a possibility that errors will appear as one did in this instance. The dropped ceiling beneath the stairs is actually shown on Section E.

The drawings for the Town House are printed exactly as they were prepared for the construction job. A few errors may remain on these drawings. Any errors that do remain give you valuable experience in detecting and dealing with error.

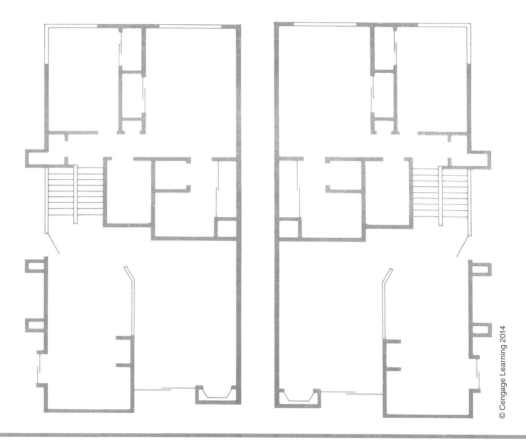

**Figure 34–4.** A reversed plan is similar to what would be seen by looking at the plan in a mirror.

At the top of the garage stairs, there is an entry area with an exterior door. A boxed number 33 refers to the plan notes printed on Drawing 2. This note indicates a stub wall dividing the entry from the dining room. Boxed note references are used throughout the Town House drawings. Reading each related note as it is encountered will help you to better understand the plan.

As you enter the dining room and kitchen, the floor material changes. The kitchen ceiling is dropped to 7'-6" as dimensioned on Section A-A, Drawing 4. The areas with dropped ceiling above are shown by light cross-hatching on the floor plan. The kitchen has base cabinets on three walls and a peninsula of cabinets separating the kitchen and dining room. Opposite the peninsula of base cabinets, there is a 4'-8" alcove with cabinets, **Figure 34–6.**

Visually walk back through the dining room along a 36"-high guard rail, plan note 13, to a two-riser stair down to the living room (see **Figure 34–7**). At the front of the large 18'-4" × 13'-4" living room there is a fireplace (see **Figure 34–8**). To the left of the fireplace is a 7'-0" × 8'-0" sliding glass door that opens onto a small courtyard. A note on the floor plan indicates a *36 MOJ RADIUS ABOVE*. The legend of symbols and abbreviations on sheet 0 shows that MOJ means ***measure on the job***. Looking at the Front Elevation, Drawing 5, you can see that this note refers to a window with a 36-inch radius at the top.

From the entry area, another stair leads up to a hall serving the bedrooms and bath (see **Figure 34–9**). Across the hall, a pair of 2'-6" × 6'-8" hollow-core doors open into Bedroom #2. By the stairs, a door opens into bathroom #2. This is a small bathroom with a tub, water closet, and lavatory. At the end of the hall, you enter the master bedroom. There is an 8'-0" × 5'-0" aluminum sliding window at the far end of the Master Bedroom. On your left as you walk into the bedroom, there is a wardrobe closet with a shelf and pole (S&P). On your right there is a master bath. The master bath

Orienting the Drawings 253

## table of contents

| | | |
|---|---|---|
| | | COVER SHEET AND INDEX |
| a | | SITE PLAN |
| b | | GENERAL NOTES |
| c | | GENERAL NOTES |
| 1 | PLAN A | FOUNDATION PLAN |
| 2 | | FLOOR PLAN |
| 3 | | FRAMING PLAN |
| 4 | | SECTIONS |
| 5 | | ELEVATIONS |
| 6 | PLAN B | FOUNDATION PLAN |
| 7 | | FLOOR PLAN |
| 8 | | FRAMING PLAN |
| 9 | | SECTIONS |
| 10 | | ELEVATION (1) |
| 11 | | ELEVATION (2) |
| 12 | PLAN C | FOUNDATION PLAN |
| 13 | | FLOOR PLAN |
| 14 | | FRAMING PLAN |
| 15 | | SECTIONS |
| 16 | | ELEVATIONS (1) |
| 17 | | ELEVATIONS (2) |
| 18 | PLAN D | FOUNDATION PLAN |
| 19 | | FLOOR PLAN |
| 20 | | FRAMING PLAN |
| 21 | | SECTIONS |
| 22 | | ELEVATIONS (1) |
| 23 | | ELEVATION (2) |
| 24 | PLAN A-B | INTERIOR ELEVATIONS - PLAN A & B |
| 25 | PLAN C-D | INTERIOR ELEVATIONS - PLAN C & D |
| 26 | | BUILDING TYPE I |
| 27 | | BUILDING TYPE II |
| 28 | | BUILDING TYPE III |
| 29 | | BUILDING TYPE IV |
| 30 | | BUILDING TYPE V |
| 31 | | BUILDING TYPE VI |
| 32 | | BUILDING TYPE VII |
| 33 | | BUILDING TYPE VII |
| 34 | | BUILDING TYPE VIII |
| 35 | | BUILDING TYPE VIII |
| 36 | | BUILDING TYPE IX |
| 37 | | BUILDING TYPE IX |
| 38 | | BUILDING TYPE X |
| 39 | | BUILDING TYPE X |
| 40 | | BUILDING TYPE XI |
| 41 | | BUILDING TYPE XII |
| 42 | | METER ENCLOSURE |
| 43 | | REC. CENTER   FOUND FLR PLAN |
| 44 | | REC CENTER   SECTION ELEV |
| D1 | | FOUNDATION DETAILS |
| D2 | | FRAMING DETAILS |
| D3 | | FRAMING DETAILS |
| D4 | | FRAMING DETAILS |

Courtesy of Berkus-Group Architects

**Figure 34–5.** Table of contents for Hidden Valley construction drawings.

**Figure 34–6.** Hidden Valley kitchen as seen from the entry.

**Figure 34–8.** End wall of living room.

**Figure 34–7.** Hidden Valley stairs.

**Figure 34-9.** View from Hidden Valley living room.

has a 42" × 60" tub, a vanity (called a *pullman* here), a large wardrobe closet, and a toilet room.

To thoroughly understand the plan, you should read all the plan notes on Drawing A-2 and locate the features to which they refer. Then locate each of the details, indicated by a circle. The numerals above the horizontal line in these circles indicate the detail being referenced. The letter and numeral below the horizontal line indicate on which drawing the detail appears. As you find these plan notes and details, refer to the framing plans, building sections, and elevations to thoroughly understand each.

### USING WHAT YOU LEARNED

Finding your way around a large set of drawings that covers several buildings begins with finding the particular drawing sheets that pertain to the building you are working on. Phase I of Hidden Valley includes 11 buildings housing 55 dwelling units. For the purposes of construction, each dwelling unit can be treated as a separate building, which has its own unit number. On what drawing sheet would you expect to find the floor plan for dwelling unit number 40, shown in **Figure 34–2**?

**Figure 34–2** shows unit 40 is plan C1. Refer to the Table of Contents for the drawings, Figure 34–5, to see that the Plan C floor plan is on Sheet 13.

## Assignment

Refer to the Town House drawings in your textbook packet and **Figures 34–1** and **34–2** to complete the assignment.

1. How many buildings are included in the first phase?
2. What plans are included in building Type VI?
3. In the first phase, find Building II. On which side (compass direction) is the courtyard?
4. How thick is the concrete slab for the garage floor in Plan B?
5. What size are the floor joists under the dining room in Plan B?
6. In Plan B, what supports the kitchen floor joists under the back wall of the kitchen?
7. On which drawing would you find elevations of the kitchen cabinets for Plan A? For Plan B?
8. On which drawing would you find details of concrete piers under bearing posts for girders?
9. What is the height of the handrail at the dining room/living room stairs in Plan B?
10. For each of the major rooms of Plan B listed below, indicate the overall dimensions. Do not include closets, stairs, or minor irregularities. Allow for the thickness of all walls. Walls are dimensioned to the face of the framing.
    a. Living room
    b. Dining room
    c. Kitchen
    d. Deck
    e. Library
    f. Bedroom #2
    g. Master bedroom
11. What are the tread width and the riser height for the stairs between the living room and dining room in Plan B?
12. What important feature of the living room is beside the entry in Plan B?
13. How long are the studs in the partition between the master bedroom and bedroom #2 in Plan B?
14. How high above finished grade is the top of the privacy fence in front of Plan B?

# UNIT 35

# Town House Construction

## Objectives

After completing this unit, you will be able to perform the following tasks:

- Identify and explain the construction of party walls.
- Find and interpret detailed information on complex residential drawings, such as the Town House.

Multifamily buildings are constructed the same way as single-family buildings in most respects. A wall that is shared by two living units in a multifamily structure is called a *party wall* (see **Figure 35–1**). In addition to the usual requirements of a wall, a party wall provides more fire resistance and privacy.

The Town House party walls have stricter fire-resisting requirements in some places than in others. In the Town House drawings, the architect refers to party walls where the fire-resistance factor is lower. Where special fire code requirements must be met, the architect refers to area-separation walls. The terms *party wall* and *area-separation wall* are often used interchangeably. Other architects may use them with reverse meanings.

## Fire-Rated Construction

*Fire-rated construction* serves two purposes in the event of a fire. It slows the spread of fire, and it maintains structural support longer than non-fire-rated construction. Approved fire caulking or fill must be used at all utility-line fire-rated wall penetrations.

The most obvious way in which a wall, floor, ceiling, or roof can slow the spread of fire is by having a fire-resistant surface. Plaster and gypsum wallboard are fire-resistant materials. They do not burn, and they do not easily transmit the heat of fire to the framing members on the other side. Of course, the thicker the wallboard or plaster, the better it resists the flow of heat. For this reason, party walls are often required to have double thicknesses of gypsum wallboard on each side (see **Figure 35–2**).

Notice that the wallboard in **Figure 35–2** is indicated as type x. This is a special fire-rated wallboard. Although all gypsum plaster is noncombustible, standard wallboard breaks down and crumbles in the high heat of a fire. Fire-rated wallboard holds up much longer in a fire.

Fire-rated construction is frequently used to separate garages and mechanical rooms from living spaces. Party walls are also fire rated to prevent a fire from spreading between housing units. To completely separate the housing units, a fire-rated party wall should extend all the way from the foundation to the roof.

Most building codes allow an alternative to extending the party wall through the roof. The fire-rated construction may end at the bottom of the roof as long as the roof is of fire-rated construction (see **Figure 35–3**). This usually means that the roofing material resists fire for as long as the party wall does. For example, if the party wall is required to be a one-hour code wall (it resists fire for one hour), the roof must be covered only with material that also resists fire for one hour.

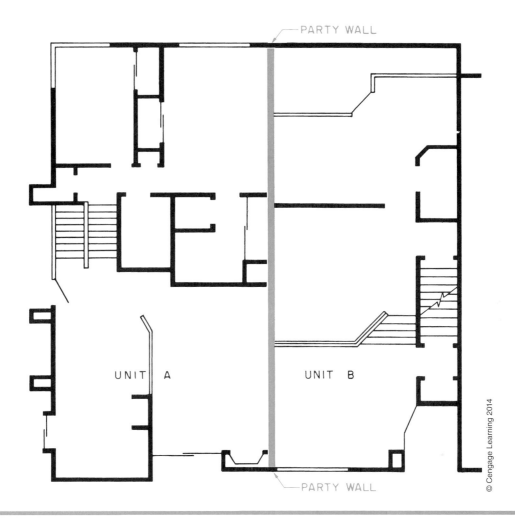

**Figure 35–1.** A party wall separates two or more units.

Fire-rated walls must also prevent fire from spreading vertically inside the wall. If left open, the spaces between the studs in a frame wall act like chimneys, allowing flame to spread very quickly from one level to another. To prevent vertical flame spread, stud spaces are not permitted to be more than one story high. The spaces between the studs are closed off with **firestops** at each level (see **Figure 35–4**). Instead of wood firestops, the wall cavity can be blocked off with fire-resistant fiberglass insulation as shown in **Figure 35–2.**

Openings are usually avoided in fire-rated walls. Where it is necessary to include a door, it is made of fire-resistant material. The fire rating of the door must comply with the building code. Fire-rated doors are often allowed to have a slightly lower rating than the walls in which they are installed. Doors in fire walls are equipped with a self-closing mechanism.

## Sound Insulation

To provide privacy between the housing units, party walls should not allow the sound from one unit to be heard in the next unit. The measurement of the capability of a building element to reduce the passage of sound is its *sound transmission classification* (STC) (see **Figure 35–5**).

Sound is transmitted by vibrations in any material: solid, liquid, or gas. To slow the passage of sound, a party wall must reduce the flow of vibrations. The materials used in the construction of most walls vibrate relatively well. Also, they transmit these vibrations to the air inside the wall. The air carries the sound to the other side, where it is transmitted to the air on the opposite side of the wall. The electrical installer may be required to utilize fiberglass or some other sound attenuation material around flush device boxes. Raceways may require installations that reduce sound transmission,

Town House Construction

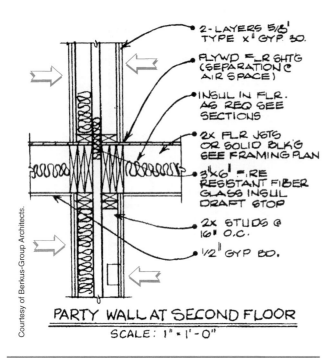

such as expansion joints, flexible raceway, or raceway offsetting through the party wall.

The sound transmission classification of a wall can be improved greatly by not allowing the studs to contact both surfaces. This may be accomplished in one of two ways. One method is to attach clips made for sound insulation to the studs and then fasten the wallboard to these clips. The clips absorb the vibrations. Using the clips and sound-deadening fiberboard results in an STC rating of 52.

An STC of 45 is achieved without clips by using 2 × 4 studs and 2 × 6 plates. The studs are staggered on opposite sides of the wall so that no studs contact both surfaces (see **Figure 35–6**). The STC can be increased to 49 by including fiberglass insulation.

**Figure 35–2.** Party walls often have double layers of gypsum wallboard on one or both sides.

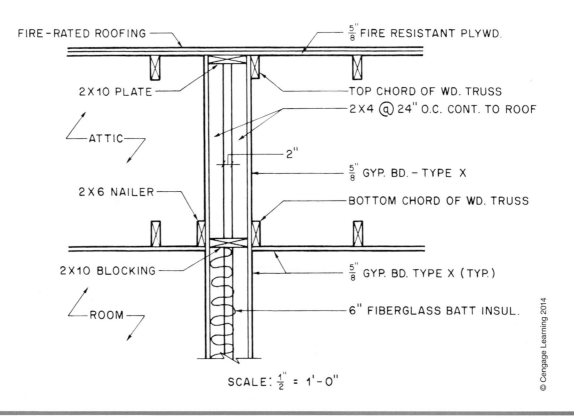

**Figure 35–3.** Fire-rated party wall in attic space.

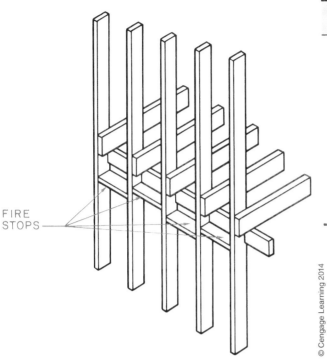

| STC RATING | EFFECTIVENESS |
|---|---|
| 25 | Normal speech can be understood quite easily. |
| 35 | Loud speech can be heard, but not understood. |
| 45 | Must strain to hear loud speech. |
| 48 | Some loud speech can barely be heard. |
| 50 | Loud speech cannot be heard. |

**Figure 35–5.** Sound transmission classes.

**Figure 35–4.** Firestops are installed between studs to prevent vertical drafts inside the wall.

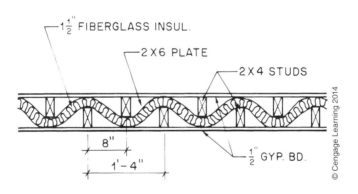

**Figure 35–6.** Typical sound-insulated wall.

### USING WHAT YOU LEARNED

The building code requires some provision for preventing the vertical spread of fire in the cavity of a frame wall that is more than one story high. The party wall in the living room of Plan A is such a wall. What provision is made for resisting the vertical spread of fire in this wall?

The logical place to begin is by looking at the floor plan to identify the wall in question. There is a callout on the party wall for detail 21 on Sheet D3. That detail has to do with construction at the roof and does not show any fire blocking, so we'll have to continue our search. Next, look at the First Floor Framing Plan on Sheet 3. It shows that there is a Detail 23 on Sheet D3. That detail shows a 3″ × 6″ fire-resistant fiberglass Insulation draft stop in the wall cavity at the floor level.

Town House Construction 261

## Assignment

Refer to the Town House drawings in your textbook packet to complete the assignment.

1. List the type, thickness, and number of layers of wallboard at each of the following locations in the Town House:
    a. Area separation wall in Plan B, dining room
    b. Party wall in Plan B, kitchen
    c. Party wall in Plan B, garage
    d. Area separation wall in Plan B, bedroom #2
    e. Area separation above ceiling in Plan B, master bedroom
    f. Party wall in Plan A, living room
    g. Garage ceiling under nook in Plan B
2. What is used to stop the vertical spread of fires inside the party walls in the Town House?
3. How is fire prevented from spreading over the top of the area separation walls where they meet the Town House roof?
4. What is the total thickness of the party wall at the library in Plan B?
5. What is done to stop the transmission of sound through the air space in the party wall of the library in Plan B?
6. What is the STC rating of the party wall of the library in Plan B?
7. In Plan B, what is the dimension from the floor to the bottom of the enclosure for the fluorescent light above the master bathroom vanity cabinet and lavatory?
8. On Plan B, on the exterior of the building below the library window there is a sloping shelf. At what angle does that shelf slope?
9. On Plan B, what is the total thickness of the wall separating the master bedroom from the next living unit?
10. What size are the floor joists in the Plan B Library?

# Unit 36
# Plumbing, Heating, and Air Conditioning

All residences have a plumbing system that consists of a water supply system; a water distribution system; and a *drain, waste, and vent (DWV)* system. The water supply system provides a source of water to the structure, and the water distribution system provides hot and/or cold water to each of the fixtures requiring water in the structure. After the water is used, the drain, waste, and vent system is used to dispose of the water into a municipal sewage system or an onsite disposal system.

Even though installing gas piping is not defined as plumbing, plumbers sometimes install the piping. The gas piping system supplies fuel gas to the gas-fired fixtures, which might include water heaters, furnaces, ranges, and clothes dryers.

## Plumbing Materials

The materials most often used for plumbing are copper, plastic, cast iron, and black iron. A brief description of each is given in the list that follows.

- *Copper* is frequently used for plumbing because it resists corrosion. However, it is relatively expensive. Copper pipes and fittings may be threaded or smooth for soldered joints.
- *Plastic* materials are used in the water supply, water distribution, and DWV systems. These materials are lightweight, noncorrosive, and easily joined. While plastics are allowed in each area of a plumbing system, the plastics used in a water distribution system must have a minimum temperature rating of 180°F (82°C). Plastics are not suitable for some applications where high strength is required.
- *Cast iron* is commonly used where DWV piping passes through the foundation and outside the building, where it is buried. Cast iron is strong and has excellent resistance to corrosion. Cast iron is not generally used for water supply or water distribution systems in residential construction.
- *Black iron* is used almost exclusively for gas piping. Black iron pipes and fittings are threaded, so they can be screwed together. Brass fittings are frequently used to join black iron pipe.

## Fittings

A wide assortment of fittings is used for joining pipe, making offsets at various angles, controlling the flow of water, and gaining access to the system for service. Most fittings are made of the same materials as the pipe. Plumbers

## Objectives

After completing this unit, you will be able to perform the following tasks:

○ Explain the basic principles of plumbing design.

○ Identify the plumbing symbols used on drawings.

must be familiar with all types of fittings so they can install their work according to the specifications of the designer (see **Figure 36–1(A)–(E)**).

- *Couplings* (see **Figure 36–1(A)**) are used to join two pipes in a straight line. Couplings are generally used only where a single length of pipe is not long enough.
- *A union* (see **Figure 36–1(B)**) allows piping to be disconnected without having to cut the pipe. A union consists of three parts, with one part being attached to each pipe and a nut to secure the connection. Then the two parts of the union are screwed together. When it becomes necessary to disconnect the pipe, the two halves of the union are unscrewed.
- *Elbows* (see **Figure 36–1(C)**) are used to make changes in direction of the piping. Elbows turn either 90°, 45°, or 22.5°.
- Some fittings have a hub on each end to accept the outside diameter of the pipe. Others, called *street fittings*, have a hub on one end, and the other end is the same as the outside diameter of the pipe. Street fittings can be joined directly to other fittings, with no pipe between them.
- *Tees and wyes* (see **Figure 36–1(D)**) have three connections to allow a second pipe to join the first from the side. Tees have a 90° side connection. A sanitary tee has a curve in the side connection to help direct the flow. Wyes have a 45° side connection.
- **Cleanouts** (see **Figure 36–1(E)**) allow access to sewage plumbing for cleaning. A cleanout consists of a threaded opening and a matching plug. When cleaning is necessary, the plug is removed and a drain cleaning cable, also known as a *snake* or *auger*, is run through the line. Cleanouts are installed in each straight run of DWV at the base of drainage stacks, where pipe changes direction more than 45°, and several other areas dictated by plumbing codes.
- *Valves* are used to stop, start, or regulate the flow of water. The faucets on a sink or lavatory are a type of valve. A valve can be used to isolate one part of a system from the rest. Every building must have an isolation valve, and most plumbing fixtures must have an isolation valve located near the fixture. Bathtubs and showers do not typically require separate

isolation valves. Some valves are made with a port through which the piping on that side of the valve can be drained after the valve is closed. These are called *stop-and-waste valves* (see **Figure 36–1(F)**).

## Design of Supply Piping

In most communities, water is distributed through a system of water mains under or near the street. When a new house is constructed, the municipal water department taps (makes an opening in) this main.

**Figure 36–1(A).** Coupling.

**Figure 36–1(B).** Union.

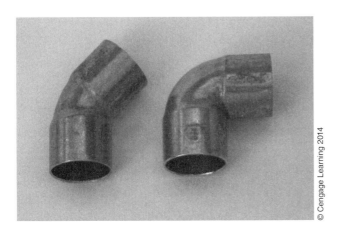

**Figure 36–1(C).** 45° and 90° elbows.

**Figure 36–1(D).** Tee.

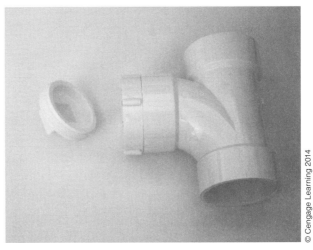

**Figure 36–1(E).** Sanitary tee with cleanout.

**Figure 36–1(F).** Stop-and-waste valve.

The supply piping from the municipal tap to the house is installed by plumbers who work for the plumbing contractor.

The main supply pipe entering the house must be larger in diameter than the individual branches installed from the main to each point of use. There are two basic reasons for this. First, water develops friction as it flows through pipes, and the greater size reduces this friction in the long supply line. Second, when more than one fixture is used at a time, the main supply must provide adequate flow for both. Generally, the main supply pipe for a one- or two-family house is ¾-inch or 1-inch pipe.

At the point where the main supply enters the building, a water meter is installed. The water meter measures the amount of water used. The municipal water department relies on this meter to determine the proper water bill for the building. The main water shutoff valve is located near the water meter.

The cold-water distribution piping continues throughout the house, which also provides the water supply to a water heater. The piping system exiting the water heater creates the hot-water distribution system serving the entire house. The size of each pipe providing water to each fixture is dictated by the specific fixture requirements and the plumbing code. **Figure 36–2** shows the water distribution system for a house.

When a valve is suddenly closed at a fixture, the water tends to slam into the closed valve. This causes a sudden pressure buildup in the pipes and may cause the pipes to *hammer* (a sudden shock in the

### GREEN NOTE

*Americans use large quantities of water inside their homes. The average family of four can use 400 gallons of water every day, and, on average, approximately 70 percent of that water is used indoors.*

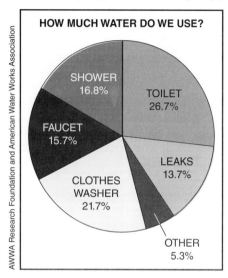

*The bathroom is the largest consumer of indoor water. The toilet alone can use 27 percent of household water. Almost every activity or daily routine that happens in the home bathroom uses a large quantity of water, such as:*

- *Older toilets use between 3.5 and 7 gallons of water per flush. However, WaterSense-labeled toilets require 75 to 80 percent less water.*
- *A leaky toilet can waste about 200 gallons of water every day.*
- *A bathroom faucet generally runs at 2 gallons of water per minute. By turning off the tap while brushing your teeth or shaving, a person can save more than 200 gallons of water per month.*

*Outside the bathroom, there are many opportunities to save water. Here are some common water efficiency measures, along with a few solutions to those problems you may not have known existed:*

- *High-efficiency washing machines can conserve large amounts of water. Traditional models use between 27 and 54 gallons of water per load, but new, energy- and water-conserving models (front-loading or top-loading, non-agitator ones) use less than 27 gallons per load.*
- *Washing the dishes with an open tap can use up to 20 gallons of water, but filling the sink or a bowl and closing the tap saves 10 of those gallons.*
- *Keeping a pitcher of water in the refrigerator saves time and water instead of running the tap until it gets cold.*
- *Not rinsing dishes prior to loading the dishwasher could save up to 10 gallons per load.*

supply piping). For quick-closing valves, water hammer arresters are required by most plumbing codes. The water hammer arrester has a piston that transmits shock waves from the system to a gas-filled chamber (see **Figure 36–3**). When a valve is suddenly closed, the gas chamber acts as a shock absorber. Although water cannot be compressed, gas can be. When the pressure tends to build up suddenly, the gas in the arrester compresses and cushions the resulting shock.

## Drainage Waste and Vent System

The main purpose of a drainage system is to remove wastewater and solids from a building. A drain is installed at each fixture, and all individual drains are connected to eventually create a building drain that exits the building. The building drain connects to a building sewer, which conveys the wastewater and solids to a municipal sewer, septic tank, or other approved point of disposal. The purpose of a vent is to allow air circulation within the system to equalize positive and negative pressures within the piping. Each fixture must be protected by a vent to ensure safe operation of the drainage system. A vent can terminate independently through a roof or be connected with other individual vents to create a branch vent before terminating through a roof. A venting device known as an *air admittance valve (AAV)* is accepted by some codes. When a plumbing fixture is operated

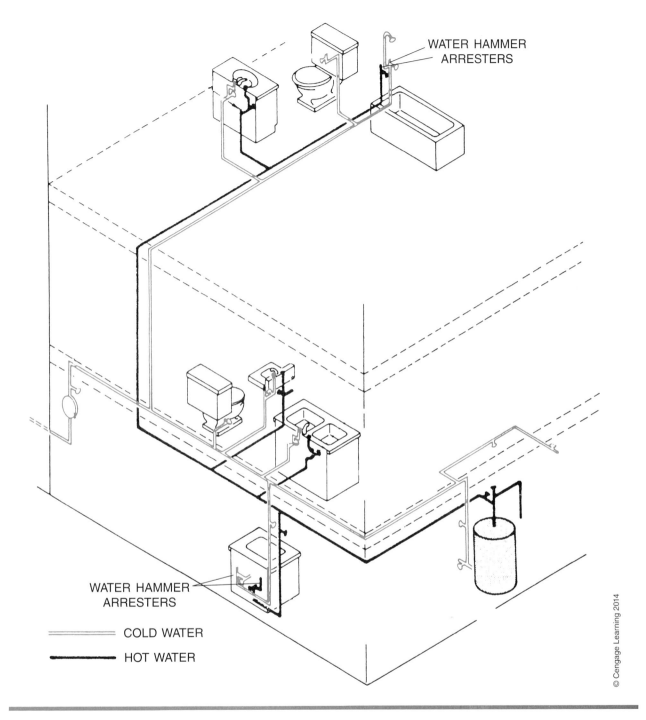

**Figure 36–2.** Hot- and cold-water piping.

and water drains out of the system, negative pressure causes the valve to open, allowing air to enter as needed to equalize the pressure. When the flow stops, gravity closes the valve, preventing the escape of sewer gases through the valve. An AAV eliminates the need for venting through a roof.

## Traps

The sewer contains foul-smelling, germ-laden gases that must be prevented from entering the house. If wastewater simply emptied into the sewer from the pipe, this sewer gas would be free to enter the building. To prevent this from happening, a trap is installed at each fixture. A **trap** is a fitting that naturally fills with water to prevent sewer gas from entering the building (see **Figure 36–4**). Not all traps are easily seen. A **water closet** (toilet) creates its own trap by nature of its design (see **Figure 36–5**).

## Vents

As the water rushes through a trap, it is possible for a siphoning action to be started. (The air pressure entering the fixture drain is higher than that on the other side of the trap. This forces the water out of the trap.)

To prevent DWV traps from siphoning, a vent is installed near the outlet side of the trap. The vent is an opening that allows air pressure to enter the system and break the suction at the trap (see **Figure 36–6**). Because the vent allows sewer gas to pass freely, it must be vented to the outside of the building. Unless protected by an AAV, all the fixtures are usually vented into one main vertical pipe, through the roof (see **Figure 36–7**).

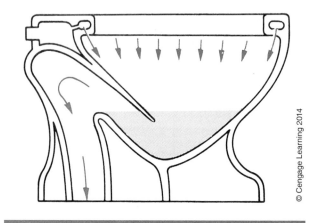

**Figure 36–5.** A water closet has a built-in trap.

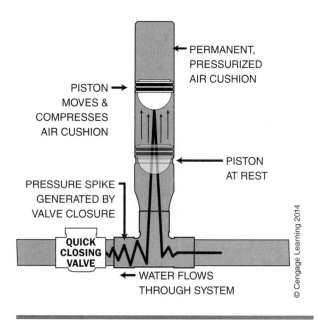

**Figure 36–3.** Water Hammer Arrester.

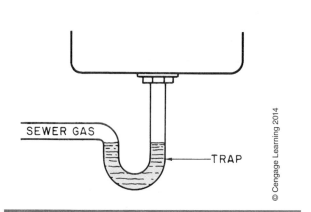

**Figure 36–4.** A trap fills with water to prevent sewer gas from entering the building.

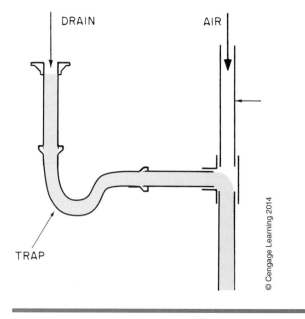

**Figure 36–6.** Venting a trap allows air to enter the system and prevents siphoning.

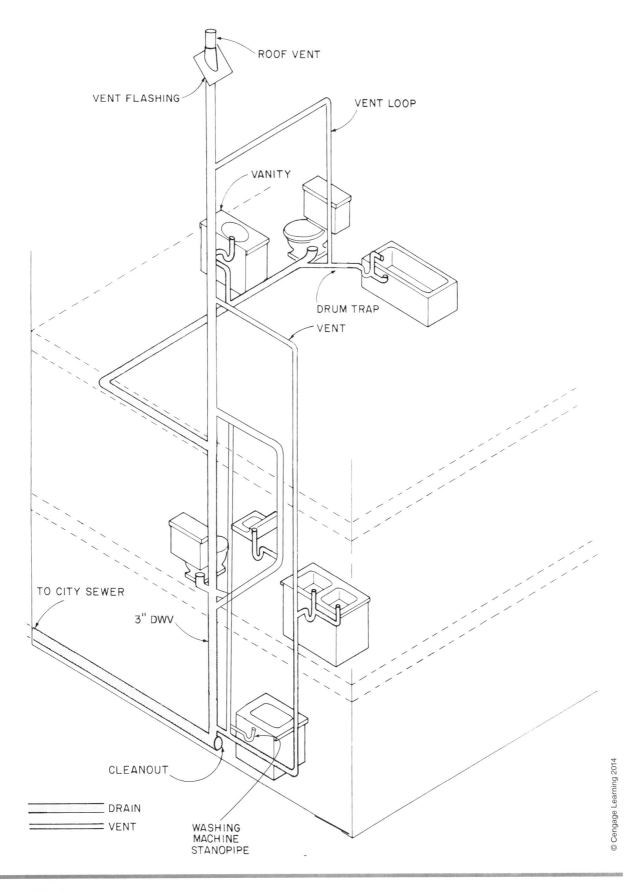

**Figure 36-7.** DWV system.

Plumbing, Heating, and Air Conditioning 269

## Plumbing Plans

For residential construction, the architect does not usually include a plumbing plan with the set of working drawings. The floor plan shows all the plumbing fixtures by standard symbols. These symbols are easily recognized, because they resemble the actual fixture. The dimensions of the fixtures are provided by the manufacturer on rough-in sheets (see **Figure 36–8**).

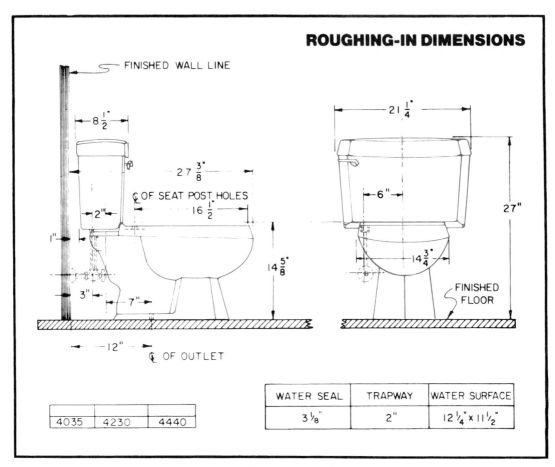

**Figure 36–8.** Typical manufacturer's rough-in sheet.

If the building and the plumbing are fairly simple, plumbers may prepare estimates and bids and complete the work from the symbols on the floor plan only. For more complex houses, the plumbing contractor usually draws a plumbing isometric (see **Figure 36–9**), or a special plumbing plan. The drawing set that accompanies this textbook includes a plumbing plan and details for Plan A of the Town House. This sheet includes more details than would normally be found on a plumbing plan for a single-family housing unit. The extra details are included here to help you understand the plumbing plan.

Plumbing plans show each kind of piping by a different symbol. Common plumbing symbols are shown in the Appendix. It will help you understand the plumbing plan if you trace each kind of piping from its source to each fixture. For example, trace the gas piping for the Town House. The gas lines can be recognized by the letter G in the piping symbol. The gas supply is shown as a broken line until it is inside the garage. Broken lines are used to indicate that the pipe is underground or concealed by construction. Although it is not noted on this plan, the plumbing contractor should know that the building code requires the gas line to be run in a sleeve where it passes through the foundation and the concrete slab (see **Figure 36–10**). Just inside the garage wall, the broken line changes to a solid line. At this point, a symbol indicates that the solid line (exposed piping) turns down or away. Here, the gas piping runs above the concrete slab and along the garage wall. A callout on this line indicates that the diameter of the pipe is ¾ inch. At the back of the garage, the gas line has a tee. Both of the outlets of this tee are ½ inch in diameter. One side of the tee supplies the *forced-air unit (FAU)*. The other side of the tee continues around behind the FAU to another tee and then to the water heater. The side outlet of the second tee supplies a log lighter in the fireplace. This branch is shown on the first floor plumbing plan. Notice that the log-lighter branch is reduced further to ¼-inch diameter.

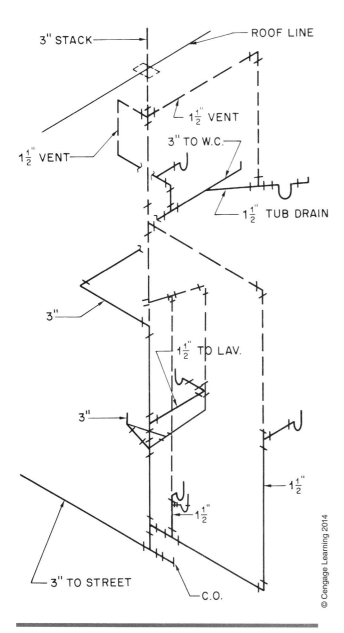

**Figure 36–9.** Single-line isometric of system shown in Figure 36–8.

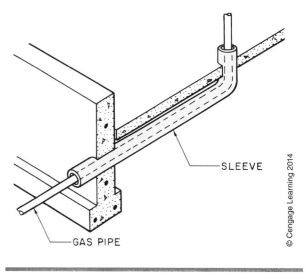

**Figure 36–10.** Sleeve for running gas piping under and through concrete. Local codes specify the design of sleeves used for gas piping. This figure shows only the basic concept.

You should trace each type of piping in a similar manner to be sure you understand it. Using colored pencils to trace the different types of piping may eliminate some of the confusion on crowded drawings.

Refer to the details on the drawing for clarification of the complex areas. As you trace each line, look for the following:

○ Kind of plumbing (hot water, cold water, waste)
○ Diameter
○ Fittings
○ Exposed or concealed
○ Where line passes through building surfaces

## Heating, Ventilating, and Air Conditioning

Plans for residential construction do not usually include sheets for *HVAC (heating, ventilating, and air conditioning)*, but the floor plan often includes a small amount of basic information—for example, where major equipment such as the furnace, air-handling unit, or air-conditioning unit is to be located and where diffusers (conditioned air outlets) and air returns are to be placed. The HVAC contractor installs the equipment according to building codes and industry practices without the aid of formal drawings.

Many HVAC systems are of the forced-air type. When heat is called for, these systems provide some means of forcing air over a surface heated by gas, oil, or electricity. That surface might be a combustion chamber, a liquid-filled coil, or an electric heating element. When cooling is called for, the air is forced over a cold coil from a refrigeration unit or a heat pump. The conditioned air is then forced through sheet metal or plastic ducts. The ducts carry the air to openings called diffusers throughout the house. The air is finally picked up by the return grille and returned to the forced-air unit (see **Figure 36–11**).

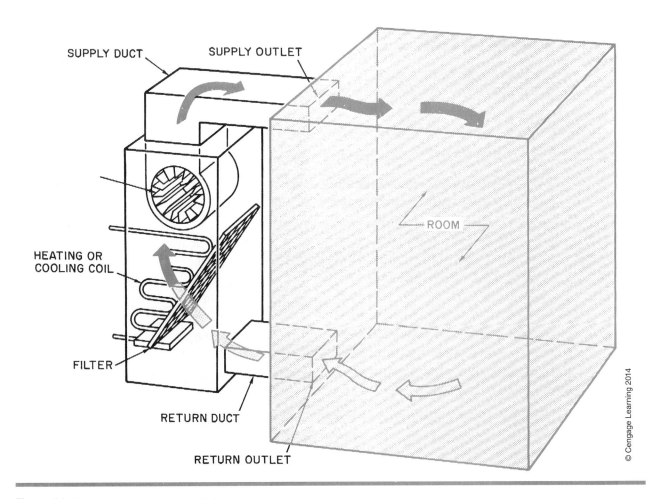

**Figure 36–11.** The air cycle in a forced-air system.

Other types of heating systems include electric resistance heat at the point of use; radiant floor heat, hot water circulated through tubing in or under the floor; and hot water circulated through baseboard heating units. With any of these systems, cooling can only be provided by a separate air-conditioning system.

>
> **USING WHAT YOU LEARNED**
>
> Plumbers are not the only professionals who need to know where the plumbing will be installed. Carpenters, too, need to understand the plumbing drawings so that they can coordinate their work with the plumbers and ensure that they do not obstruct piping and related systems. This is especially true for larger DWV piping. Referring to the Town House drawings in your textbook packet, what size is the waste piping into which the toilet in the Master Bath empties?
>
> Refer to the First Floor Plan on the Plumbing Drawing, P1. The Master Bath is clearly labeled and the toilet is at the location of the callout for Detail 4. The heavy line going from the tub area to that toilet is waste piping. A callout in the tub area indicates that this pipe and the one to the toilet in Bathroom #2 are to be 3 inches. This can also be seen on Detail 4, which indicates that the toilet empties into a 4" × 3" closet bend, an elbow intended for use on toilets. This allows the toilet to connect to the 4-inch fitting but reduces it's diameter to 3 inches where it connects to the pipe.

## Assignment

Refer to the Town House Plan A drawings in your textbook packet to complete the assignment.

1. What size pipe supplies the washing machine?
2. What size is the cold-water supply to the water heater?
3. What size is the cold-water branch to the lavatory in Bathroom #2?
4. At what point does the ¾-inch cold-water branch to the kitchen reduce to ½ inch for the hose bib?
5. Does the cold-water supply turn up or down as it leaves the bathroom area to supply the kitchen area?
6. List each of the fittings that water will pass through after it drains out of the master bathroom lavatory.
7. List each of the fittings that water will pass through to flow from the main shutoff at the building line to the shutoff on the supply side of the water heater.
8. What size is the waste piping from the water closet in the master bath?
9. What size is the waste piping from the kitchen sink?
10. What size is the waste piping from the washing machine?
11. Where is the air-conditioning compressor located?
12. Where is the tubing that connects the air-conditioning compressor to the forced-air unit?
13. What size tubing connects the compressor to the forced-air unit?
14. How many diffuser outlets supply conditioned air to the rooms?
15. Where are the sizes of the diffuser outlets given?

# UNIT 37
# Electrical

## Objectives

After completing this unit, you will be able to perform the following tasks:

- Identify the electrical symbols shown on a plan.
- Explain how the lighting circuits are to be controlled.

## Current, Voltage, Resistance, and Watts

To understand the wiring in a building, you should know how electricity flows. Electricity is energy. To do any work (turn a motor, light a lamp, or produce heat) the electrical energy must have movement. This movement is called *current*. The amount of current is measured in **amperes**, sometimes called *amps*. A single household-type light bulb requires a current of slightly less than 1 ampere. An electric water heater might require 50 amperes.

The amount of force or pressure causing the current to flow affects the amount of current. The force behind an electric current is called *voltage*. If 115 volts causes a current flow of 5 amperes, 230 volts will cause a current flow of 10 amperes.

The ease with which the current is able to flow through the device also affects the amount of current. The ease or difficulty with which the current flows through the device is called the *resistance* of that device. As the resistance goes up, the current flow goes down. As the resistance goes down, the current flow goes up.

The amount of work the electricity can do in any device depends on both the amount of current (amps) and the force of the current (**volts**). Electrical work is measured in **watts**. The number of watts of power in a device can be found by multiplying the number of amperes by the number of volts. Stated another way, the current flowing in a device can be found by dividing the number of watts by the voltage. For example, how much current flows through a 1,500-watt heater at 115 volts? 1,500 divided by 115 equals about 13 amperes. **Figure 37–1** shows the current, wattage, and voltage of some typical electrical equipment.

| DEVICE | AMPERES | VOLTS | WATTS |
|---|---|---|---|
| Ceiling light fixture | 1.3 | 115 | 150 |
| Vacuum Cleaner | 6.1 | 115 | 700 |
| Radio | 0.4 | 115 | 4 |
| Clock | 0.4 | 115 | 4 |
| Dishwasher | 8.7 | 115 | 1,000 |
| Toaster | 13 | 115 | 1,500 |
| Cook Top | 32 | 230 | 7,450 |
| Oven | 29 | 230 | 6,600 |
| Clothes Dryer | 25 | 230 | 5,750 |
| Washing Machine | 10 | 115 | 1,150 |
| Garbage Disposal | 7.4 | 115 | 850 |

© Cengage Learning 2014

**Figure 37–1.** Current, voltage, and power ratings of some typical electrical devices.

### GREEN NOTE

*Electricity is a form of energy and almost everything electrical about a home has an impact on the energy use of the home and its livability. One of the first electrical systems we see in any home is its lighting and lighting plays a major role in electrical energy use. Lighting can also impact other green aspects of the home. Lamps, commonly called light bulbs can have lifetimes of as low as 750 hours for standard incandescent lamps to as high as 50,000 hours for light-emitting diode / LEDs.*

*The recycling requirements for electric lamps is another factor to consider. All fluorescent lamps, including tubular and compact fluorescents (CFLs) as well as high-density discharge lamps contain mercury and should not be discarded into landfills. Lamps that contain mercury should not be discarded with standard residential and commercial waste. The EPA provides guidelines for disposal of these lamps.*

*Lighting can be made greener by paying attention to how much light is needed and when it is needed. Dimmers reduce energy consumption and well-planned lighting will not provide more light than is needed in a space. Motion sensors can be used effectively to turn lights off when no one is in the room.*

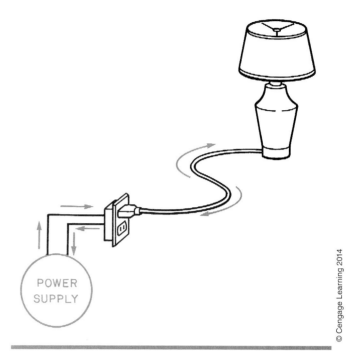

**Figure 37–2.** A complete circuit includes a path from the supply to the device and back again.

## Circuits

In order for current to flow, it must have a continuous path from the power source, through the electrical device, and back to its source. This complete path is called a *circuit* (see **Figure 37–2**).

Many circuits include one or more switches. A switch allows the continuous path to be broken (see **Figure 37–3**). By using two 3-way switches, the circuit can be controlled from two places (see **Figure 37–4**). When the circuit is broken by a switch or a broken wire, or for any other reason, it is said to be *open*.

Any material that carries electric current is called a conductor. In **Figure 37–2**, each of the wires is a conductor. When two or more wire conductors are bundled together, they make a cable (see **Figure 37–5**).

In larger buildings, the wiring is frequently installed by pulling individual wires through steel or plastic pipes, called *conduit*. In houses, it is more common to use cables containing the needed wires plus one ground conductor. The ground conductor does not normally carry current. The *ground,* as it is usually abbreviated, connects all the electrical devices in the house to the ground. If, because of some malfunction, the voltage reaches a part of the device that someone might touch, the ground protects him or her from a serious shock. The current that might otherwise flow through the person follows the ground conductor to the earth. The earth actually carries this current back to the generating station.

Additional protection against serious shock can be provided by using a *ground-fault circuit interrupter (GFCI* or *GFI)*. A GFCI is a device that measures the flow of current in the hot (supply) conductor and the neutral (return) conductor. If a faulty device allows some of the current to flow through another path, such as a person, rather than the neutral conductor, the GFCI stops all current flow immediately. GFCIs are so effective that the *National Electric Code®* requires their use on circuits for outlets installed outdoors, in kitchens, in bathrooms, in garages, and near any other water hazards.

The electrical service entrance is discussed in Unit 11. The service feeder cable ends at a distribution panel. From the distribution panel, the electrical system is split up into several branch circuits (see **Figure 37–6**). Each branch circuit includes a circuit breaker or fuse.

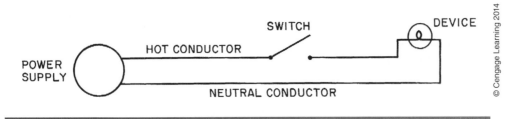

**Figure 37–3.** A switch is used to break (or open) the circuit.

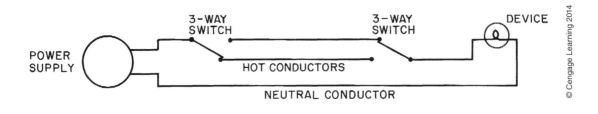

**Figure 37–4.** Three-way switches allow a device to be controlled from two locations. Notice that if either switch is activated, the device will be energized.

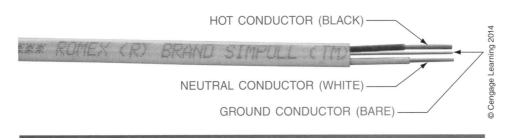

**Figure 37–5.** This cable has two circuit conductors and one ground conductor.

The circuit breaker or fuse opens the circuit if the current flow exceeds the rated capacity of the circuit. Branch circuits for special equipment such as water heaters and air conditioners serve that piece of equipment only. Branch circuits for small appliances and miscellaneous use may serve several outlets. Branch circuits for lighting are restricted to lighting only, but a single circuit may serve several lights. Lighting circuits also include switches to turn the lights on and off.

The National Fire Protection Association publishes the *National Electrical Code®*, which specifies the design of safe electrical systems. Electrical engineers and electricians must know this code, which is accepted as the standard for all installations. The following are among the items it covers:

○ Kinds and sizes of conductors
○ Locations of outlets and devices
○ Over current protection (fuses and circuit breakers)
○ Number of conductors allowed in a box
○ Safe construction of devices
○ Grounding
○ Switches

The specifications for the structure indicate such things as the type and quality of the equipment to be used, the kind of wiring, and any other information that is not given on the drawings. However, electricians must know the *National Electrical Code®* and any state or local codes that apply because specifications sometimes refer to these codes.

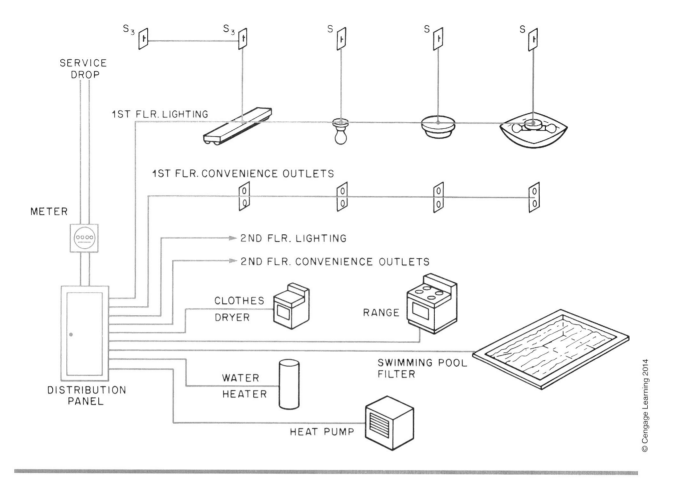

**Figure 37–6.** The electrical service is split up into branch circuits at the distribution panel.

## Electrical Symbols on Plans

The drawings for residential construction usually include electrical information on the floor plans. Only the symbols for outlets, light fixtures, switches, and switch wiring are included. The exact location of the device may not be dimensioned. The position of the device is determined by the electrician after observing the surrounding construction. It should also be noted that all wiring is left to the judgment of the electrician and the regulations of the electrical codes. Switch wiring for light fixtures is included only to show which switches control each light fixture. Switch wiring is shown by a broken line connecting the device and its switch (see **Figure 37–7**).

In rooms without a permanent light fixture, one or more convenience outlets may be split wired and controlled by a switch. In split wiring, one-half of the outlet is always hot; the other half can be opened by a switch (see **Figure 37–8**). The most common electrical symbols are shown in the Appendix at the back of this book.

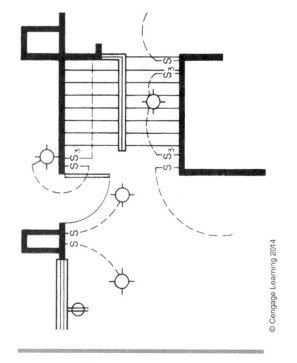

**Figure 37–7.** Switch legs on a plan.

Electrical 277

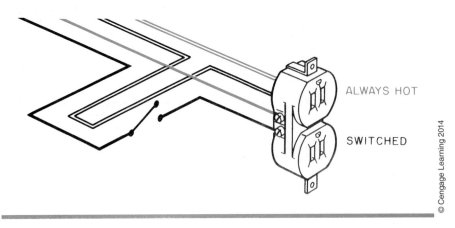

**Figure 37–8.** Split-wired outlet.

### USING WHAT YOU LEARNED

It is helpful for workers in all aspects of building construction to be familiar with the limited electrical wiring information that is typically included on the drawings for residential buildings. Estimators need to know how many of each kind of device will be used; carpenters need to be alert to possible problems with installing electrical equipment in the spaces they construct; and plumbers often need to connect piping to appliances that also use electricity.

In Floor Plan A of the Town House drawings in your textbook packet, what is controlled by the two wall switches at the top of the garage stairs? One is a three-way switch controlling the light fixture over the stairs. The other controls half of a split-wired outlet in the living room. The switches and lines indicating what they are connected to are clearly shown on the floor plan.

## Assignment

Refer to the drawings for the Town House, Floor Plan A drawing in your textbook packet, including the garage.

1. How many light fixtures are shown on the floor plan, including the garage?
2. How many switches are shown on the floor plan, including the garage?
3. How many duplex outlets are shown?
4. Briefly describe the location of each split-wired outlet and the switch or switches that control each.
5. List five pieces of equipment shown on the floor plan that probably require separate branch circuits.
6. How many outlets are to have ground-fault circuit interrupters included in their circuits?
7. What is the location of the switch or switches that control the light over the stairs to the bedroom level?
8. What is the location of the switch or switches that control the light fixture over the stairs to the garage?
9. Where are the smoke detectors located?
10. Where is each of the two telephone outlets?

# Test

**A.** For each of the symbols in Column I, indicate the object in Column II it represents.

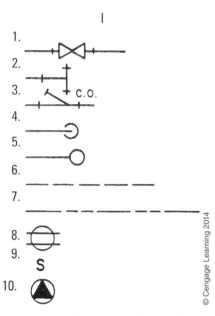

| I | II |
|---|---|
| 1. | a. Electrical outlet for special equipment |
| 2. | b. Pipe turning down or away |
| 3. | c. Switch wiring |
| 4. | d. Convenience outlet |
| 5. | e. Valve |
| 6. | f. Pipe turning up or toward |
| 7. | g. Hot-water piping |
| 8. | h. Wall switch |
| 9. | i. Wye fitting with a cleanout |
| 10. | j. Tee fitting |

**B.** Refer to the illustrated Hidden Valley drawing index and partial site plan to answer the following questions.

1. In which building is unit 56 located?
2. On which drawing sheet is the foundation plan for unit 56?
3. On which drawing sheet are the foundation details for Plan B units?
4. List all the drawing sheets that are not needed to construct Building IV.
5. Of the buildings shown, which would be built first?

**C.** Refer to the Town House drawings to answer these questions.

1. In Plan B, which rooms have their floors at the same elevation as the kitchen?
2. In Plan B, what separates the living room and dining room?
3. In Plan B, as you climb the stairs from the first floor, what room do you enter?
4. In Plan B, as you climb the stairs from the garage, what room do you enter?
5. In Plan B, what is the distance from the bottom of the garage stairs to the finished floor in the library?
6. In Plan B, what is the width in the living room area, measuring to the inside face of the framing?
7. What size are the anchor bolts in the foundation of Plan B?
8. In the front part of the foundation plan for Plan B, there is a 24-inch square by 12-inch deep pad for a pier that supports a wood beam. What does this beam support?

279

9. In Plan B, what supports the floor joists under bedroom #2 at the end nearest the front of the building?
10. In Plan B, what prevents the party wall in the nook area from wracking?
11. In Plan B, where the bedroom floor joists rest on the dining room–kitchen partition, how is the blocking between the joists fastened to the top plate of the wall?
12. What is the wall surface material at the area separation wall in the dining room of Plan B?
13. What is the total thickness, including surface material, of the party wall in the garage stairway in Plan B?
14. What size lumber (thickness × width × length) would be ordered for the joists in the Plan A kitchen floor?
15. In Plan A, what is the finished size (thickness × width × length) of the floor joists under bedroom #2?
16. What is the length of the cantilevered part of the floor joists under bedroom #2 of Plan A?
17. What material is used for the wood sill on the foundation of Plan A at the exterior corner of the kitchen?
18. In Plan A, what kind and size of fasteners are used to anchor the railing outside the dining room entrance?
19. In Plan A, how far is the air-conditioner compressor pad from the forced-air unit?
20. In Plan B, how many three-way switches are required?
21. In Plan A, how many ceiling light fixtures are there, including luminous soffits?
22. In Plan A, how many plumbing vents penetrate the roof?
23. What diameter of pipe supplies the hose bib at the front of the kitchen?
24. In Plan A, how many cleanouts are included in the waste piping?
25. In Plan A, where is the hot-water shutoff closest to the shower in bathroom #2?
26. In Plan A, where is the forced-air unit located?
27. How many return air grilles are there on Plan A?
28. What kind of fuel does the FAU use?
29. What is the major difference between units 38 and 42 in Building IV? (Refer to the partial site plan on page 282.)
30. What is the thickness and depth of the header over the Plan A garage door?
31. How much is the concrete slab thickened for a haunch at the entrance to the garage in Plan A? Give the thickness or width and depth of the haunch without the floor.

# table of contents

|     |          |                                  |
|-----|----------|----------------------------------|
|     |          | COVER SHEET AND INDEX            |
| a   |          | SITE PLAN                        |
| b   |          | GENERAL NOTES                    |
| c   |          | GENERAL NOTES                    |
| 1   | PLAN A   | FOUNDATION PLAN                  |
| 2   |          | FLOOR PLAN                       |
| 3   |          | FRAMING PLAN                     |
| 4   |          | SECTIONS                         |
| 5   |          | ELEVATIONS                       |
| 6   | PLAN B   | FOUNDATION PLAN                  |
| 7   |          | FLOOR PLAN                       |
| 8   |          | FRAMING PLAN                     |
| 9   |          | SECTIONS                         |
| 10  |          | ELEVATION (1)                    |
| 11  |          | ELEVATION (2)                    |
| 12  | PLAN C   | FOUNDATION PLAN                  |
| 13  |          | FLOOR PLAN                       |
| 14  |          | FRAMING PLAN                     |
| 15  |          | SECTIONS                         |
| 16  |          | ELEVATIONS (1)                   |
| 17  |          | ELEVATIONS (2)                   |
| 18  | PLAN D   | FOUNDATION PLAN                  |
| 19  |          | FLOOR PLAN                       |
| 20  |          | FRAMING PLAN                     |
| 21  |          | SECTIONS                         |
| 22  |          | ELEVATIONS (1)                   |
| 23  |          | ELEVATION (2)                    |
| 24  | PLAN A-B | INTERIOR ELEVATIONS-PLAN A & B   |
| 25  | PLAN C-D | INTERIOR ELEVATIONS-PLAN C & D   |
| 26  |          | BUILDING TYPE I                  |
| 27  |          | BUILDING TYPE II                 |
| 28  |          | BUILDING TYPE III                |
| 29  |          | BUILDING TYPE IV                 |
| 30  |          | BUILDING TYPE V                  |
| 31  |          | BUILDING TYPE VI                 |
| 32  |          | BUILDING TYPE VII                |
| 33  |          | BUILDING TYPE VII                |
| 34  |          | BUILDING TYPE VIII               |
| 35  |          | BUILDING TYPE VIII               |
| 36  |          | BUILDING TYPE IX                 |
| 37  |          | BUILDING TYPE IX                 |
| 38  |          | BUILDING TYPE X                  |
| 39  |          | BUILDING TYPE X                  |
| 40  |          | BUILDING TYPE XI                 |
| 41  |          | BUILDING TYPE XII                |
| 42  |          | METER ENCLOSURE                  |
| 43  |          | REC. CENTER    FOUND FLR PLAN    |
| 44  |          | REC. CENTER    SECTION ELEV      |
| D1  |          | FOUNDATION DETAILS               |
| D2  |          | FRAMING DETAILS                  |
| D3  |          | FRAMING DETAILS                  |
| D4  |          | FRAMING DETAILS                  |

Courtesy of Berkus Group Architects

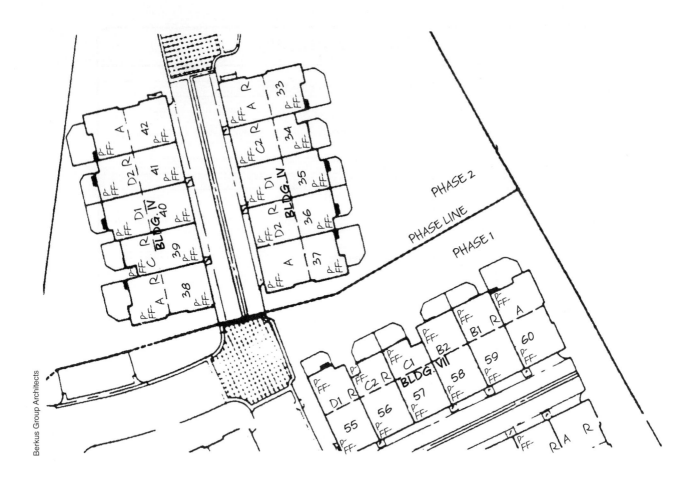

# HEAVY COMMERCIAL CONSTRUCTION

Part IV presents a thorough examination of the information found on prints for heavy commercial construction. The materials and methods used for large buildings are different from those used in light-frame construction. Also, the drawing set for a commercial project usually includes many more sheets than are found in the drawing sets for smaller buildings. To understand these drawings and make practical use of them, you need to understand the organization of the drawings and how the heavier materials are described on the drawings.

Part IV also discusses the drawings used for air conditioning and heating, plumbing, and electrical systems in much more detail than earlier sections of the textbook. For those who work in the mechanical or electrical trades, the importance of understanding the drawings for these systems is obvious. However, all the trades work in the same spaces, and the work of one trade affects the other trades. Estimators, superintendents, inspectors, and many other construction professions also require an understanding of all the construction trades.

The School Addition project in your textbook packet drawings is an excellent building for inclusion in this book. It uses structural steel framing, reinforced concrete foundations, and a varied assortment of other materials typically associated with heavy construction. Unlike most large commercial buildings, however, the School Addition is small enough, so that most of the

drawings can be packaged with this book. Some drawings have been eliminated, either because they pertain to other work done on the existing school building or because they do not provide new opportunities for the study of commercial construction drawings. All the School Addition drawings are numbered as they were in the complete set, and the textbook makes frequent reference to the drawing numbers. The Material Keynotes, which are explained in this part of your textbook, are printed in Appendix A at the end.

# UNIT 38
# Heavy Commercial Construction

## Architectural Style

Large buildings, such as the School Addition described in the drawings in your textbook packet, are usually designed in a different style from one- and two-family homes or even larger light-frame buildings, such as the Town House.

Most larger commercial buildings are too large to employ the same construction methods as those used with homes and smaller buildings and have different architectural styles than they do (see **Figure 38–1**). For example, commercial buildings usually have a larger roof area, so it is not practical to slope the roof. This means that not only does the flat roof change the appearance of the building, but it necessitates the use of different roof framing and roof covering materials. Commercial buildings are designed to be as maintenance free as possible. The wood or vinyl molding that decorates the exterior of many houses would require too much maintenance for a commercial office building. This is not to say that a commercial building cannot be architecturally pleasing. In fact, because more money is generally available to design and construct large commercial buildings, they often tend to be extremely attractive and viewed as works of art. However, the aesthetic qualities in a commercial building are generally built into the overall design of the building, rather than applied after the shell is erected.

Larger buildings require greater strength. The weight of the materials required to build the School Addition is many times the weight of the materials to build even a large house. Also, the external loads (wind, snow, and so forth) on a larger building are greater than those on a smaller building. All these greater loads require stronger structural members, more precisely engineered construction methods, and fastenings that transmit the loads from one structural member to the next. This usually dictates the use of structural steel and reinforced concrete for the structural elements of the building—the frame and foundation.

## Structural Steel

Steel is made by alloying small amounts of carbon and other elements with iron. The amount of carbon in the steel determines its toughness. Other elements, such as chromium, copper, nickel, or titanium, are used to produce specific properties. Alloying chromium with the steel, for example, makes stainless steel. Structural steel is mild steel (a small amount of carbon is used) that is rolled into specific shapes used for construction. The most common

## Objectives

After completing this unit, you will be able to perform the following tasks:

○ Describe the major differences between light-frame construction and heavy commercial construction.

○ Describe common structural steel shapes and their designations.

○ Explain the differences between foundations for heavy structures and those for light-frame buildings.

**Figure 38–1.** Generally, medium to large commercial buildings are very functional in design.

structural shapes are shown in **Figure 38–2**. W-shapes are used for most beams and columns in steel-frame buildings. The designator for a structural steel member includes a symbol or capital letter to indicate the basic shape, one or more numbers to indicate the size, and a weight-per-foot designation. For example, drawing S200 for the School Addition (see **Figure 38–3**), shows a W6 × 25 column. This is a *wide*-flange shape, with a *6*-inch web, weighing *25* pounds per foot. Standard dimensions for structural steel shapes are published in the *American Institute for Steel Construction (AISC) Manual of Steel Construction.*

Structural steel members are most often joined by welding. Bolts are sometimes used to attach dissimilar materials, such as wood and steel framing members. Where bolts are to be used, they are usually shown by either symbols or a drawing of a bolt on the details (see **Figure 38–4**). Detail J-14 of the drawings for the School Addition (see **Figure 38–4**) introduces a technique that is often used on drawings for commercial construction. Instead of describing all the information on the detail drawing itself, many of the drawing call-outs refer to a material legend. The *material legend*, which is a running list on the right side of most of the school drawings, identifies all the material items in the building by a legend number. Roof Edge Detail J-14, the drawing shown in **Figure 38-4**, shows four items identified only as 5.10D. The material legend on Sheet A504 identifies these as anchor bolts.

## Welding Symbol

Welds are depicted in drawings by the use of standard welding symbols (see **Figure 38–5**). The American Welding Society (AWS) has developed a standard system for welding symbols. Each part of the welding symbol has a specific meaning. The arrow indicates what steel parts are to be welded. The term *welding symbol* should not be confused with *weld symbol*. The welding symbol is the complete symbol, including arrow, reference line, and any information added to it. The weld symbol indicates the type of weld to be made. The type of weld used most often in structural steel for construction is called a *fillet weld* and is indicated by a triangle. Other types of welds include plug welds, groove welds, and butt welds (see **Figure 38–6**). The reference line is always drawn parallel to the bottom edge of the drawing. The lower side of the reference line is called the *arrow side*. Any information, such as the type of weld, that is shown below the reference line applies to the side of the assembly to which the arrow points. The upper side of the reference line is called the *other side*. Information above the reference line applies to the side of the assembly away from the arrow (see **Figure 38–7**). The tail of the arrow is used to specify welding processes, electrode materials, and any other specifications that cannot be shown elsewhere on the welding symbol.

| Shape | Letter Designation | Size Designation |
|---|---|---|
|  Wide Flange | W | depth of web × weight per foot |
| 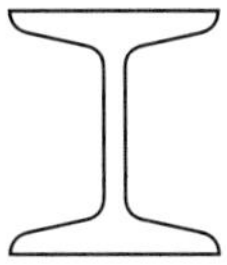 Standard Beam | S | depth of web × weight per foot |
|  Channel or Miscellaneous Channel | C or MC | depth × weight per foot |
| 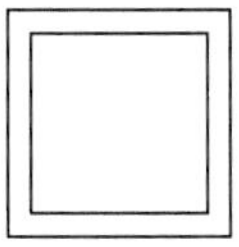 Structural Tube | TS | width × depth |
| 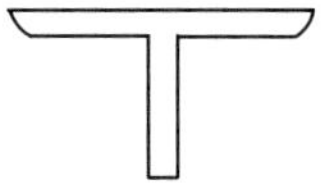 Tee | T | depth × weight per foot |
| 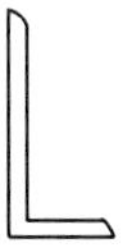 Angle | L | vertical leg × horizontal leg × thickness |

**Figure 38–2.** Common structural shapes and their designations.

## Shop Fabrication

Many structural subassemblies are more easily or more accurately fabricated in a shop environment. For example, if a special beam is to be installed with a welded subassembly on its ends, the subassembly would probably be fabricated in the steel shop so the beam can be delivered to the site ready for erection. A large structural steel frame might require as many *fabrication drawings* (sometimes called *shop drawings*) as structural drawings. Usually, the field contractors do not deal with shop drawings, and they are not covered in this book. However, it is important to recognize notes

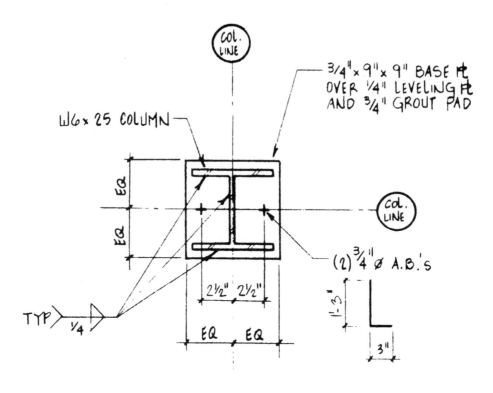

**Figure 38–3.** Structural steel designation on a drawing.

on the structural drawings that indicate a difference between field erection and shop fabrication. One example of such information is the field weld symbol (see **Figure 38–8**). The field weld symbol is a small flag at the break in the arrow of the welding drawing. This symbol indicates the weld is to be done in the field, not in the fabrication shop.

## Foundations for Commercial Construction

Houses and small frame buildings usually have very simple foundation systems. The superstructure does not impose a heavy load on the foundation, so a simple *haunched slab* or *inverted-T foundation* is all that is required to support the building in most soil conditions. As buildings become larger and more complex, they place a greater load on the foundation system. Also, larger buildings tend to be designed with much of the load being transmitted through a relatively small number of columns, instead of being spread over a sill member by closely spaced studs. To support these loads, the foundations of heavy buildings are carefully engineered by structural engineers and are more heavily reinforced than the foundations for smaller buildings.

Generally, continuous wall foundations are similar to those for smaller buildings. They consist of a *spread footing* that supports a continuous concrete wall. Because of the greater size of the structure, the footing and wall may be more massive for a heavy building

288 UNIT 38

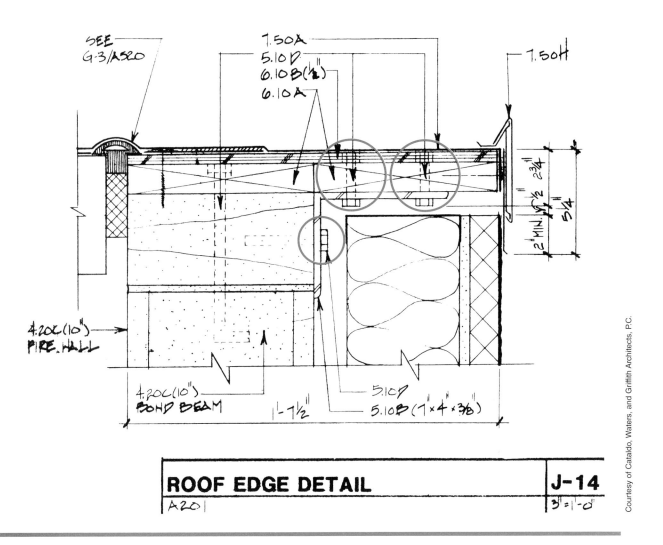

**Figure 38–4.** Bolts are sometimes used for fastening structural members.

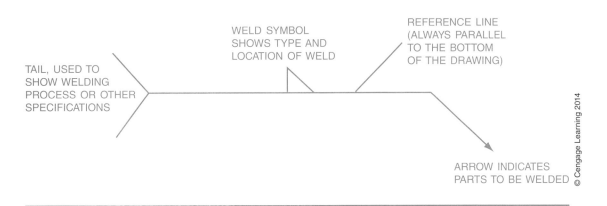

**Figure 38–5.** The basic welding symbol.

Heavy Commercial Construction 289

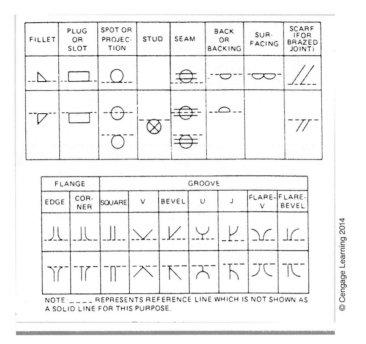

**Figure 38–6.** Weld symbols used to show various types of welds.

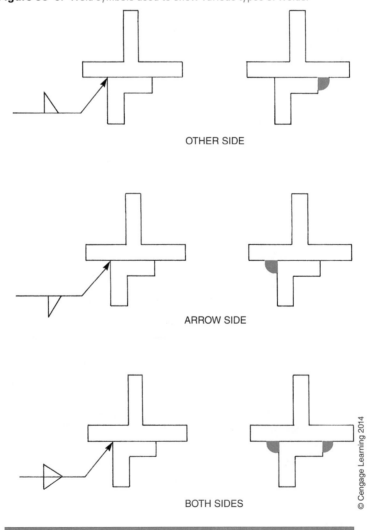

**Figure 38–7.** Arrow-side, other-side significance.

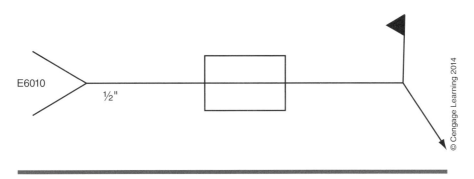

**Figure 38–8.** This can be recognized as a field weld symbol by the flag. It is a ½-inch plug weld on both sides of the part and is to be done with E6010 electrode or filler metal.

and will probably require more and larger-diameter reinforcing steel. The overall layout of the foundation is shown on a foundation plan (see **Figure 38–9**). The dimensions and the reinforcement of the footings and walls are planned by engineers to carry the necessary loads, so it is likely that there will be several detail drawings to further describe the foundation systems at different points in the building (see **Figure 38–10**). Some of the reinforcement may be bars that protrude only a short distance out of the footing and into the wall. Reinforcing bars used in this way are called *dowels* and are used to keep the wall from shifting on the footing

Column footings are located by the column centerlines. All these footings must be centered under the columns they support unless specifically noted otherwise. The outline of the footing is shown on the foundation plan using dashed lines. This indicates that the footing is actually hidden from view by the concrete slab. Column footing dimensions are usually shown in a schedule (see **Figure 38–11**). There are usually several column footings of the same size. The mark-schedule system allows the drafter to show all the similar footings with a single entry on the schedule.

*Piles* are long poles or steel members that are driven into the earth to support columns or other grade beams. Pile foundations support their loads by means of several piles in a cluster, topped with a pile cap made of reinforced concrete. Pile foundations are not as common as spread footings and continuous walls. *Grade beams* are reinforced concrete

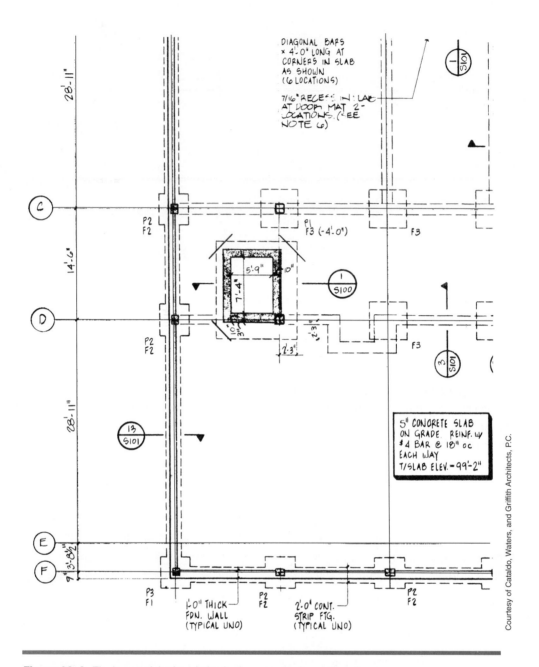

**Figure 38-9.** The layout of the foundation is shown on the foundation plan.

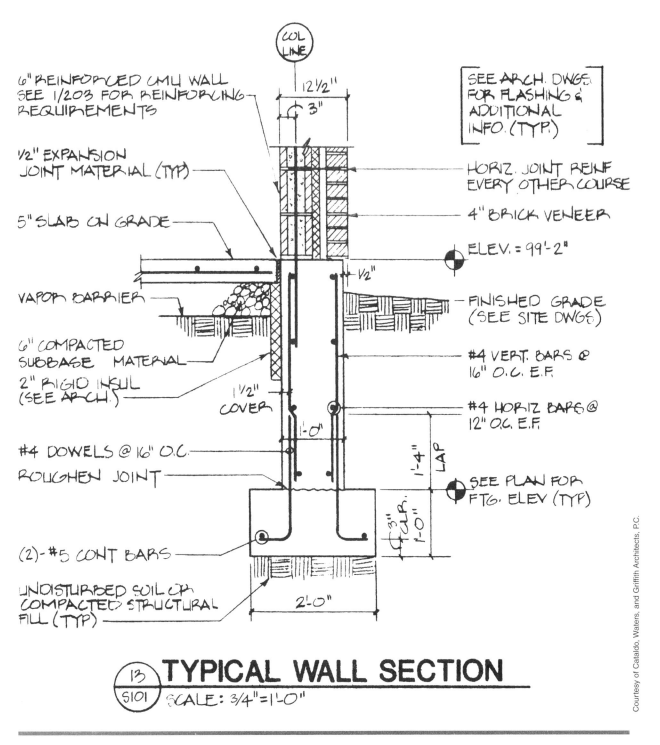

**Figure 38–10.** Foundation detail.

## FOOTING SCHEDULE

| MARK | SIZE | REINF. | REMARKS |
|---|---|---|---|
| F1 | 3'-0" x 3'-0" x 12" | (4) #5 BARS EA. WAY | |
| F2 | 4'-0" x 4'-0" x 12" | (5) #5 BARS EA. WAY | |
| F3 | 4'-6" x 4'-6" x 12" | (5) #6 BARS EA. WAY | |
| F4 | 3'-0" x 4'-6" x 24" | (5) #5 SHORT BARS (7) #5 LONG BARS | OVERPOUR EXISTING BUILDING FOOTING (SIM. TO 14/S101) |

Courtesy of Cataldo, Waters, and Griffith Architects, P.C.

**Figure 38–11.** Footing schedule.

### USING WHAT YOU LEARNED

Just as it is important to understand the descriptions and sizes of all of the framing materials for a home construction project, it is also important to understand how the materials for a heavy project are indicated on drawings and in specifications. As stated in this unit, heavy construction projects usually involve a lot of structural steel.

On the second floor of the School Addition, classrooms 203 and 204 are separated by a folding partition. What structural shape and size is the roof beam above the folding partition? Begin by finding the folding partition on the second floor plan, which is C-14 on Sheet A102. Notice a callout for A-3/A504.

That detail shows the folding partition suspended from a partial drawing of a structural shape identified only as 5.10A, which would be a note on the Material Legend on the right side of the sheet. 5.10A simply says it is a steel beam. Based on where it is located (overhead on the second floor), it is most likely the roof beam in question. The Roof Framing Plan would be among the structural drawings, those with the letter S preceding the sheet number. Searching the structural drawings, we find that S201 is titled "Roof Framing Plan & Details." Looking in the corresponding area of the Roof Plan, we find a W18 × 40 beam with a callout referring to 5/S201. Detail 5 is a Folding Partition connection detail and it includes a W18 beam, so that it confirms what we found on the Roof Framing Plan.

# Assignment

1. List five features of the School Addition that would be very different on a light-frame building, and explain the differences.
2. Describe each of the structural steel designations listed below:
   W14 × 82
   MC12 × 45
   L6 × 3½ × 5/16
3. What is the structural steel shape and depth of the steel beam shown supporting the roof bar joists in detail 3/S202?
4. Draw a sketch showing where the welds would be made for each of the following symbols:

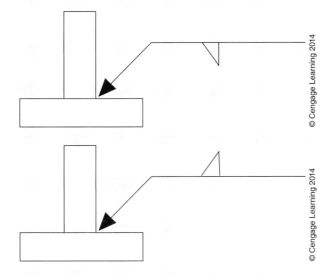

5. Describe the information given in the welding symbol on detail 3/S202.
6. Based on the information shown on Sheet S100, would you say that the School Addition has a pier foundation, grade beams, or a continuous wall foundation with spread footings?
7. How many pieces of reinforcing steel are shown in the footing on Wall Section 13/S101? Describe each, including its size and purpose.
8. How many 4'-6" × 4'-6" × 12" column footings are shown on the foundation plan for the School Addition?
9. There are several pairs of dashed (hidden) lines running vertically through the interior of the foundation plan. A note on one of these pairs of lines indicates that the slab is to be thickened. Why is it thickened at these points? How much thicker is it at these points than the surrounding area? What additional reinforcement is provided in the thickened portion of the slab?

# UNIT 39
# Coordination of Drawings

## Objectives

After completing this unit, you will be able to perform the following tasks:

- Interpret the indexes and keys normally found on drawings for commercial construction.
- Use the number system for referencing details, sections, and other drawings.
- Explain the use of a grid system to locate columns and piers.
- Use the plans, elevations, and details to describe the layout of the rooms and spaces in a large commercial building.

## Indexes

A set of working drawings for a large commercial building might fill more than 100 sheets. The School Addition in your textbook packet, although small by comparison with many large buildings, is part of a renovation and expansion program that spans three buildings and is described on 66 drawing sheets. A key plan in the lower right corners of sheets A101 and A102 shows the addition in relationship to the rest of the Junior/Senior High School building. Finding one's way around these 66 sheets would be a difficult task without some kind of an index. As with most commercial construction drawing sets, the cover page of the drawing set includes the index (see **Figure 39–1**).

It is not reasonable to include all 66 drawing sheets, plus the cover page, in your textbook packet because that would add the equivalent of more than two hundred pages to the book. Therefore, the packet only includes the drawings that are necessary to completely understand the additions to the Junior/Senior High School. The Index of Drawings, found in **Figure 39–1,** lists all the original drawing numbers. The drawing numbers that are underlined indicate the drawings contained in the drawing packet. Parts of other sheets are included as illustrations in this textbook. Drawings for civil (site work) that are not needed for an understanding of the structure and drawings for work on other school property are not included.

The *Drawing Index* includes the number of the drawing and the drawing title. *Drawing numbers* are made up of one or two letters and numerals. The letters indicate the basic type of construction shown on the drawing. Most architects use the letter *A* to indicate architectural drawings, *S* to indicate structural drawings, *C* for civil, and so on. The numbers indicate the sequence of the drawings in that section. Some architects indicate levels of drawings by using a series within a range to designate a level. For example, 100 through 199 might be reserved for foundations, 200 through 299 might be floor plans and elevations showing the overall superstructure, and 300 through 399 might be roof framing. This system is usually quite easy to spot by studying the index briefly.

## Material Keying

In light-frame construction of small buildings, it is customary for callouts on the drawings to label the material used. On larger projects, however, it is common for a material keying system to be used so that the drawings themselves

| | GENERAL CONSTRUCTION CONTRACT |
|---|---|
| A000 | MASTER KEYNOTES |
| G001 | SITE PLAN (JR/SR HIGH SCHOOL) |
| G002 | KEY PLANS (JR/SR HIGH SCHOOL) |
| G003 | KEY PLANS (BERNE ELEMENTARY & WESTERLO ELEMENTARY |
| C100 | SITE DEMOLITION PLAN (JR/SR HIGH SCHOOL & BERNE ELEMENTARY |
| C101 | LAYOUT PLAN (JR/SR HIGH SCHOOL) |
| C102 | GRADING PLAN (JR/SR HIGH SCHOOL) |
| C103 | PLANS & DETAILS (JR/SR HIGH SCHOOL) |
| C201 | PARTIAL SITE PLAN, SOUTH (JR/SR HIGH SCHOOL) |
| C202 | PARTIAL SITE PLAN, NORTH (JR/SR HIGH SCHOOL) |
| C203 | PLAN AND DETAILS - SEWAGE DISPOSAL (JR/SR HIGH SCHOOL & BERNE ELEMENTARY) |
| C204 | PLAN AND DETAILS; NEW WELL (JR/SR HIGH SCHOOL & BERNE ELEMENTARY) |
| C205 | DRAINAGE PROFILES AND DETAILS (JR/SR HIGH SCHOOL & BERNE ELEMENTARY) |
| C206 | DETAILS SEWAGE DISPOSAL (JR/SR HIGH SCHOOL & BERNE ELEMENTARY) |
| C207 | PLANS AND DETAILS (JR/SR HIGH SCHOOL & BERNE ELEMENTARY) |
| C208 | SECTIONS AND DETAILS (JR/SR HIGH SCHOOL & BERNE ELEMENTARY) |
| A001 | ROOM FINISH SCHEDULE, DOOR SCHEDULE AND DETAILS (JR/SR HIGH SCHOOL) |
| A101 | FIRST FLOOR PLAN (JR/SR HIGH SCHOOL) |
| A102 | SECOND FLOOR PLAN (JR/SR HIGH SCHOOL) |
| A103 | CAFETERIA & BOILER ROOM PLAN, DEMOLITION PLAN & DETAILS (JR/SR HIGH SCHOOL) |
| A104 | FLOOR PLAN & DETAILS (BUS GARAGE) |
| A201 | EXTERIOR ELEVATIONS & BUILDING SECTIONS (JR/SR HIGH SCHOOL) |
| A301 | STAIR PLANS & SECTIONS (JR/SR HIGH SCHOOL) |
| A302 | STAIR DETAILS (JR/SR HIGH SCHOOL) |
| A303 | RAMP & ELEVATOR (JR/SR HIGH SCHOOL) |
| A501 | WALL SECTIONS (JR/SR HIGH SCHOOL) |
| A502 | WALL SECTIONS (JR/SR HIGH SCHOOL) |
| A503 | PLAN DETAILS (JR/SR HIGH SCHOOL) |
| A504 | WALL DETAILS (JR/SR HIGH SCHOOL) |

**Figure 39–1.** Index from the cover page of school drawings.

| | |
|---|---|
| A510 | WINDOW TYPES AND DETAILS (JR/SR HIGH SCHOOL) |
| A511 | EXTERIOR ELEVATIONS (WESTERLO ELEMENTARY) |
| A512 | WINDOW TYPES & DETAILS (WESTERLO ELEMENTARY) |
| A520 | ROOF PLAN & DETAILS (JR/SR HIGH SCHOOL) |
| A610 | INTERIOR ELEVATIONS & TOILET ROOM FLOOR PLAN (JR/SR HIGH SCHOOL) |
| S100 | FOUNDATION PLAN & NOTES (JR/SR HIGH SCHOOL) |
| S101 | FOUNDATION DETAILS (JR/SR HIGH SCHOOL) |
| S200 | SECOND FLOOR FRAMING PLAN & DETAILS (JR/SR HIGH SCHOOL) |
| S201 | ROOF FRAMING PLAN & DETAILS (JR/SR HIGH SCHOOL) |
| S202 | STEEL FRAMING DETAILS (JR/SR HIGH SCHOOL) |
| S203 | MISCELLANEOUS STEEL DETAILS & MASONRY REINFORCING DETAILS (JR/SR HIGH SCHOOL) |
| | ROOF RECONSTRUCTION CONTRACT |
| A521 | ROOF PLAN & DETAILS (JR/SR HIGH SCHOOL) |
| A522 | ROOF PLAN & DETAILS (BERNE ELEMENTARY) |
| A523 | ROOF PLAN & DETAILS (BERNE ELEMENTARY) |
| | ASBESTOS ABATEMENT CONTRACT |
| ASB-1 | ASBESTOS ABATEMENT - PLANS AND DETAILS (JR/SR HIGH SCHOOL) |
| ASB-2 | PARTIAL PLAN - BOILER ROOM (JR/SR HIGH SCHOOL) |
| ASB-201 | ASBESTOS ABATEMENT - PLANS AND DETAIL (WESTERLO ELEMENTARY) |
| | PLUMBING CONTRACT |
| P-1 | PARTIAL PLAN - ADDITION, FIRST FLOOR AND SECOND FLOOR (JR/SR HIGH SCHOOL) |
| P-2 | PARTIAL PLAN - CAFETERIA EXPANSION; DETAILS (JR/SR HIGH SCHOOL) |
| | HEATING, VENTILATING & AIR CONDITIONING CONTRACT |
| H-1 | FIRST & SECOND FLOOR PLAN ADDITION; SYMBOL LEGEND (JR/SR HIGH SCHOOL) |
| H-2 | PARTIAL PLAN BOILER ROOM; REMOVALS AND NEW WORK (JR/SR HIGH SCHOOL) |
| H-3 | PARTIAL PLAN - CAFETERIA EXPANSION; DETAILS (JR/SR HIGH SCHOOL) |
| H-4 | SCHEDULES AND DETAILS (JR/SR HIGH SCHOOL) |

Cataldo, Waters, and Griffith Architects, P.C.

**Figure 39–1.** *(continued)*

| H-5 | DETAILS (JR/SR HIGH SCHOOL) |
|---|---|
| H101 | PARTIAL PLAN - BOILER ROOM; DETAILS; SCHEDULES (WESTERLO ELEMENTARY) |
| H201 | PARTIAL PLAN, BOILER ROOM, DETAILS, SCHEDULES (WESTERLO ELEMENTARY) |
| H202 | DETAILS AND SCHEDULES (WESTERLO ELEMENTARY) |
| | **ELECTRICAL CONTRACT** |
| E-1 | FIRST & SECOND FLOOR PLAN, ADDITION LIGHTING (JR/SR HIGH SCHOOL) |
| E-2 | FIRST & SECOND FLOOR PLAN, ADDITION UTILITIES (JR/SR HIGH SCHOOL) |
| E-3 | PARTIAL PLANS CAFETERIA & COMPUTER LABS (JR/SR HIGH SCHOOL) |
| E-4 | BASEMENT PLAN, PART PLANS BOILER ROOM, DETAILS (JR/SR HIGH SCHOOL) |
| E-5 | FIRST FLOOR PLAN, DETAILS (JR/SR HIGH SCHOOL) |
| E-6 | SECOND FLOOR PLAN, DETAILS (JR/SR HIGH SCHOOL) |
| E-7 | SCHEDULES (JR/SR HIGH SCHOOL) |
| E101 | GROUND FLOOR PLAN; DETAILS; SCHEDULES (WESTERLO ELEMENTARY) |
| E102 | FIRST FLOOR PLAN; DETAILS; SCHEDULES (WESTERLO ELEMENTARY) |
| E103 | SECOND FLOOR PLAN; DETAILS; SCHEDULES (WESTERLO ELEMENTARY) |

Cataldo, Waters, and Griffith Architects, P.C.

**Figure 39–1.** *(continued)*

are not cluttered by excessive notation. The following, which is reprinted from the cover page of the School Addition drawings, explains the system:

### MATERIAL KEYING SYSTEM

A KEYNOTING SYSTEM IS USED ON THE DRAWINGS FOR MATERIALS REFERENCES AND NOTES. REFER TO THE KEYNOTE LEGEND ON THE DRAWINGS FOR THE INFORMATION WHICH RELATES TO EACH KEYNOTE SYMBOL ON THE RESPECTIVE DRAWING.

EACH KEYNOTE SYMBOL CONSISTS OF A NUMBER FOLLOWED BY A LETTER SUFFIX. THE NUMBER RELATES TO THE SPECIFICATION SECTION WHICH GENERALLY COVERS THE ITEM THAT IS REFERENCED, AND THE LETTER SUFFIX COMBINED WITH THE NUMBER CREATES A KEYNOTE SYMBOL WHICH IDENTIFIES THE SPECIFIC REFERENCE NOTATION USED ON THE DRAWING. THE LETTER SUFFIX DOES NOT RELATE TO ANY CORRESPONDING REFERENCE LETTER IN THE SPECIFICATION.

THE ORGANIZATION OF THE KEYNOTING SYSTEM ON THE DRAWINGS, WITH THE KEYNOTE REFERENCE NUMBERS RELATED TO THE SPECIFICATIONS SECTIONS NUMBERING SYSTEM, SHALL NOT CONTROL THE CONTRACTOR IN DIVIDING THE WORK AMONG SUBCONTRACTORS OR IN ESTABLISHING THE EXTENT OF WORK TO BE PERFORMED BY ANY TRADE.

As the material keying system reflects the numbering of the C.S.I. Master Format, you may find it helpful to keep a bookmark in your text at **Figure 33–1**. The entire Master Format is too lengthy to include in this book, but that figure includes all of the Division numbers.

**Figure 39–2** illustrates how this keynote system works. Notice that many of the callouts on this detail are a decimal number followed by a letter. That is the *keynote symbol*. Referring to the Index of Drawings (see **Figure 39–1**), we find that Sheet A000 is Master Keynotes. The complete legend of Master Keynotes, as it appeared on Sheet A000, is reprinted on pages 297 through 299. The keynotes that pertain to a particular drawing sheet are also copied on the right end of the drawing sheet. In **Figure 39–2** the keynotes on the right side of the figure pertain to the drawings that were together on Sheet A302, where J-14 was located. The top right callout on J-14 has a keynote symbol, 3.30A. Referring to the material keying legend, 3.30A is cast-in-place concrete. This keynote

**Figure 39–2.** Material keynote system.

replaces the customary callout on the drawing that would indicate that the landing and tread surfaces are cast-in-place concrete. More specific information, such as the grade of concrete and type of finish, would be indicated in Division 3 of the specifications. We know this because virtually all large construction projects use the C.S.I. Master Format to organize the specifications and Division 3 of the C.S.I. format is Concrete. Moving counterclockwise around the drawing, the next keynote symbol is 5.50E and is referenced to the stair stringer that is *beyond* the main part of the drawing. According to the material keying legend, 5.50E is covered in Division 5, Section 5.50 of the specifications, Miscellaneous Metals, and is an MC12 × 10.6 stringer. (Unit 40 explains the meaning of this structural steel designation.)

Many of the material keynotes are referenced on several of the drawing sheets. In that case, they are reproduced on each of those sheets, as well as on the master legend on Sheet A000. All of the whole sheets with this book include the material keying legends as they were intended by the architects.

## Reference Symbols Used on Drawings

A large building such as the School Addition requires a large number of sections and details to describe all aspects of the structure. The symbols for section views and the symbols for detail drawings are similar in appearance and format but have one significant difference. While both commonly use a circle as a symbol, the circle on a plan or other large-scale drawing to indicate existing detail contains the detail number above a horizontal line. Below the horizontal line is the sheet number on which the detail appears. Some element of the symbol, such as a rectangular flag, indicates that it is a detail drawing symbol. Section view symbols use the same scheme for specifying section number and sheet number but use a pointer instead of a rectangular flag to indicate the direction in which the section is viewed. **Figure 39–3** is reprinted from the cover page for the School Addition. In this case, Section B-2, which would be found on Sheet A100, is drawn with a view of the building from the top of the page toward the

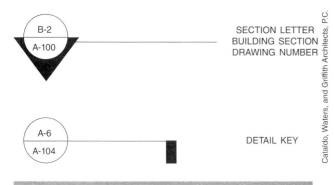

**Figure 39–3.** Detail and section-view symbols.

**Figure 39–4.** Elevation reference symbol.

bottom. (The Index of Drawings shows that the A series of drawings includes A000 and A101 through A610, but no A100. Apparently, A100 is just a hypothetical sheet for the sake of explaining how symbols are used.)

An example from the drawings that are used with this textbook can be seen on the First Floor Plan, Sheet A101. At the extreme right and left ends of the floor plan there is a section-view symbol, with its pointers directed toward the top of the drawing (north, according to the symbol in the lower right corner of the drawing). This symbol refers to Section View J-14 on Sheet A201. Section J-14 is in the top right of Sheet A201 and is a full cross-section view of the building as seen by a person standing on the south side of the building and looking north. The detail drawing symbol also can be found on the First Floor Plan, located just above the section-view symbol at the left end of the building. The direction of the rectangular flag indicates that it is also drawn as though viewed while facing north. That detail, which is at a larger scale (¾" = 1'-0") than the building sections (⅛" = 1'-0"), is number A-4 on Sheet A502. Detail A-4 includes further detail symbols, such as one about halfway up the wall referring to E-2 on A504, an expansion joint detail, at 3" = 1'-0".

Interior elevations, which are often used to show the locations of fixtures, built-in furniture, and special architectural details, are usually referenced with their own type of symbols on floor plans. The symbol for an interior elevation also shows the direction from which the elevation is viewed, and it includes the number of the elevation and the sheet on which that elevation drawing appears. The symbol in **Figure 39–4** references elevation drawing L-10, which would be found on Sheet A103. The direction the triangular shape of the symbol points is the direction in which the elevation is viewed.

## Structural Grid Coordinate

Buildings with columns of reinforced concrete or structural steel as a main element of the building frame are laid out on a structural grid. The plans are drawn with a system of horizontal and vertical reference lines, so that all columns and details can be referenced to this grid. Vertical lines are numbered, and horizontal lines are lettered; or the system is reversed, with horizontal lines numbered and vertical lines lettered (see **Figure 39–5**). Columns are located by grid lines through their centers. The locations of major architectural components, such as walls, are located by dimensions that are referenced to the grid lines. The dimensions from the grid lines to the faces of exterior walls are usually given at the corners of the building. Interior partitions and other architectural elements may be located by dimensioning to the faces of these exterior walls, which are referenced to the structural grid.

## Mental Walk-Through

As with each of the buildings discussed earlier in this textbook, one of the first steps you should take to familiarize yourself with the School Addition is to take a *mental walk-through*. Start at the front entrance to the first floor, in the lower right corner of the First Floor Plan. Enter door 101, and step into a stairwell marked stair #1. At the far end of the stairwell is door 102, which opens into a corridor. If you turn right in the corridor, you can walk up a ramp into the existing building, 10 inches higher than the first floor of the addition. Across the main corridor of the addition are the doors into two toilet rooms. A symbol in front of those doors indicates that Elevation A-8 on A610 will show you what the toilet room doors look like. As you continue your walk-through, you will come to many more elevation symbols. Look at each of the elevations on your walk-through to help you form a mental picture of the addition. You will also see several detail and section symbols. It will not be necessary to check each of these to form a mental picture of the addition, but you should look up a couple of each, just to make sure you are comfortable locating them.

Coordination of Drawings  301

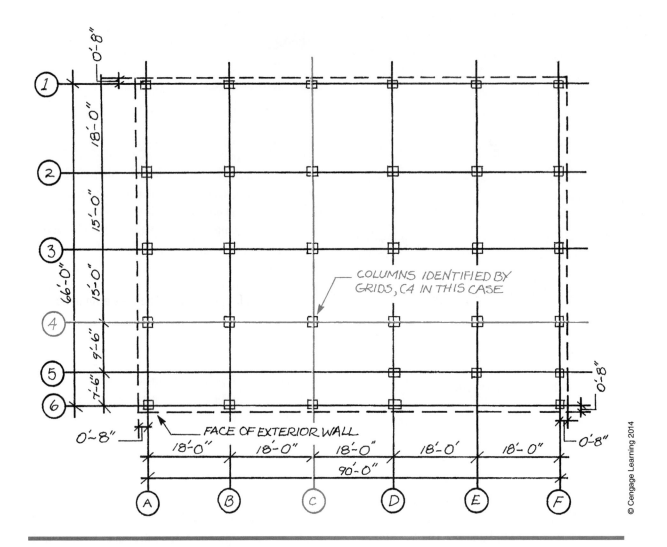

**Figure 39–5.** Structural grid coordinate system.

Moving to the west in the main corridor, you come to a door (103) on your left. This is the entrance to a storage room. A dimension inside the storage room indicates that it is 13'-8½" wide. The exterior wall (south wall) of the storage room is on grid line F (see the left end of the floor plan). The wall with the door is on grid line D, which is 32'-7½" from grid line F. You could determine the actual face-to-face dimension between the north and south walls by looking at the details and sections, but that is not necessary for this mental picture of the building. Notice that where the east and west walls of the storage room intersect with the wall of the corridor, there are two heavy dashed crosses, indicating columns. Most walls are on column lines. These are on vertical grid lines 6 and 7.

Now walk through the janitor closet (JAN. CL.), each of the classrooms, and all corridors and stairwells.

Only a few highlights are described in this textbook, but after you have followed the walk-through described here, you should take a second walk-through, noticing and studying everything until you can visualize it.

Along both sides of the corridor are several references to Material Keynote 10.50A. Find that keynote on the legend at the right end of the drawing sheet or on the master keynote legend.

Through door 105 you enter classroom 103. On the floor plan, room 103 is only separated from room 104 by a single dashed line. That warrants investigation. A detail symbol directs you to A-6 on Sheet A504. That is a large-scale section view of a track detail. Looking again at the floor plan, you see a material keynote reference near the lower end of the dashed line: 10.62A. According to the legend, 10.62A is a folding partition, so rooms 103 and 104 are separated by a folding partition.

Across the corridor from classroom 103 is stair #2, which goes to the second floor. At the extreme left end of the corridor is a large rectangular area identified by keynote 14.20A, an elevator. Take the elevator to the second floor.

When you step off the elevator, you are on the second floor. Your walk-through is continued on Sheet A102, Second Floor Plan. The second floor of the addition is almost identical to the first floor. It would be foolhardy to take for granted that the two levels are identical, but a quick survey shows that they have the same rooms in approximately the same layout. The rooms on the second floor are numbered in the 200 series to indicate that they are on the second floor.

### USING WHAT YOU LEARNED

When working on projects that are based on a structural grid, and that is nearly all large buildings, it is vital to understand the grid and to be able to identify it quickly on the drawings. The School Addition is based on a grid with horizontal grid lines A through F and vertical grid lines 1 through 9. Major architectural and structural features are dimensioned at these grid lines.

For example, how far is the centerline of the folding partition in separating Classrooms 103 and 104 from the western most grid line? This question can be answered by studying the first floor plan. The westernmost grid line is line 1, as numbered at the top left corner of the floor plan. The folding partition is shown by a broken line, with a callout for Detail A-6/A504 which confirms that this line is the partition. The partition aligns with the grid line near door 106 in the north wall of classroom 104. At the top of the drawing we see that this is grid line 3. Grid line 2 is 14'-2½" from grid line 3 and grid line 1 is also 14'-2 ½" from grid line 2, so the distance from grid line 1 to grid line 3 is 28'-5". That is the dimension from the western grid line to the centerline of the partition.

## Assignment

Refer to the drawings of the School Addition included in your textbook packet to complete this assignment.

1. List the sheet numbers for all the structural drawings for the addition.
2. On what sheet would you look to find wall details?
3. On what sheet would you find a framing plan for the roof?
4. What is indicated by Keynote 10.80C?
5. What keynote number is used to indicate plywood?
6. On the First Floor Plan, at about the midpoint of the north wall, a circle is drawn over a triangular flag. The notation inside the circle is J-6/A201. Explain what this symbol indicates and where further information can be found.
7. According to the building elevations and several wall details, there is a precast concrete sill or band that is about waist high on the exterior walls. What is the dimension from the top of the foundation to the bottom of this sill or band?
8. What is the height of the concrete block wall on the north wall of the first floor corridor?
9. What is the material on the lower surface of the canopy over the entrance at door 101?
10. What are the dimensions of the liquid marker board and tackboards on the west wall of classroom 107?
11. What are the dimensions of the mirrors in the toilet rooms?
12. Describe the location of column D3, relative to the floor plans.
13. What is the center-to-center spacing between columns F7 and F9?
14. What are the overall outside dimensions of the precast concrete band, such as is used near the base of typical exterior walls?
15. There is a door from a classroom (112) in the existing building into the storage room in the northeast corner of the addition. The door is made of what material?
16. Describe the differences between the rooms adjacent to the elevator on the first floor and on the second floor.

Coordination of Drawings

# UNIT 40
# Structural Drawings

## Objectives

After completing this unit, you will be able to perform the following tasks:

○ Describe the footings for columns and walls, including dimensions and reinforcement.

○ Interpret the information found on a foundation plan, including dimensions of foundation walls, reinforcement of foundations, and the locations of the various elements.

○ Describe each column, beam, and lintel shown on the structural drawings.

○ Interpret structural details and sections.

## Foundations for Commercial Buildings

The foundations for large, commercial buildings perform the same functions as those for light-frame buildings. The foundation supports the loads (weight) imposed on the superstructure and spreads those loads over a large enough area so that the earth can support it uniformly. In most commercial construction, the foundation system is composed of spread footings and stem walls and pads that act as footings under columns. These are the same elements that are found in most one- and two-family homes. The biggest difference in the foundation for a commercial building and that for a small house is the thickness of the concrete and the amount of reinforcing steel. There are apt to be more detail drawings for the foundation of a larger building, because of the need to describe the different sizes and shapes of footings and the reinforcement at many points in the foundation. The same knowledge is required to read the foundation plan for a large building as for a smaller one.

The drawings consist of a plan that shows the layout of the foundation and the major dimensions and detail drawings that describe reinforcement, expansion joints, and variations in design at special locations. The placement of column footings is indicated by referencing the structural grid coordinates. The structural grid indicates where the centerline of the column will be, but it is very important that the footing be placed accurately on this center. If the column does not rest on the center of the footing, the footing may tip due to the uneven pressure on the soil below (see **Figure 40–1**).

## Structural Steel Framing

A structural steel building frame consists of **columns** (vertical members) and *beams* (horizontal members.) The largest beams, called *girders,* attach to the columns (see **Figure 40–2**). *Joists* are intermediate beams and are supported at their ends by the girders. *Lintels* are the beams that support the weight above an opening, such as a door, window, or nonstructural panel. If the first floor is a concrete slab, it is described on the foundation plans and details. The second and higher floors are described on framing plans, as is the roof frame. Many buildings have several floors that are framed alike. In this case a note might indicate that a framing plan is typical.

Framing plans for structural steel are drawn on the structural grid coordinate system (see **Figure 40–3**). Beams are shown by a single-line symbol with an accompanying note to indicate the structural shape to be used. The lines indicating beams stop short of the symbol for the girder or column when the beam is framed into the supporting member and does not continue over it. The

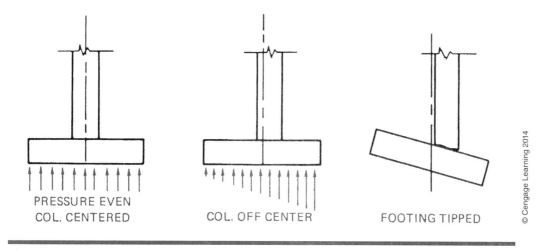

**Figure 40–1.** The column load must be centered on the footing.

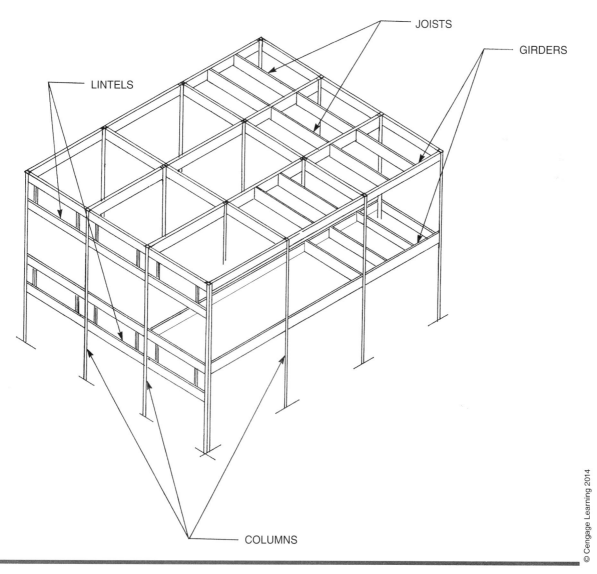

**Figure 40–2.** Major parts of a building frame.

Structural Drawings 305

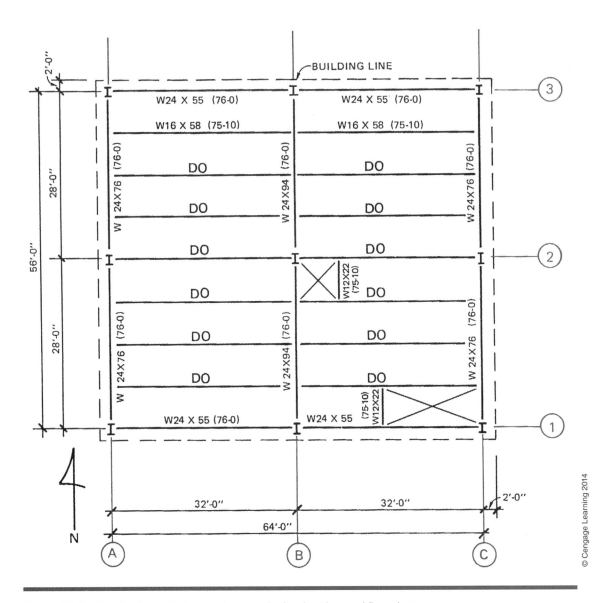

**Figure 40–3.** Structural coordinates are shown on the framing plans and floor plans.

abbreviation *do,* which stands for *ditto,* indicates that the specification for the first member in a series is to be repeated for all members. A number in parentheses at the end of the designation is used to indicate the elevation of the top of the member. This may be the elevation or the distance above or below the floor line (see **Figure 40–4**).

Joists are frequently *open-web steel joists,* sometimes called *bar joists* (see **Figure 40–5**). Bar joists are manufactured in H, J, and K series, depending on the grade of steel used and their strength requirements. K series are stronger than H series, and H series are stronger than J series. A bar joist designation includes a number to indicate depth, a letter to indicate strength series, and a number to indicate chord size. For example, the designation 16H6 indicates a 16-inch-deep H series joist with number 6 chords.

The actual lengths of members are not shown on the general contract drawings. This information is shown on shop drawings that are drawn by the steel fabricator some time after the construction drawings are completed. It is an easy matter, however, to find the span of the member by looking at the framing plan. Connections are shown on details and sections.

Lintels are sometimes categorized as *loose steel.* In masonry construction, a steel lintel is placed above each opening to support the weight of the masonry above the opening (see **Figure 40–6**). These lintels are

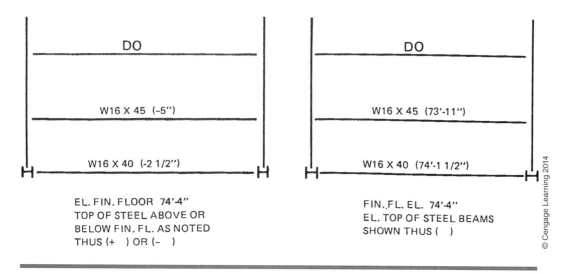

**Figure 40–4.** Notations on the drawings indicate the relative elevations of beams.

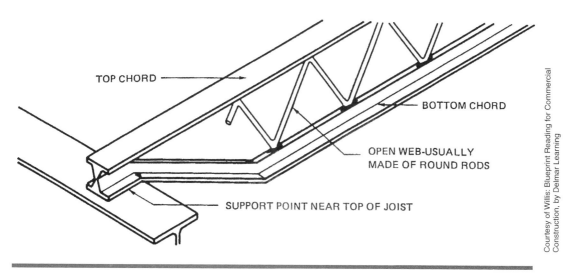

**Figure 40–5.** Open-web steel joists (often called bar joists).

not attached to the steel building frame, which is why they are called loose steel. If there are several lintels alike, they are normally shown on the plans by a symbol and are then described more fully in a lintel schedule (see **Figure 40–7**).

## Masonry Reinforcement

Masonry materials have great compressive strength. That is, they resist crushing quite well, but they do not have good tensile strength. The **mortar** in masonry joints is especially poor at resisting the forces that tend to pull it apart, such as a force against the side of a wall or a tendency for the wall to topple. Masonry joint reinforcement is done by embedding specially made welded-wire reinforcement in the joints (see **Figure 40–8**). Greater strength can be achieved by building the masonry wall with reinforcement bars in the cores of the masonry units, then filling those cores with concrete. Concrete used for this purpose is called *grout*. Reinforcing steel is also embedded in masonry *walls* to tie structural elements together. For example, it is quite common for rebars to protrude out of the foundation and into the exterior masonry walls. The strength of a masonry wall can be increased considerably by the use of bond beams (see **Figure 40–9**). A *bond beam* is made by placing a course of U-shaped masonry units at the top of the wall. Reinforcing steel is placed in the channel formed by the U shape; then the channel is filled with grout. The result is a reinforced beam at the top of the wall.

Structural Drawings 307

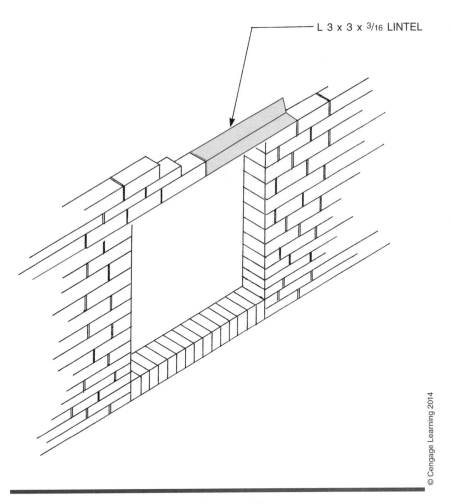

**Figure 40–6.** A lintel may be considered loose steel if it is not attached to the frame.

| LINTEL SCHEDULE | | | | |
|---|---|---|---|---|
| MARK | MATERIAL | TYPE | MAS. OPNG | REMARKS |
| L1 | WT 8 x 13 | ⊥ | SEE ARCH. | ATTACH ONE END TO COL. SEE 1/S200 |
| L2 | WT 8 x 13 | ⊥ | " | END BEARING BOTH ENDS |
| L3 | WT 8 x 13 | ⊥ | " | CONTINUOUS W/ CENTER SUPPORT ON T.S. 4x4 |
| L4 | WT 4 x 9 | ⊥ | " | AT ALL BELOW WINDOW UNIT VENTILATOR OPENINGS |
| L5 | (2) ∠ 6x4x5/16 W/ ¼" x 9½ CONT. BOT. PL | ⊥⊥ | " | LINTEL IN FIRE WALL WELD BOT. PL TO ANGLES. |
| L6 | (2) ∠ 6x4x5/16 | ⊥⊥ | " | IN EXIST. BLDG. WALL |
| L7 | (3) ∠ 4x3½x5/16 | ⊥⊥⊥ | " | AT CAFETERIA HVAC UNIT WALL OPENINGS COORDINATE WITH G-7/H3 AND A103 (2-LOCATIONS) |

**Figure 40–7.** Lintel schedule.

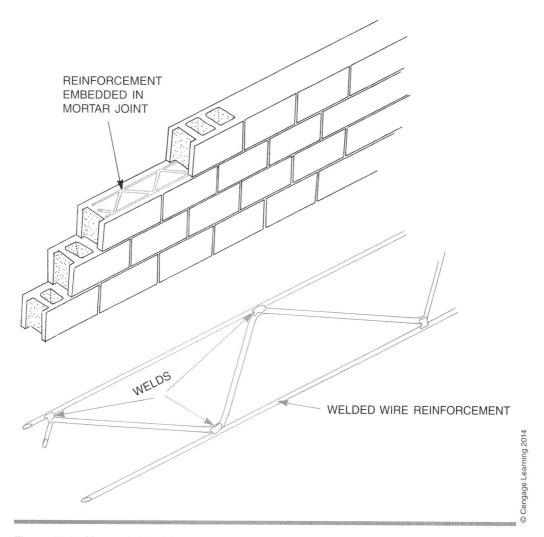

**Figure 40–8.** Masonry joint reinforcement.

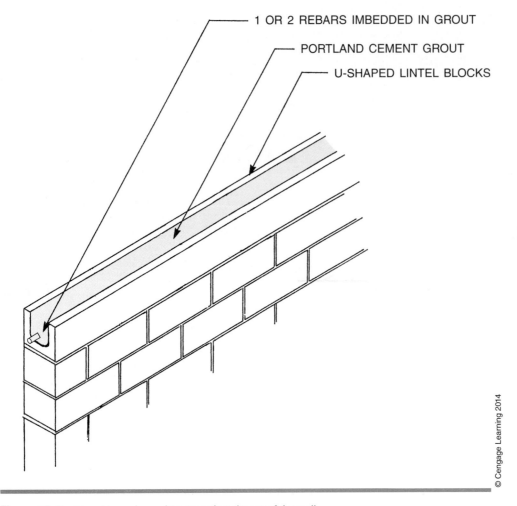

**Figure 40–9.** A bond beam is used to strengthen the top of the wall.

 **USING WHAT YOU LEARNED**

The structural drawings for the School Addition in your textbook packet contain information about how the new addition is to be anchored to the existing building. If this anchoring is not done correctly, the risk that the buildings will separate over time increases. At the ground level, there will be a ramp down from the existing building to the addition. How is that ramp anchored to the existing building?

Being at the ground level, the ramp is shown on the Foundation Plan on Sheet S100. Section View 11/S101 shows a cross section of the ramp. At the far right of that section view is a callout that reads #4 × 1′-6″ DOWELS @12″ O.C. DRILLED & GROUTED INTO EXIST. SLAB 6.

## Assignment

Refer to the drawings of the School Addition in your textbook packet to complete this assignment.

1. What are the dimensions of the footing for the northwestern most column?
2. What is the width and depth of the footing at the west end of the addition?
3. What is the elevation of the top of the floor in the elevator pit?
4. How many lineal feet of #5 reinforcement bars are needed for the footing under the east wall of the addition?
5. What size and kind of material is used to prevent the foundation wall from moving on the footings? How much of this material is needed for the west end of the addition?
6. How many pieces of what size reinforcing steel are to be used in the footing for the column between the entrances to classrooms 103 and 104?
7. How closely is the vertical reinforcement spaced in a typical section of the foundation wall?
8. What is the overall length of each piece of rebar used to secure the interior masonry partitions to the concrete slab?
9. What is the spacing of the dowels used to secure the exterior masonry walls to the foundation?
10. Describe how the corridor walls are secured to the columns.
11. How many pieces of W18 × 50 steel are used in the construction of the addition?
12. What size and shape structural steel supports the north ends of the floor joists under classroom 203?
13. What size and shape structural steel supports the ends of the girder in the second floor at the front (side with the door) of the elevator shaft?
14. What is the nominal depth of the joist in the second floor corridor?
15. What is the elevation of the top of the second floor girder between D5 and D6, relative to the second finish floor elevation of 110'-3"?
16. What is the shape and size of the member that supports the masonry above the windows in classroom 108?

# UNIT 41

# HVAC & Plumbing Drawings

## Objectives

After completing this unit, you will be able to perform the following tasks:

- Identify and briefly describe the major pieces of HVAC equipment to be used in a building.
- List the sizes of pipes and fittings shown on mechanical plans, and explain the major functions of each.
- Describe the sizes and shapes of air-handling ducts shown on mechanical drawings.

Heating, ventilating, air conditioning, and plumbing are often referred to as the mechanical systems. For residential construction, these systems are rarely covered in depth with drawings. The subcontractors who will install these systems can do most of the work with the very minimal information shown on the floor plans. If additional drawings are necessary, those subcontractors will prepare their own. In commercial construction, however, such as the school addition, the mechanical systems are more complex and the drawing set usually includes drawings for *HVAC (heating, ventilating, and air conditioning)* and for plumbing. The original School Addition drawings include two plumbing drawings, identified by the P prefix in the sheet number; and five HVAC drawings, identified by an H prefix. The School Addition drawings included in your textbook contain only one plumbing drawing and one HVAC drawing.

## Heating, Ventilating, and Air Conditioning Plans

The HVAC system for a building larger than a single-family house is designed by an engineer who specializes in this work. The drawings for this work are usually drawn on a basic outline of the floor plan for the building. This allows the mechanical drawings to show where HVAC and plumbing equipment is located in relation to rooms, walls, floors, and so on. The coordination of structural, mechanical, and electrical work is often an issue of some concern on construction projects, and so it is important to understand where and how pipes will pass between beams or through walls.

Most commercial buildings are heated by burning fuel (natural gas or fuel oil) and distributing the resulting heat in the form of warm air, hot water, or steam. Warm-air heating systems require ducts throughout the building to carry the warm air to the occupied spaces and to return the cool air to the source. A warm-air system has the advantage of allowing for inexpensive air conditioning. The same ducts can be used to distribute either warm air in winter or cool air in summer. The ducts, however, require much more space in ceilings and walls and are harder to route through and between structural elements than are pipes. When hot water or steam is used to transport the heat to the occupied space, two pipes are required: one to supply the hot water or steam and one to return the cooler medium to its source. The basic type of system is readily apparent to anyone looking at the HVAC plans. The plans for an air-handling system include ductwork. The plans for a hydronic system include more pipes and fittings.

## Unit Ventilators

At first look, the heating plans for the School Addition might appear foreign and difficult to understand, but like most of the drawings you have studied before this point, these drawings are easy to understand when taken one step at a time. Notice that each of the rooms in the addition has a rectangle on the outside wall with a callout of UV-1, UV-2, or something similar. These indicate that this building is heated by unit ventilators. **Figure 41–1** is the Schedule of Unit Ventilators. (In the actual drawing package it was H-7 on Sheet H-4.) A *unit ventilator* is a unit that combines a means of mixing room air and outside air to ventilate the room while also warming it. The warmth is provided by the hot water that is piped in from the boiler.

**Figure 41–2** explains the principles of a unit ventilator. The heating controls include a function that controls the balance of recirculated inside air and fresh outside air. During the day, when the building is in use, the damper is opened to allow more fresh air to be circulated through the room. At night, the damper is closed, so that less cool air enters the room, resulting in an energy savings. The amount of air that is drawn in from the outside when the dampers are closed as much as possible is indicated on the drawings and on the schedule as *CFM (cubic feet per minute)*. Other information on the Schedule of Unit Ventilators describes such things as the total amount of air that passes through the unit, the water temperature intended by the designer, and the speed of the fan.

This school does not include air conditioning. If it did, the unit ventilators would be supplied with cold water when cooling is called for. The same system of balancing inside and outside air would be used for energy conservation and fresh air, but instead of passing the air over heated pipes, it would be passed over cold pipes. The Schedule of Unit Ventilators would include information for air conditioning.

The unit ventilators are only a few feet long; yet the plans show a continuous string of equipment of some sort along the exterior walls of the classrooms. Although not labeled anywhere on the plans, the equipment is bookcases; the Construction Notes refer to them. The manufacturers of unit ventilators also manufacture bookcases and filler pieces that are designed to fit with the ventilators and fill the rest of the wall space. These bookcases have a false back, with space for pipes to and from the ventilators.

In some spaces, such as stairways and storage closets, the heaters are not labeled as *UV (unit ventilators)*. These are *cabinet unit heaters (CUH)* that mount on the surface of the wall and do not have ventilation capability and *fan coil units (FCU)* that are intended to provide more heat than ventilation. **Figure 41–3** shows the schedules for the units marked as CUH, UH, and FCU.

| UNIT VENTILATOR SCHEDULE | | | | | | | | | | | | | H-7 |
|---|---|---|---|---|---|---|---|---|---|---|---|---|---|
| MARK | MAKE | MODEL | FAN | | | | V/PH/HZ | HEATING | | | | | |
| | | | TOT CFM | MIN O.A. | RPM | HP | | HOT WATER | | | | | |
| | | | | | | | | EAT | EWT | GPM | MBH | ΔP | |
| UV-1 | AAF | AV-4000 | 1000 | 270 | 650 | 1/8 | 120/1/80 | 50 | 180 | 2.0 | 36 | 0.99 | |
| UV-2 | AAF | AV-5000 | 1250 | 270 | 650 | 1/8 | 120/1/60 | 50 | 180 | 2.0 | 42 | 1.27 | |
| UV-3 | AAF | AV-3000 | 750 | 250 | 650 | 1/8 | 120/1/60 | 40 | 180 | 2.0 | 28 | 0.79 | |
| UV-4 | AAF | AH-3000 | 750 | 270 | 650 | 1/8 | 120/1/60 | 40 | 180 | 2.0 | 36 | 0.79 | |
| UV-5 | AAF | AH-3000 | 750 | 270 | | 1/2 | 120/1/60 | 40 | 180 | 2.0 | 36 | 0.79 | |
| | | | | | | | | | | | | | |

| | | | |
|---|---|---|---|
| TOT CFM | total cubic feet per minute through the unit | EWT | entering water temperature |
| MIN O.A | minimum outside airflow into the unit | GPM | gallons per minute of water through the unit |
| RPM | revolutions per minute of the fan | MBH | thousands of BTUs per hour (a measure of heat output) |
| HP | fan motor horsepower | | |
| V/PH/HZ | fan motor volts, phase rating (1 or 3), and hertz (frequency or cycle) | ΔP | delta pressure (drop in water pressure through the system) |
| EAT | entering air temperature | | |

Cataldo, Waters, and Griffith Architects, P.C.

**Figure 41–1.** Unit ventilator schedule.

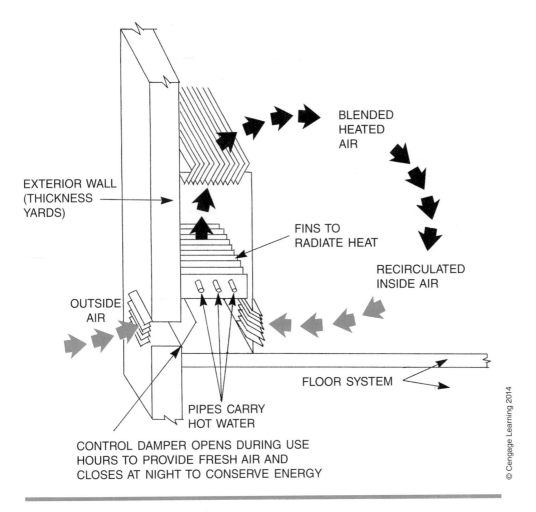

**Figure 41–2.** Operation of a unit ventilator.

### CABINET UNIT HEATER SCHEDULE

| MARK | MAKE | MODEL | STYLE & ARRANGEMENT | STEAM | | WATER | | | | CFM | RPM | EAT | MBH | ELECTRICAL | |
|------|------|-------|---------------------|-------|------|-------|------|-----|------|-----|------|-----|------|------------|-----|
| | | | | PSIG | lB/HR | EWT | ΔT | GPM | ΔP" | | | | | V/ø/HZ | HP |
| CUH–1 | STERLING | RWI–1130–04 | RECESSED WALL, INVERTER | | | 180 | 34 | 2.0 | 0.44 | 420 | 1050 | 60 | 33.8 | 120/1/60 | 1/10 |

### UNIT HEATER SCHEDULE

| MARK | MAKE | MODEL | TYPE | WATER | | | MBH | CFM | EAT | LAT | MOTOR | | |
|------|------|-------|------|-------|-----|-----|-----|-----|-----|-----|-------|------|--------|
| | | | | ENT. | GFM | ΔP | | | | | RPM | H.P. | V/ø/HZ |
| UH–1 | MODINE | HS–18L | HORIZONTAL | 180 | 1.1 | 0.4 | 9.4 | 364 | 60 | 84 | 1550 | 16MHP | 120/1/60 |

### FAN COIL UNIT SCHEDULE

| MARK | MAKE | MODEL | CFM & TOT./F.A. | EXT. S.P. | ELECTRICAL | | HEATING | | | | |
|------|------|-------|-----------------|-----------|------------|------------|---------|-----|-----|-----|-----|
| | | | | | WATTS | VOLTS/PH/HZ | EAT | EWT | GPM | MBH | ΔP |
| FCU–1 | AAF | SFG–TFA–3000 | 350 | 0 | 110 | 120/1/60 | 50 | 180 | 1.0 | 18 | 0.1 |

**Figure 41–3.** Schedules of cabinet unit heaters, unit heaters, and fan cabinet unit heaters.

## Heating Piping

Once the basic type of heating system has been determined, the next step is to understand the piping that carries water to and from the heating units. It is easy to spot the pipes on the heating plan. A solid line indicates supply pipes, and a dashed line indicates return pipes. Pipe sizes are indicated by callouts throughout the plans. Fittings such as valves, elbows, and tees are represented by symbols. **Figure 41–4** shows the most common symbols for mechanical systems. There are standard piping symbols, but each drafter uses a few special representations of his or her own, so a symbol legend is included with the drawings.

Pipes are easier than ducts to coordinate with structural elements because the direction of the pipe can be changed easily and frequently. Pipes might run horizontally in a ceiling for some distance, then loop around a column and drop down to a unit ventilator. It is not practical to show the layout of all pipes with separate plan and riser views. The common method for showing pipes where both horizontal and vertical layout is involved is to use isometric schematics. Plumbing isometrics are discussed in Unit 35. Sheet H-1 of the School Addition drawings combines plan views and isometric schematics. The long runs of pipes are shown in plan view to indicate where the pipes are in relation to the walls. Where the pipes need to drop down from the ceiling to the level of the ventilators, they are shown in isometric views (see **Figure 41–5**).

For the sake of explanation, the following discussion traces the hot-water supply piping (broken line) in room 109, the storage room on the first floor, E-7 on Sheet H-1. Most of what is discussed here is also shown in **Figure 41–5,** but we will start closer to the center of the building, where the branch tees to the right (north) off the main supply running from top to bottom of the drawing. See the symbol legend A-11 on Sheet H-1 to interpret the symbols as you trace the piping. It is recommended that you use a colored pencil to trace the piping as you read this.

The first thing we encounter is a valve, an elbow toward the top of the drawing, and then another elbow back to the right. Construction Note 13 indicates that the pipes pass through a structural steel member. The line is broken for clarity through the area where ductwork is shown that would make reading the piping drawing difficult. Pick the line up again in the storage room 109. Next we come to a tee where the pipe drops down to the level of the unit ventilator. If we continued on to the right or north, we would come to Construction Note 2, which tells us that the symbol indicates a riser to the second floor. The riser is not shown in the wall, where it would actually be positioned, because doing so would bury the riser symbol in the wall with material symbols and other lines. The pipe going down is actually shown at an angle up and to the right. This is the isometric portion of the drawing. At the level of the unit ventilators, the pipes are arranged in a special way to eliminate noises that might result from expansion and contraction as the water temperature changes (see **Figure 41–6**). The ends of the pipes are shown with break lines, which indicate that the pipes actually continue to their obvious destinations: fittings on the unit ventilators.

## Air-Handling Equipment

The unit ventilators introduce fresh air into the School Addition. A separate ventilation system removes stale air from each room of the school. All the rooms in the addition except the smallest closets and the stairways are provided with a system of ducts and a fan to ensure air circulation. A *louvered grille* in the room receives the air and channels it into a duct that carries it to the roof, where a fan exhausts it to the outside.

**Figure 41–7** explains the designations for outlet and inlet grilles. The ducts that carry the air to the roof vent are shown by their outlines and do not indicate their sizes. A square or rectangle with a diagonal line through it indicates where the duct rises to the next floor or the roof. Several notes indicate that ducts rise up to EF-# on the roof. *EF* is an abbreviation for exhaust fan. The number indicates which of the several exhaust fans on the roof the duct is connected to. Some of the exhaust fans occupy several square feet of space on the roof, so they cannot be placed too close together.

Trace the flow of air as it is exhausted from the janitor's closet on the first floor. The path starts with a louver marked RB (return)—8 × 6 (cross-sectional size in inches)—75 (cubic-feet-per-minute nominal airflow). Between the janitor's closet and the storage room, the air enters a 6 × 6 duct, which goes up through a chase (space between two walls) to the second floor. At the second floor it is joined by the exhaust air from the second floor janitor's closet. On the Second Floor Plan

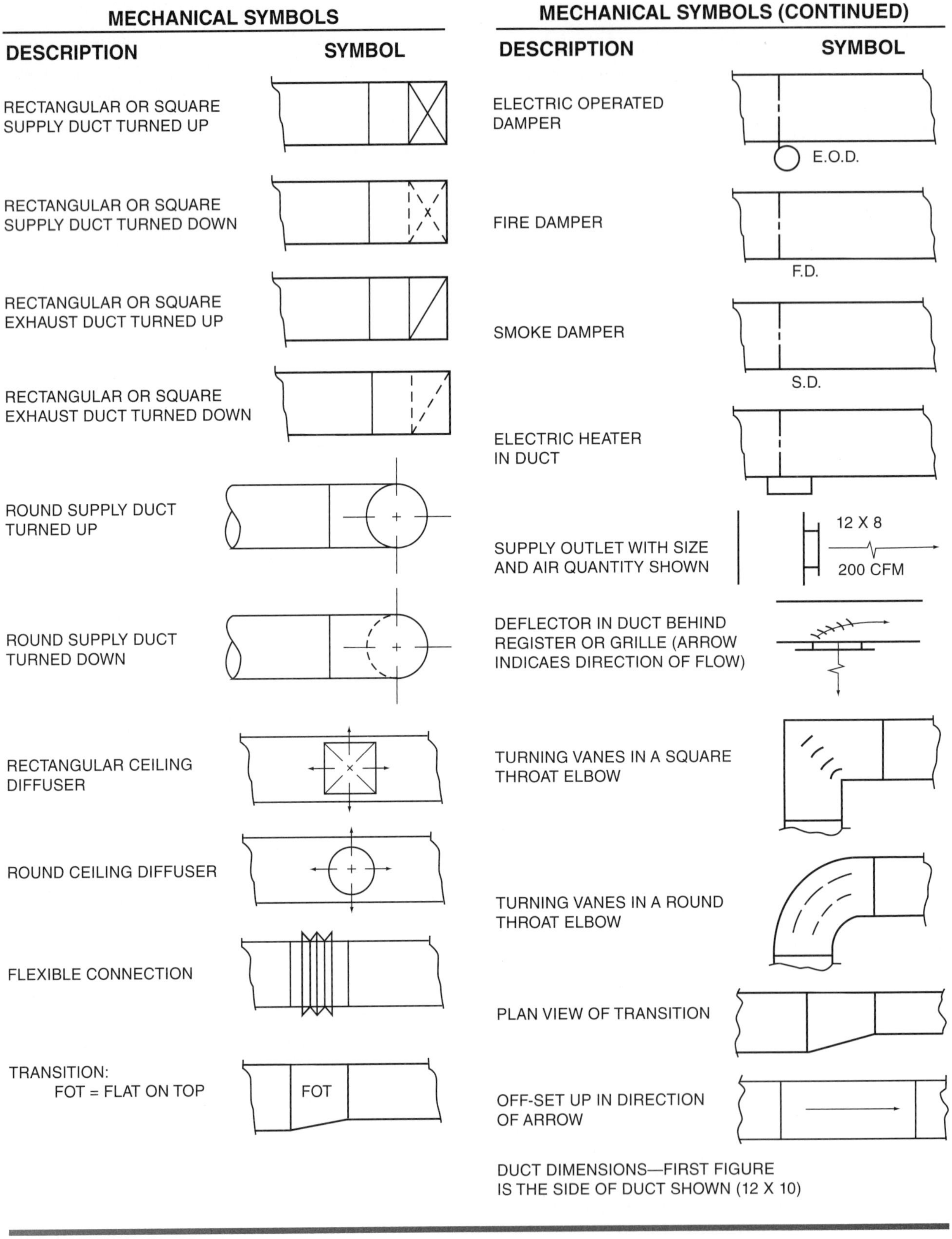
MECHANICAL SYMBOLS
DESCRIPTION
SYMBOL
RECTANGULAR OR SQUARE SUPPLY DUCT TURNED UP
RECTANGULAR OR SQUARE SUPPLY DUCT TURNED DOWN
RECTANGULAR OR SQUARE EXHAUST DUCT TURNED UP
RECTANGULAR OR SQUARE EXHAUST DUCT TURNED DOWN
ROUND SUPPLY DUCT TURNED UP
ROUND SUPPLY DUCT TURNED DOWN
RECTANGULAR CEILING DIFFUSER
ROUND CEILING DIFFUSER
FLEXIBLE CONNECTION
TRANSITION: FOT = FLAT ON TOP
FOT
MECHANICAL SYMBOLS (CONTINUED)
DESCRIPTION
SYMBOL
ELECTRIC OPERATED DAMPER
E.O.D.
FIRE DAMPER
F.D.
SMOKE DAMPER
S.D.
ELECTRIC HEATER IN DUCT
SUPPLY OUTLET WITH SIZE AND AIR QUANTITY SHOWN
12 X 8
200 CFM
DEFLECTOR IN DUCT BEHIND REGISTER OR GRILLE (ARROW INDICAES DIRECTION OF FLOW)
TURNING VANES IN A SQUARE THROAT ELBOW
TURNING VANES IN A ROUND THROAT ELBOW
PLAN VIEW OF TRANSITION
OFF-SET UP IN DIRECTION OF ARROW
DUCT DIMENSIONS—FIRST FIGURE IS THE SIDE OF DUCT SHOWN (12 X 10)
© Cengage Learning 2014

**Figure 41–4.** Mechanical and plumbing symbols.

## MECHANICAL SYMBOLS (CONTINUED)

| DESCRIPTION | SYMBOL |
|---|---|
| ACCOUSTICAL LINING INSIDE INSULATION | |
| BRANCH TAP IN DUCT | |
| SPLITTER FITTING WITH DAMPER | S.D. |
| VOLUME-DAMPER | V.D. |
| BACKDRAFT DAMPER | BDD |
| ACCESS DOOR IN DUCT 10" X 10" SIZE | 10 x 10 A.D. |
| PNEUMATIC OPERATED DAMPER | P.O.D. |
| THREE-WAY VALVE | |
| PRESSURE REDUCING VALVE | |
| PRESSURE RELIEF VALVE OR SAFETY VALVE | |
| SOLENOID VALVE | |
| PIPE TURNED UP (ELBOW) | |
| PIPE TURNED DOWN (ELBOW) | |
| TEE (OUTLET UP) | |
| TEE (OUTLET DOWN) | |

## MECHANICAL SYMBOLS (CONTINUED)

| DESCRIPTION | SYMBOL |
|---|---|
| BACKFLOW PREVENTER | B.F.P. |
| UNION | |
| REDUCER | |
| CHECK VALVE | FLOW |
| GATE VALVE | OR |
| GLOBE VALVE | OR |
| BALL VALVE | |
| BUTTERFLY VALVE | |
| DIAPHRAGM VALVE | |
| ANGLE GATE VALVE | |
| ANGLE GLOBE VALVE | |
| PLUG VALVE | |
| LOW PRESSURE STEAM | LPS |
| LOW PRESSURE CONDENSATE | LPC |
| PUMPED CONDENSATE | PC |
| FUEL OIL SUPPLY | FOS |
| FUEL OIL RETURN | FOR |
| HOT WATER SUPPLY | HWS |
| HOT WATER RETURN | HWR |
| COMPRESSED AIR | A |
| REFRIGERANT SUCTION | RS |

**Figure 41–4.** *(continued)*

## MECHANICAL SYMBOLS (CONTINUED)

| DESCRIPTION | SYMBOL |
|---|---|
| REFRIGERANT LIQUID | —— RL —— |
| REFRIGERANT HOT GAS | —— RHG —— |
| CONDENSATE DRAIN | —— CD —— |
| FUEL GAS | —— G —— |
| CHILLED WATER SUPPLY | —— CWS —— |
| CHILLED WATER RETURN | —— CWR —— |

## PLUMBING SYMBOLS

| DESCRIPTION | SYMBOL |
|---|---|
| METER | —— M —— |
| SPRINKLER PIPING | —— S —— |
| SPRINKLER HEAD | (half-filled circle) |
| FLOOR DRAIN | F.D. |
| CLEAN-OUT | C.O. |
| TUB | (rectangular tub symbol) |
| TANK-TYPE WATER CLOSET | (tank-type toilet symbol) |
| WALL-MOUNTED LAVATORY | (wall-mounted lavatory symbol) |
| URINAL | (urinal symbol) |
| SHOWER | (square shower symbol) |
| WATER HEATER | WH |
| MANHOLE | MH |

## PLUMBING SYMBOLS (CONTINUED)

| DESCRIPTION | SYMBOL |
|---|---|
| WALL HYDRANT | (wall hydrant symbol) |
| YARD HYDRANT | Y.H. |
| FLUSH VALUE WATER CLOSET | (flush valve toilet symbol) |
| COUNTER-TYPE LAVATORY | (counter lavatory symbol) |
| KITCHEN SINK (DOUBLE BOWL) | (double bowl sink symbol) |
| SOIL, WASTE OR DRAIN LINE | ———————— |
| PLUMBING VENT LINE | — — — — — |
| COLD WATER (DOMESTIC) | ———————— |
| HOT WATER (DOMESTIC) | – – – – – |
| HOT WATER RETURN (DOMESTIC) | — — — — |
| FIRE LINE | —— F —— |
| FUEL GAS LINE | —— G —— |
| ACID WASTE LINE | —— AW —— |
| VACUUM LINE | —— V —— |
| COMPRESSED AIR LINE | —— A —— |
| BACKFLOW PREVENTER | —— BFP —— |
| GATE VALVE | —⧖— OR —⋈— |
| GLOBE VALVE | —●— OR —⬤⋈— |
| CHECK VALVE (ARROW INDICATES DIRECTION OF FLOW) | —↗N— |

**Figure 41–4.** *(continued)*

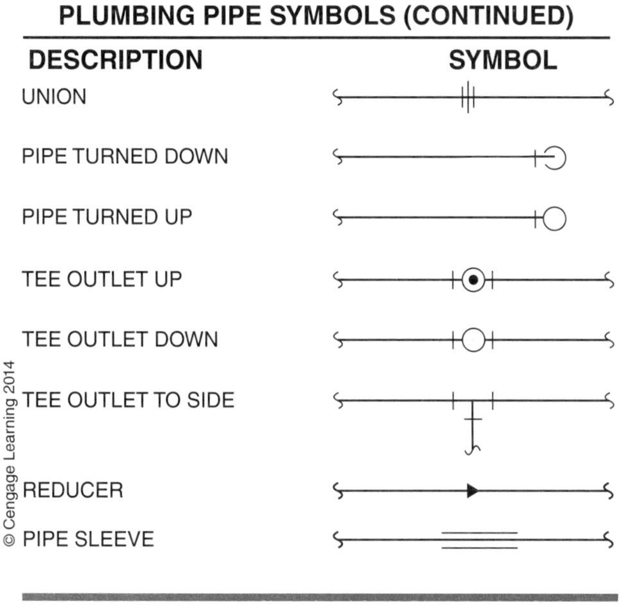

PLUMBING PIPE SYMBOLS (CONTINUED)
DESCRIPTION
SYMBOL
UNION
PIPE TURNED DOWN
PIPE TURNED UP
TEE OUTLET UP
TEE OUTLET DOWN
TEE OUTLET TO SIDE
REDUCER
PIPE SLEEVE



**Figure 41–4.** *(continued)*

2
15
TYP
84
5
72
FILLER
3
4
6X6 UP
1/2"
2
FCU-1
100CFM
RB-8X6-100
36
STORAGE
109
3/4" TO
UV1-1
1/2" TO
FCU-1
1"



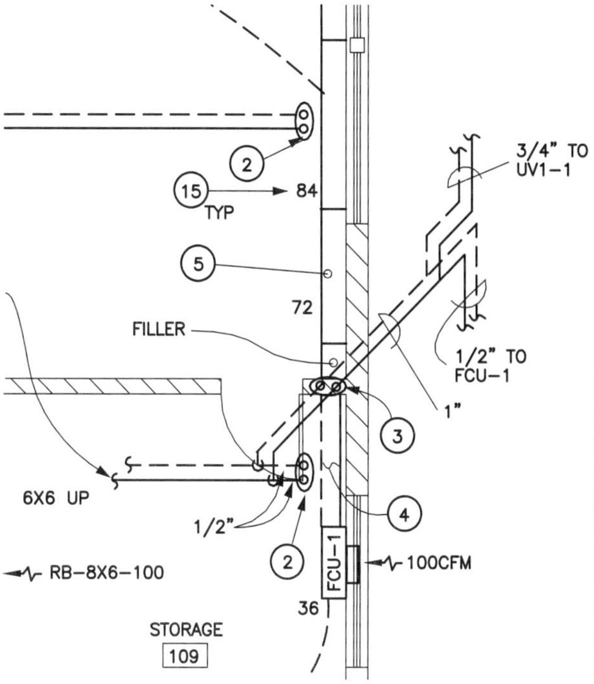

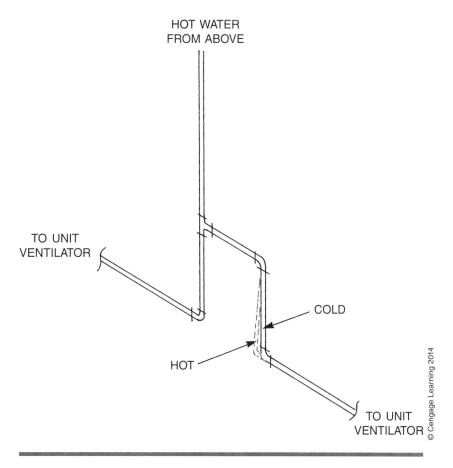

**Figure 41–5.** This isometric drawing shows how the heating pipes drop down from the ceiling to the level of the ventilators.

HOT WATER
FROM ABOVE
TO UNIT
VENTILATOR
COLD
HOT
TO UNIT
VENTILATOR



**Figure 41–6.** Noises are eliminated by allowing pipes to expand and contract.

there is also a symbol indicating a fire damper designated as W/ Access Door At Floor.

This damper (metal gate) closes off the duct if it gets hot in a fire, **Figure 41–8**. This is necessary because the building code requires fire-rated construction, and an open duct is a great place for fire to spread. The access door at the floor allows for maintenance and repairs to the damper. Notes at both levels indicate that the duct between floors is 6 × 6. Above the second floor louver, the duct is 8 × 8; and at the second floor ceiling, it is offset an unspecified distance into the storage room, where it turns up to EF-4 on the roof. The reason for the offset is because it would otherwise be too close to the other duct such that there would not be enough room for both exhaust fans on the roof.

## Plumbing

The plumbing for the School Addition consists of *DWV (drainage, waste, and vent)* piping to remove effluent (wastewater), hot- and cold-water distribution, and fixtures. The DWV system is more complex for a building this size than for a house, because the large flat roof collects quite a bit of water in a rainstorm and there must be a way to get rid of it. Therefore, the Second Floor Plumbing Plan shows roof drains, marked RD, and storm drain piping, indicated by a heavy line. Callouts along the storm drain indicate the size of the pipe and its slope and give information about how the storm drain is to be coordinated with structural components of the building. On the First Floor Plumbing Plan the storm drain is represented by a heavy dashed line, indicating that it is below grade.

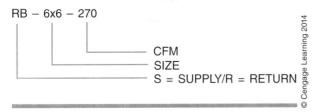

**Figure 41–7.** Explanation of grille designations.

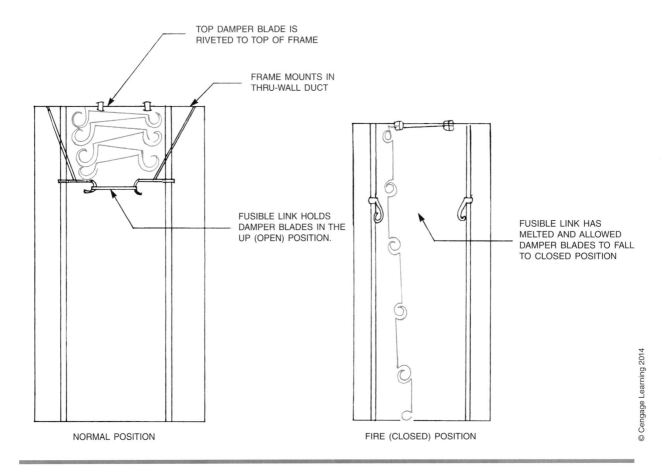

**Figure 41–8.** Fire damper for ductwork.

Notice that the *storm drain* and the *sanitary drain* are separate systems. The storm drain only needs to carry rainwater away from the building. The sanitary drain must connect to a sewer or septic system. Although the storm drain and the sanitary drain pipes cross near the janitor's closet on the first floor, they are not connected.

In the region where this building was constructed, drainage pipes outside the building lines are installed by the site contractor, not the plumbing contractor. Therefore, the plumbing drawings only show the storm drain and sanitary drain to a point 5′-0″ outside the building lines. A note where each line passes through the building wall indicates the *invert elevation* (elevation of the bottom of the inside of the pipe).

The hot- and cold-water piping is similar to that found in homes, except that the hot water is a two-pipe system. A return hot-water pipe allows the hot water to circulate continuously, ensuring that hot water is readily available at the point of use. The hot-water supply and hot-water return can be distinguished from each other by the number of short dashes in the symbol. These are explained on the symbol legend in the drawing packet. The hot-water return, which is usually smaller than the supply, maintains circulation to a point close to where most of the hot water will be used. It is often not practical to continue a hot-water return pipe to every fixture in an area with several fixtures like the School Addition toilet rooms. **Figure 41–9** shows riser diagrams for the fixture connections in the School Addition. (A riser diagram shows the vertical arrangement of the piping.) Notice that there is no hot-water return shown on these diagrams, because the return is connected to the supply before the supply reaches the first fixture.

A plumbing riser diagram is only used to indicate design intent and does not indicate a piping route. The sizes indicated by the engineer are minimum sizes. Company preferences and local codes can dictate that larger sizes be used. Notice that Riser 1 and Riser 4 in **Figure 41–9** are identical. This does not mean that they will always be installed the same way. The notes on Riser 1 are different from the notes on Riser 4. The notes on both risers are relevant for Risers 1 and 4, and the vacuum breakers required on S-1 in Riser 1 are also required on S-1 fixtures in Riser 4. The note to route all piping in the casework in Riser 4 is also relevant for the piping in Riser 1.

It is important for plumbers to know and comply with local codes. In Riser 3, the vent serving MB-1 is offset 90° below the fixture. That is an illegal installation by most codes, and the plumber would install the **vent pipe** in accordance with local codes and not use the riser diagram as a guide for routing the pipe. Where the pipes seem to connect but are actually different piping systems, you can outline the pipe route with different color pencils to clarify the different piping systems.

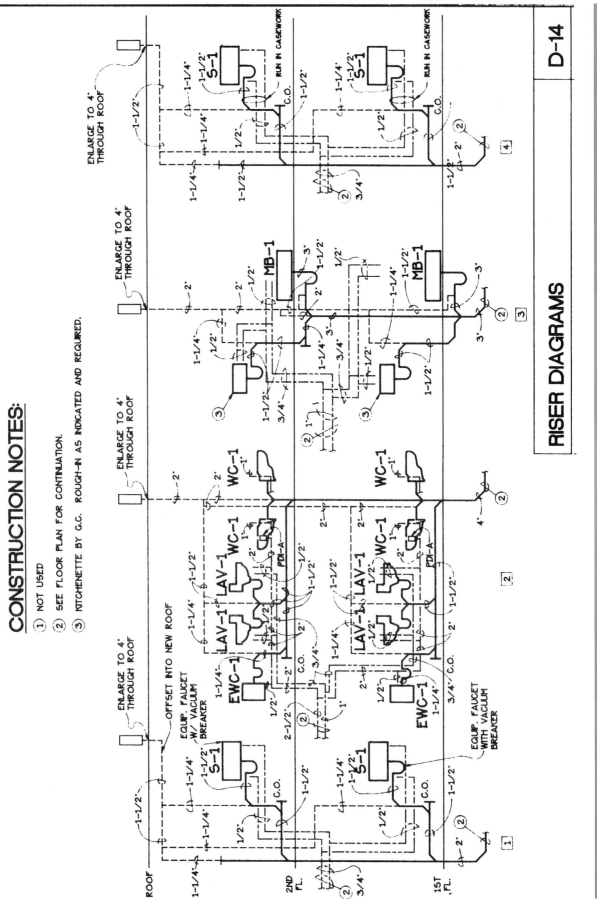

**Figure 41-9.** Riser diagrams for school fixtures.

### USING WHAT YOU LEARNED

Even for workers not directly involved in installing the HVAC system in a building, it is useful to understand where piping, ductwork, and other mechanical equipment will be installed. For example, iron workers who install the structural steel need to know where pipes run through bar joists or around steel beams. As an exercise in tracing piping, list all the pipe sizes through which hot water passes from the time it enters the School Addition until it reaches the unit ventilator in Classroom 104. The easiest way to do this is to start at the unit ventilator and work back to where the piping enters the building. Find the Unit Ventilator (UV-2) in Classroom 104 on the First Floor Plan E-7. There are actually two UVs in this classroom, but both have the same size supply pipes. The pipes at this point are near the top of the classroom and labeled as ¾". Follow the pipes to where they turn east (notice that the North Arrow point to the right) to follow the hall toward the existing building. At this point, they are labeled as 1" pipes. After they pass the tees to Classroom 107, they are 1¼" pipes. Beyond the tees, entering Classroom 103, they are 1½" pipes, and, after the next tees, entering storage room 101, they are 2". They are 2" all the way to where they enter the addition. So, in the order the water passes through the pipes, they are 2", 1½", 1¼", 1", and ¾".

## Assignment

Refer to the drawings of the School Addition in your textbook packet to complete this assignment.

1. How many unit ventilators are shown on the first floor?
2. How is the machinery room on the first floor behind the elevator heated?
3. List in order the sizes of pipe that the water flows through, starting from the point where the water enters the addition to get to the heating unit in Classroom 210 and then back to the point where it leaves the addition.
4. What are the dimensions of the grille where stale air is vented out of Classroom 209?
5. List in order the sizes of the ductwork that exhaust air passes through from the point where the air leaves Classroom 209 until it is outside the building.
6. List in order the sizes of the ductwork that exhaust air passes through from the point where the air leaves the girls' toilet room on the first floor until it exits the building.
7. How many gallons per minute of water are expected to flow through the heating unit in stair #2?
8. Why are two roof drains shown on the First Floor Plumbing Plan of this two-story building?
9. What is designated as EWC-1 on the plumbing plans?
10. What is the vertical distance between the storm drain and the sanitary drain at the point where they cross?
11. What is the diameter of the pipe that is used for the storm drain where it goes from the second floor to the first floor?
12. What size pipe is used for the domestic hot-water return line?
13. Where does the domestic hot-water return line tee out of the domestic hot-water supply line?
14. Where would you shut off the cold-water supply to the kitchenette unit in Room 101 without shutting off the cold water to the toilet rooms on the same floor?

# UNIT 42
# Electrical Drawings

## Objectives

After completing this unit, you will be able to perform the following tasks:

- Explain the information found on a lighting plan.
- List the equipment served by an individual branch circuit using electrical plans, riser drawings, and schedules.
- Explain the information on a schematic diagram.

## Electrical Drawings

Electrical drawings deal predominantly with circuits. Knowledge and understanding of the basic circuits used in buildings is necessary for those working with and interpreting electrical drawings. The four basic methods of showing electrical circuits are:

1. Plan views
2. Single-line diagrams
3. Riser diagrams
4. Schematic diagrams

## Plan Views

The electrical floor plan of a building shows all the exterior walls, interior partitions, windows, doors, stairs, cabinets, and so on along with the location of the electrical items and their circuitry.

**Power Circuits**

The *power circuit* electrical floor plan shows electrical outlets and devices and includes duplex outlets, specialty outlets, telephone, fire alarm, and the like. The conventional method of showing power circuits is to use long dash lines when specified to be installed in the slab or underground and to use solid lines when concealed in ceilings and walls. Short dash lines indicate exposed wiring. Solid lines are often used in commercial installations when the raceway (conduit) is to be exposed (surface mounted). Slash marks through the circuit lines are used to indicate the number of conductors. Full slash marks are the circuit conductors, a longer full slash is the neutral, and half slash marks are the ground wires. When only two wires are required, no slash marks are used. The typical wire size for commercial construction is No. 12 AWG (American wire gauge). The arrows indicate "home runs" to the designated panel. The panel and circuit number designations are adjacent to the arrowheads. **Figure 42–1** lists the typical circuiting symbols. The electrical installer will normally group home runs in a raceway with combinations similar to the following:

- Three-phase systems with three circuit wires and a common neutral in a raceway
- Single-phase systems with two circuit wires and a common neutral in a raceway

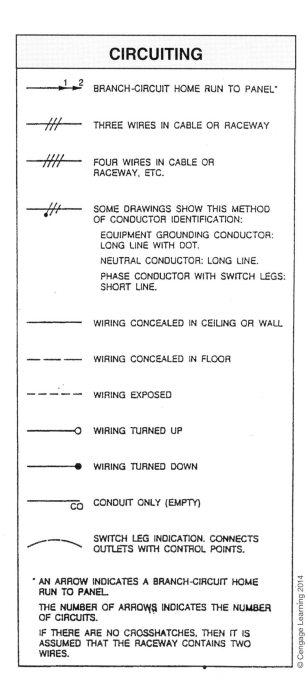

**Figure 42–1.** Circuiting symbols.

**Note:** To reduce the harmonic problems caused by solid-state devices (nonlinear loads), many electrical designers are requiring separate neutrals with each phase conductor.

Some equipment, such as television, fire alarm, clock, PA, and sound, might not be connected with wiring lines on the plan. This indicates that wiring for that equipment is not part of the general circuit wiring and is probably not to be fed through the same panels as the rest of the electrical equipment. For example, on the School Addition the PA system and the clocks are fed from the sound system console and do not involve the general circuit wiring.

### Lighting Circuits

There is usually a *lighting circuit* electrical floor plan for each level or major space within a building. This floor plan shows light fixtures, emergency lighting, security lighting, and special lighting control (photocell, motion detector, etc.). Typically included is a reflected ceiling plan (see **Figure 42–2**). That is, it is a plan view but shows what is on the ceiling as though it were reflected onto the floor plan (see **Figure 42–2**). The reflected ceiling plan shows each light fixture with a circle, square, or rectangle that approximates the shape of the fixture. The lighting circuitry on the School Addition is shown with three types of wiring circuit lines, solid for unswitched, dotted for switched, and line-dash-line for the motion sensor circuits. The conventional method of showing lighting circuits is the same as previously explained in the paragraph above on power circuits.

*Motion sensors* are an interesting feature of the lighting design for the School Addition. The motion sensors are shown as small circles containing the letter *M*. A motion sensor detects any movement in the area of the sensor, so that in a room controlled by a motion sensor, if there is no movement in the room for a period of time, the sensor opens its contacts, like turning a switch off. A motion detector can be used together with a switch so that the lights are controlled normally by the switch. But if someone forgets to turn the lights off, the motion detector will.

A low-voltage lighting control system requires a *relay* that is activated by the light switch. The relay opens or closes the higher-voltage fixture (see **Figure 42–3**). The relays on the School Addition are indicated by the letter *R* in a small square. Notice that every lighting switch is connected to a relay.

### Symbols

*Electrical symbols* are used to simplify the drafting and later the interpreting of the drawings. Electrical symbols *are not standardized* throughout the industry. Most drawings have a symbol legend or list. You must be knowledgeable of the symbols specifically used on each project since designers modify basic

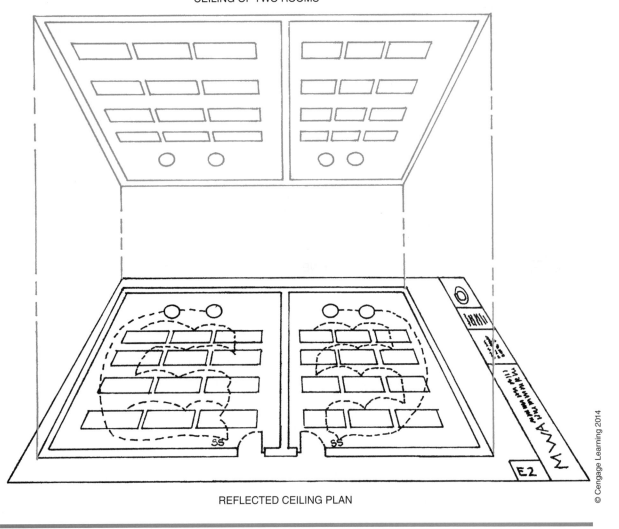

**Figure 42–2.** Reflected ceiling plan.

symbols to suit their own needs. Many symbols are similar (circle, square, etc.). The addition of a line, dot, shading, letters, numbers, and so forth gives the specific meaning to the symbol. Learning the basic form of the various symbols is the best starting point in developing the ability to interpret the drawings and their related symbol meanings. **Figure 42–4** shows common electrical symbols and abbreviations.

The School Addition drawing E-1, symbol legend A-4, contains an electrical symbol list for the project. From the symbol list, it can be seen that there are duplex receptacles, switches, telephone outlets, special-purpose outlets, and fire alarm devices mounted at various heights. Then there is the General Note in the symbol legend that specifies that all mounting heights are to be verified and "modified as directed." This is an example of why the installer must become familiar with the drawings and specifications far in advance of the scheduled time for the installation. The installer must request clarification or direction and give the designers reasonable time to clarify the questionable specified instructions.

## Single-Line Diagrams

The electrical service *single-line diagram* on the School Addition is shown on drawing E-2, Single-Line Diagram E-17. The electric power is brought into the school by way of the *service entrance section (SES)* The note "coordinate with the local utility requirements prior to rough-in" indicates that the electrical

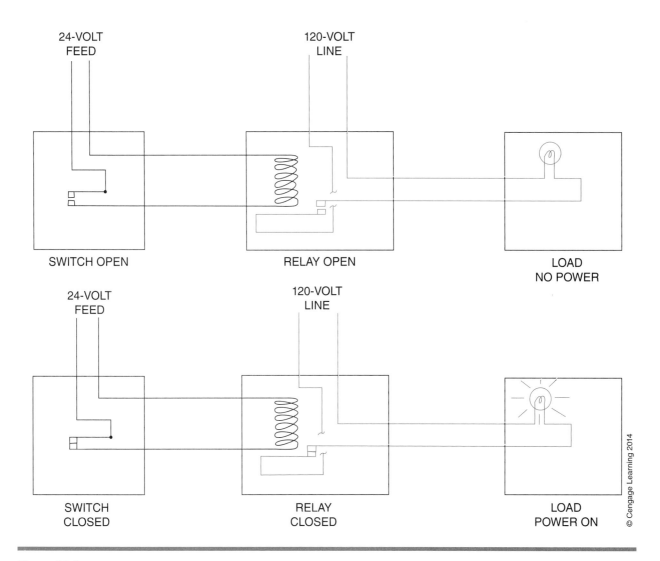

**Figure 42–3.** A relay allows a low-voltage circuit to start and stop a higher-voltage circuit.

contractor is to furnish and install the three 4-inch conduits and that the utility company wants to inspect the conduit installation prior to backfilling the trench where the conduits are installed; however, this should be verified at a preconstruction meeting with the utility company. There are times when an electrical service, or electrical distribution raceway, would require being concrete encased for protection. This type of concrete-encased underground raceway system is called a *duct bank* (see **Figure 42–5**).

The note does not specifically indicate who furnishes and installs the utility feeders in the three 4-inch conduits. This could be an expensive item and should be clarified with the utility company prior to estimating and bidding the project.

To comply with the *National Electrical Code® (NEC®)*, the service entrance section may have up to, but must not exceed, six main disconnecting units (switches or circuit breakers). Four main fused switches are shown in the existing portion of the single-line drawing. Note that this single-line drawing shows only a portion of the complete existing service entrance section. The service entrance section indicates that it is located outside by the NEMA 3R ENCLOSURE definition. NEMA, National Electric Manufacturers Association, generates specifications that are recognized as design standards for electrical boxes, devices, and equipment. NEMA 3R is a rain-tight designation. The new portion of the single-line drawing shows the electrical distribution to be installed in this

Electrical Drawings 327

## Electrical Reference Symbols

*ELECTRICAL ABBREVIATIONS*
(Apply only when adjacent to an electrical symbol.)

| | |
|---|---|
| Central Switch Panel | CSP |
| Dimmer Control Panel | DCP |
| Dust Tight | DT |
| Emergency Switch Panel | ESP |
| Empty | MT |
| Explosion Proof | EP |
| Grounded | G |
| Night Light | NL |
| Pull Chain | PC |
| Rain Tight | RT |
| Recessed | R |
| Transfer | XFER |
| Transformer | XFRMR |
| Vapor Tight | VT |
| Water Tight | WT |
| Weather Proof | WP |

*ELECTRICAL SYMBOLS*

Switch Outlets

| | |
|---|---|
| Single-Pole Switch | S |
| Double-Pole Switch | $S_2$ |
| Three-Way Switch | $S_3$ |
| Four-Way Switch | $S_4$ |
| Key-Operated Switch | $S_K$ |
| Switch and Fusestat Holder | $S_{FH}$ |
| Switch and Pilot Lamp | $S_P$ |
| Fan Switch | $S_F$ |
| Switch for Low-Voltage Switching System | $S_L$ |
| Master Switch for Low-Voltage Switching System | $S_{LM}$ |
| Switch and Single Receptacle | ⊖S |
| Switch and Duplex Receptacle | ⊜S |
| Door Switch | $S_D$ |
| Time Switch | $S_T$ |
| Momentary Contact Switch | $S_{MC}$ |
| Ceiling Pull Switch | (S) |
| "Hand-Off-Auto" Control Switch | HOA |
| Multi-Speed Control Switch | M |
| Push Button | ▫ |

**Receptacle Outlets**

Where weather proof, explosion proof, or other specific types of devices are to be required, use the upper-case subscript letters. For example, weather proof single or duplex receptacles would have the uppercase WP subscript letters noted alongside of the symbol. All outlets should be grounded.

| | |
|---|---|
| Single Receptacle Outlet | ⊖ |
| Duplex Receptacle Outlet | ⊜ |
| Triplex Receptacle Outlet | ≡⊜ |
| Quadruplex Receptacle Outlet | ⊕ |

**Figure 42–4.** Recommended electrical symbols.

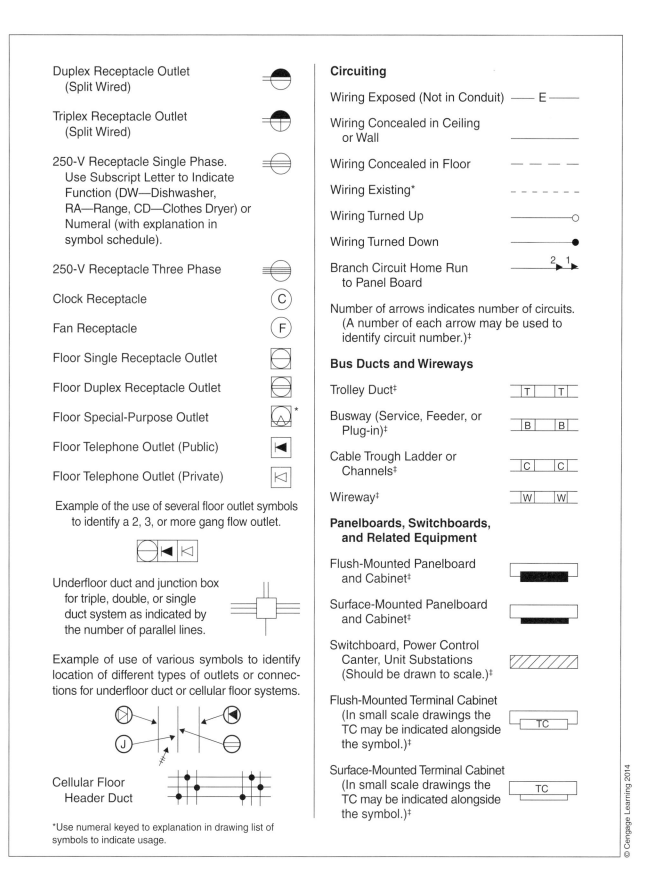

**Figure 42–4.** (continued)

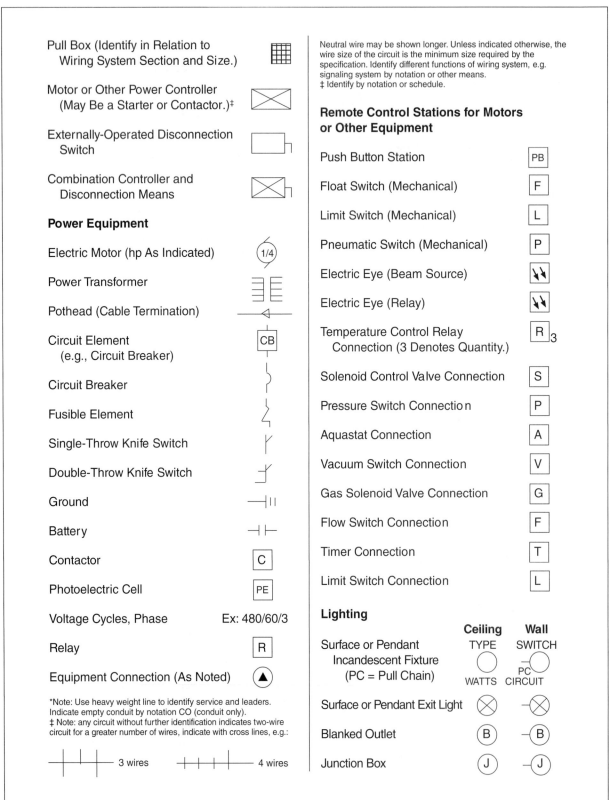

**Figure 42–4.** *(continued)*

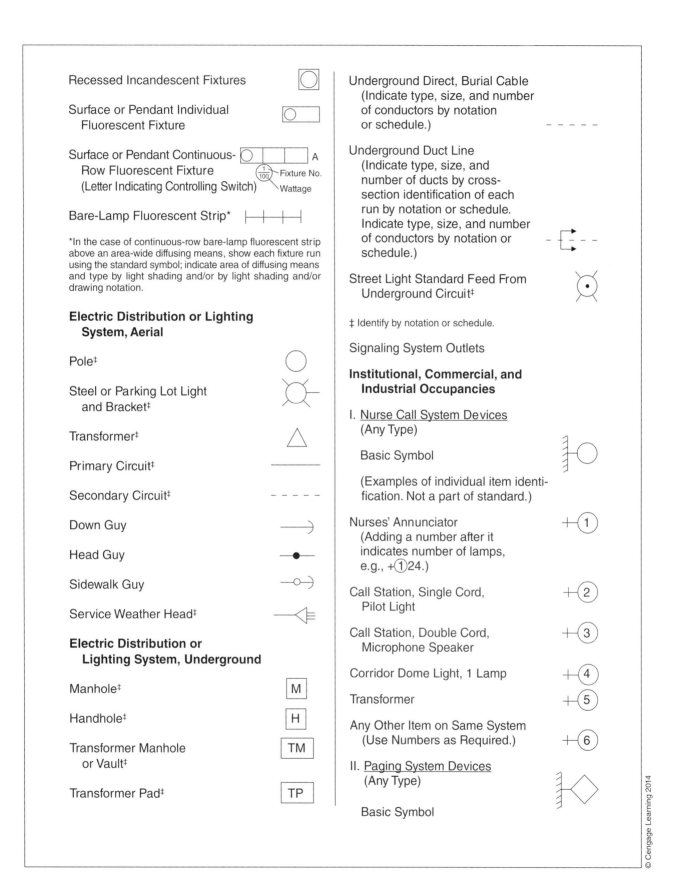

**Figure 42-4.** (continued)

(Examples of individual item identification. Not a part of standard.)

Keyboard — ◇1

Flush Annunciator — ◇2

Two-Face Annunciator — ◇3

Any Other Item on Same System — ◇4
(Use Numbers as Required.)

III. <u>Fire Alarm System Devices</u>
(Any Type) Including Smoke and
Sprinkler Alarm Devices

Basic Symbol

(Examples of individual item identification. Not a part of standard.)

Control Panel — □1

Station — □2

10" Gong — □3

Presignal Chime — □4

Any Other Item on Same System — □5
(Use Numbers as Required.)

IV. <u>Staff Register System Devices</u>
(Any Type)

Basic Symbol

(Examples of individual item identification. Not a part of standard.)

Phone Operators' Register — ◇1

Entrance Register (Flush) — ◇2

Staff Room Register — ◇3

Transformer — ◇4

Any Other Item on Same System — ◇5
(Use Number as Required.)

V. <u>Electric Clock System Devices</u>
(Any Type)

Basic Symbol

(Examples of individual item identification. Not a part of standard.)

Master Clock — ⬡1

12" Secondary (Flush) — ⬡2

12" Double Dial (Wall Mounted) — ⬡3

18" Skeleton Dial — ⬡4

Any Other Item on Same System — ⬡5
(Use Numbers as Required.)

VI. <u>Public Telephone System Devices</u>

Basic Symbol

(Examples of individual item identification. Not a part of standard.)

Switchboard — ◀1

Desk Phone — ◀2

Any Other Item on Same System — ◀3
(Use Numbers as Required.)

VII. <u>Private Telephone System Devices</u>
(Any Type)

Basic Symbol

(Examples of individual item identification. Not a part of standard.)

Switchboard — ◁1

Wall Phone — ◁2

Any Other Item on Same System — ◁3
(Use Numbers as Required.)

VIII. <u>System Devices</u>
(Any Type)

Basic Symbol

(Examples of individual item identification. Not a part of standard.)

**Figure 42–4.** *(continued)*

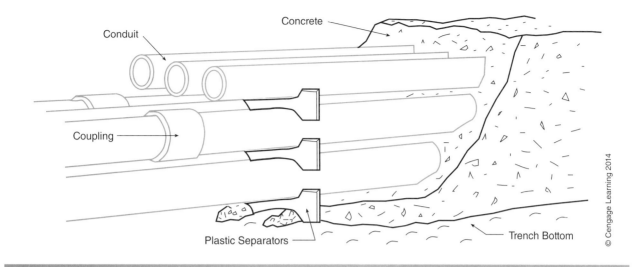

**Figure 42–5.** Duct bank system.

School Addition. The feeders for the new elevator, electrical panels L10 and L11, are fed from existing panel MDP. These new electrical loads are:

1. 60-amp circuit breaker with four #4 AWG, one #10 ground, and one #10 isolated ground in a 1¼-inch conduit feeding the elevator
2. 100-amp circuit breaker with four #1 AWG and one #6 ground in a 1½-inch conduit feeding panel L11
3. 100-amp circuit breaker with four #1 AWG and one #6 ground in a 1½-inch conduit feeding panel L10

A complete building electrical floor plan is not a part of the School Addition drawings, and the existing electrical distribution plan on drawing E-2 shows only a portion of the total electrical distribution section. This required notes 6, 7, and 8 to be added to drawing E-2, Single-Line Diagram E-17, giving the contractor the lengths of the feeders for the elevator, panels L10 and L11, which are to be included in this School Addition. In a typical design, a complete building electrical plan is provided indicating the locations of all existing and new electrical equipment. The feeder length would be determined from that drawing.

## Riser Diagrams

A *riser diagram* is so named because it usually shows the path of wiring or raceway from one level of a building to another and because the wiring rises from one floor to the next. A riser diagram does not give information about where equipment is to be located in a room or area. Riser diagrams are used because they are particularly easy to understand. Therefore, they do not require much explanation.

A *power riser diagram* (see **Figure 42–6**) shows a typical building's electrical service and related components. This figure is not the same electrical service as the School Addition Single-Line Diagram, E-17, on drawing E-2, but comparing the two diagrams shows how a power riser diagram greatly simplifies the interpretation of an installation drawing.

A *special riser diagram* is used for many systems that include:

1. Fire alarm
2. Security
3. Telephone
4. Clock
5. Signal
   ○ Bell
   ○ Call (nurse, emergency, etc.)
   ○ Water sprinkler

The *fire alarm riser diagram* (see **Figure 42–7**) shows the new School Addition fire alarm system with ¾-inch conduits to ramp area 113. The School Addition drawing E-2 shows three ¾-inch conduits with the note "Three (3) ¾-inch existing conduits from the Fire Alarm Control Panel (150 feet)." These three conduits are to be used for the School Addition fire alarm connection to the existing fire alarm system. The original fire alarm control panel was sized to accommodate this School Addition; however, you should verify that the existing *special systems* (fire alarm, security, clock, etc.) will

Electrical Drawings 333

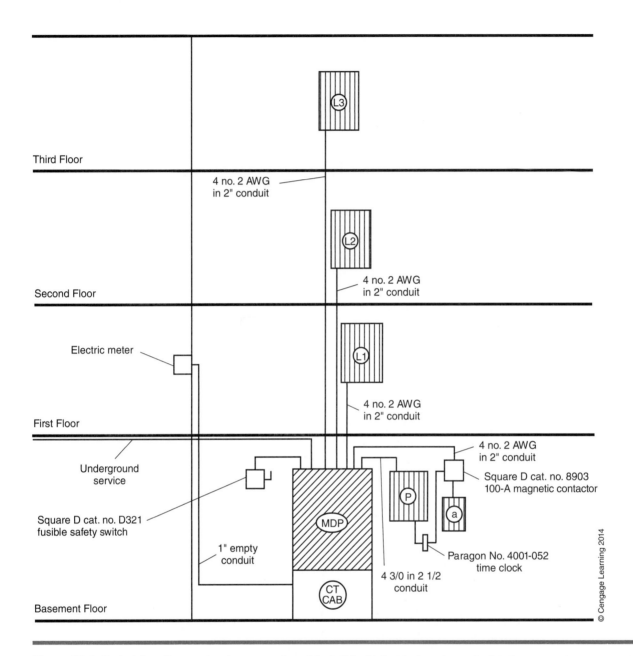

**Figure 42–6.** Typical riser diagram showing an overview of the building's electrical service and related components.

accommodate the additional requirements when adding to or modifying the existing system(s).

The *telephone riser diagram,* **Figure 42–8,** shows an existing 1¼-inch conduit, 150 feet in length, from ramp area 113 to the existing main telephone terminal cabinet. This conduit is to be extended to the telephone terminal board in room 102, First Floor Plan E-7, drawing E-2. The telephone riser diagram shows telephone conduit to be installed in the area above the suspended ceiling from the telephone terminal board to the outlet locations as shown on drawing E-2 and outlet detail A-11.

## Schematic Diagrams

A schematic wiring diagram is a drawing that uses symbols and lines to show how the parts of an electrical assembly or unit are connected. A schematic does not necessarily show where parts are actually located, but it does explain how to make electrical connections. Several of the schematics with the School Addition show connections to wires labeled G, N, and H. These stand for *ground, neutral,* and *hot.* Schematics are commonly drawn for electrical equipment that involves internal wiring—everything from washing machines

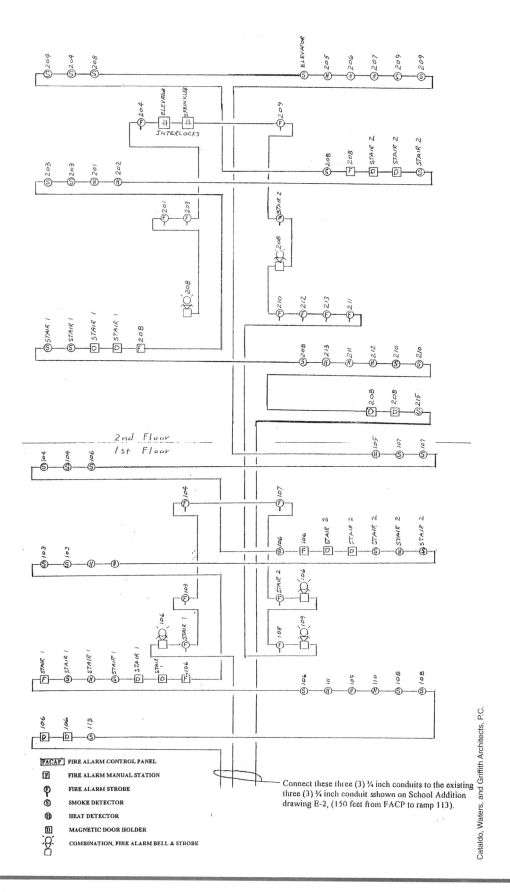

**Figure 42-7.** Fire alarm riser diagram for the school addition.

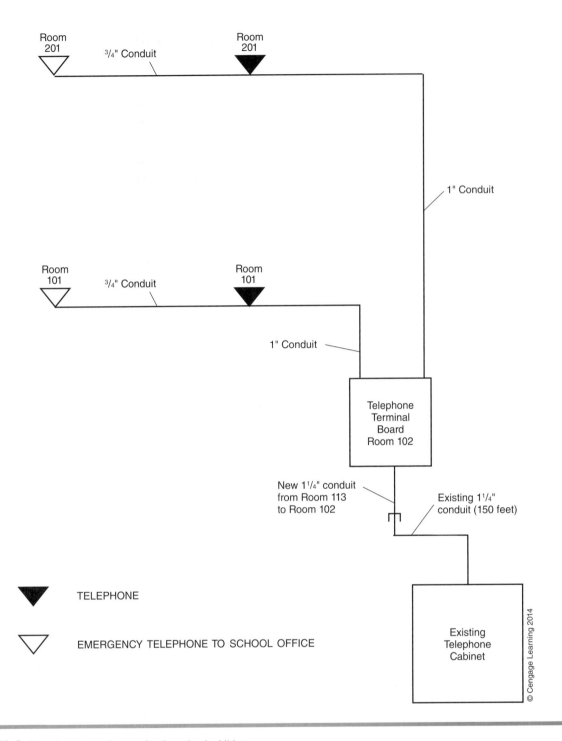

**Figure 42-8.** Telephone riser diagram for the school addition.

to computers. A basic motor control schematic (see **Figure 42-9**) shows a three-phase power source (L1, L2, and L3) through the starter contacts (M) and the overloads (OL) powering the motor. The starter control is taken from phase L1 through a stop push button, a start push button paralleled with a latching contact M to the starter coil M, and finally through the normally closed overload contacts back to phase L2. The connection label numbers (1, 2, 3, etc.), which are shown on the schematic, aid the electrician in troubleshooting. Several different labeling methods are used, but they all follow the same principles; so if you understand one method, you can understand the other methods.

## Schedules

An electrical schedule is used to systematically list equipment, loads, devices, and information. Schedules organize the information in an easily understood form and can be a valuable method for communicating the design requirements to contractors and their installers.

The *fixture schedule* (see **Figure 42–10**) lists complete information about each fixture type shown on the lighting plan. The following is a list of the kind of information that is usually shown on a fixture schedule:

○ *Mark*—The label used to indicate a fixture type. The mark is written on or next to each fixture on the plan.

○ *Make*—The identification of the manufacturer being used to establish the specific design requirements

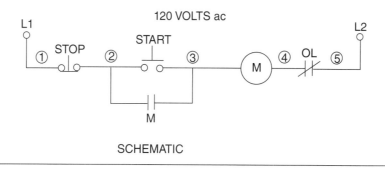

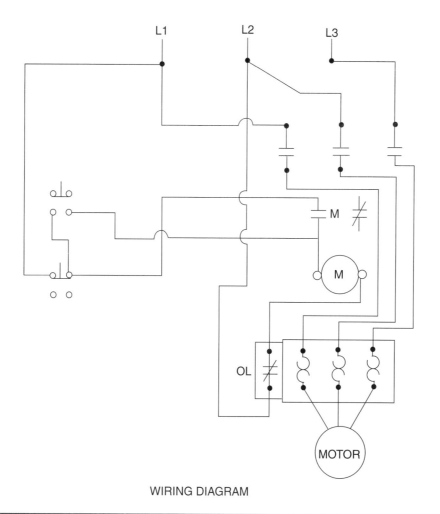

**Figure 42–9.** Schematic diagram and wiring diagram for a three-phase, AC magnetic, nonreversing motor starter.

Electrical Drawings 337

| MARK | MAKE | MODEL | VOLTS | WATTS | LAMP TYPE | LAMP NO./FIXT. | REMARKS |
|---|---|---|---|---|---|---|---|
| A | STONCO | VWXL11GC | 120 | 100 | 100WA19 | 1 | CAST GUARD. |
| FA | WILLIAMS | 1222-RWKA125 | 120 | 183 | F32T8/SP30 | 3 | SPLIT WIRE IN TANDEM PAIRS – WATTAGE IS FOR 2 FIXTURES W/ TOTAL OF 3 2-LAMP BALLASTS. ELECTRONIC BALLAST. |
| FA1 | WILLIAMS | 1222-RWKA125 | 120 | 92 | F32T8/SP30 | 3 | ELECTRONIC BALLAST. |
| FB | WILLIAMS | 2922-KA | 120 | 61 | F32T8/SP30 | 2 | ELECTRONIC BALLAST. |
| FC | KIRLIN | 96617-45-46-61 | 120 | 36 | PLC13W/27K | 2 | HPF BALLAST. |
| FC1 | KIRLIN | 96617-46-61-SM | 120 | 36 | PLC13W/27K | 2 | HPF BALLAST, SURFACE MOUNT; COLOR AS SELECTED |
| FD | WILLIAMS | 8222 | 120 | 61 | F32T8/SP30 | 2 | ELECTRONIC BALLAST. |
| FE | WILLIAMS | 1262-RWKA125 | 120 | 84 | F17T8/SP30 | 4 | ELECTRONIC BALLAST. |
| FF | TERON | EE26-P-H | 120 | 36 | PL13/27 | 2 | HPF BALLAST. |
| FG | WILLIAMS | 2122-IM | 120 | 61 | F32T8/SP30 | 2 | ELECTRONIC BALLAST. |
| FH | WILLIAMS | EPG-R272RWKA-125 | 120 | 61 | F32T8/SP30 | 2 | TO MATCH EXISTING |
| M1 | KIRLIN | SS-51277-24-43 (35W)-45-46-FR | 120 | 40 | 35W HPS | 1 | HPF BALLAST. |
| M2 | STONCO | PAR250LX | 120 | 300 | 250W HPS | 1 | HPF BALLAST EQUIP WITH MOUNTING BRACKET ARM. |
| EXITS | LITHONIA | WLES SERIES | 120 | 65 | LED | 1 | SEE SHEET E-2 FOR LOC. ARROWS & MOUNTING AS IND EQUIP W/ BATTERY BACK-UP |
| EMERG.LGT. BATT.PACK | EXIDE | B200 | 120/12 | 150 | H1212 | | |
| REMOTE LAMP | EXIDE | H1212 | 12 | 24 | H1212 | 2 | |

**Figure 42–10.** Fixture schedule.

needed to aesthetically and functionally light the various rooms or areas.
- *Volts*—It is becoming increasingly common for light fixtures to be powered by up to 277/480 volts, while the *switches* might use only 24 volts (low-voltage lighting control) to protect the person using the switch.
- *Watts*—It is necessary to know the wattage of the lamps (bulbs) in each fixture, so that the space will have the amount of illumination intended by the electrical designer.
- *Lamp Type and Quantity*—Lamp manufacturers have similar systems of designating lamp characteristics (see **Figures 42–11, 42–12,** and **42–13**).
- *Notes or Remarks*—This column is for information that does not clearly belong under the other headings.

The School Addition Fixture Schedule A-9—Floor Plans E-7 and E-14 on drawing E-1—shows type FA light fixture to be the light source for the classrooms. The School Addition fixture schedule, as shown in **Figure 42–10**, shows that the fixture is manufactured by Williams and the catalog number is 1222-RWKA125 with three F32T8/SP30 lamps rated at 120 volts with a 183-watt load. Also, the Remarks column indicates some special wiring and specific wiring and specific ballast-type requirements. The first and second floor Reflected Ceiling Plans, E-15 and E-16 on drawing E-7, indicate light fixture type FA to be a 2′ × 4′ light fixture mounted end to end.

Fifteen other light fixture types are shown on the fixture schedule and on the drawings and are identified by their corresponding mark or label. All are rated at 120 volts except one emergency battery-powered remote lamp, rated at 12 volts. This information is typically given in the specifications but is more readily presented by a schedule.

A *panel schedule* identifies the panels by their mark or label. They are shown by this same designation on the electrical floor plan. The panel schedule for the School Addition lists each of the branch circuits served by a panel, the calculated load for that branch circuit, the voltage of that circuit, and the number of poles and trip rating for each circuit breaker. A panel schedule also might include:

○ Type (surface or flush)
○ Panel main buss amperes, volts, and phases
○ Main circuit breaker/main lugs only
○ Breaker frame sizes
○ Items fed and/or remarks

On the lighting and power plans, we saw that each device is connected to a branch circuit identified by a panel number (label) and circuit number. Those numbers correspond with the numbers on the panel schedule.

When a commercial project has a kitchen, you should have a *kitchen equipment schedule*. If the drawings do not have one, it is very helpful to make one for the installer. The kitchen equipment schedule should include:

○ Equipment number or designation
○ Description of each equipment item
○ The load in horsepower or kilowatts
○ Volts
○ Wire size
○ Conduit size

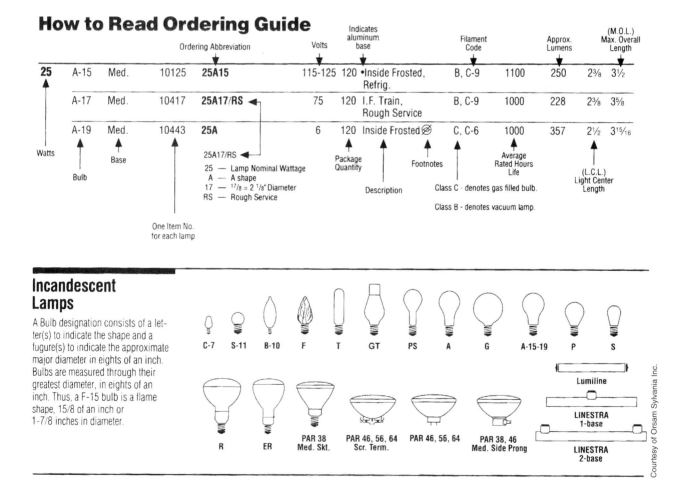

**Figure 42–11.** Incandescent lamp designations.

# How to Read Ordering Guide

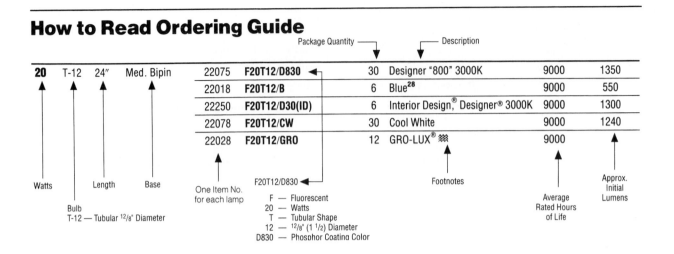

## FLUORESCENT LAMPS

The bulb shape and size of a fluorescent lamp are expressed by means of a code consisting of the letter "T" (which designates that the bulb is "tubular" in shape) followed by a number which expresses the diameter of the bulb in eighths of an inch. They vary in diameter from T-5 (5/8 inch) to T-17 (2 1/8 inches). In nominal overall length, fluorescent lamps range from 6 to 96 inches, which is always measured from back of lampholder to back of lampholder. For example, the actual overall length of the 40-watt rapid start T-12, 48 inch lamp is 47 3/4 inches. Circline lamps, which are circular, are available in four sizes: 6 1/2 inches, 8 inches, 12 inches and 16 inches outside diameter. There are also U shaped fluorescent types (Curvalume™) with T-8 and T-12 bulbs. U shaped types are measured for the distance between the ends. The overall length is measured from the face of the bases to the outside of the glass bend.

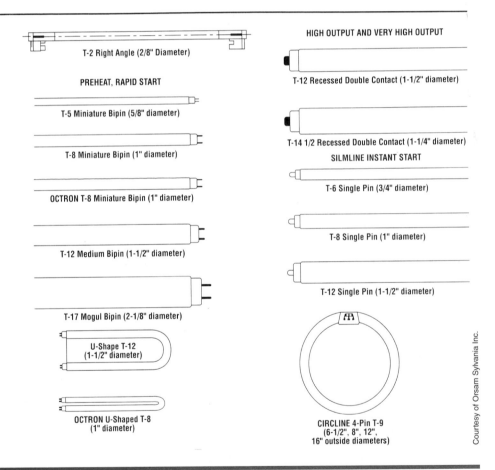

**Figure 42–12.** Fluorescent lamp designations.

340 UNIT 42

### HOW TO READ PRODUCT INFORMATION - COMPACT FLUORESCENT

| Nominal Wattage | Bulb | MOL (in) | MOL (mm) | Base | Product Number | Ordering Abbreviation | NEMA Generic Designation | Pkg Qty | Avg Rated Life (hrs) | CCT (K) | CRI | Approx Lumens Initial @25°C/77°F (@35°C/95°F) | Mean | Symbols & Footnotes |
|---|---|---|---|---|---|---|---|---|---|---|---|---|---|---|
| 13 | Twist | 4.6 | 117 | Med | 29116 | **CF13EL/MINITWIST** | | 6 | 8000 | 3000 | 82 | 800 | 640 | 2,21,28, 36,63,64 |
| 26 | D(T4) | 6.8 | 173 | G24D-3 | 20710 | **CF26DD/830/ECO** | CFQ26W/G24D/30 | 50 | 10000 | 3000 | 82 | 1800 | 1548 | 2,21,28, 34,37 |
| 32 | T(T4) | 5.5 | 140 | GX24Q-3 | 20885 | **CF32DT/E/IN/835/ECO** | CFTR32W/GX24Q/35 | 50 | 10000 | 3500 | 82 | 2328 / 2400 | 2002 / 2064 | 2,21,28, 33,35,59 |
| 40 | L(T5) | 22.6 | 573 | 2G11 | 20586 | **FT40DL/841/RS/ECO** | FT40W/2G11/RS/41 | 10 | 20000 | 4100 | 82 | 3150 | 2709 | 2,21,28 |

| | |
|---|---|
| **Nominal Wattage** | Design wattage on reference ballast. Actual wattage dependent on ballast. |
| **Bulb** | Describes the shape of the bulb. |
| **Base** | Base designations for compact fluorescent lamps are the NEMA designations. Please see page 111 for base illustrations. |
| **MOL** | Maximum overall length. The actual length of the lamp measured from the bottom of the base to the top outside edge of the glass. In many cases, the bottom of the base is the bottom of the center post of the base of the lamp. |
| **Symbols & Footnotes** | Most symbols and footnotes that apply to a specific product will appear in this space. The explanations of the symbols and footnotes are at the end of the fluorescent section. |
| **Ordering Abbreviation** | A text description of the lamp. Please see below for several examples and explanations of some of the codes. |
| **NEMA Generic Designation** | Designation assigned by NEMA (National Electrical Manufacturers Association). |
| **CCT** | Correlated Color Temperature. The degree of "whiteness" of the light. Expressed in kelvins (K). Please see page 109 for more information. |
| **CRI** | Color Rendering Index. A numbering system for rating the relative color rendering quality of a light source compared to a standard. Please see page 109 for more information. |
| **Initial & Mean Lumens** | Initial lumens are measured when the lamp has been operating for 100 hours. Mean lumens are typically measured at 40% of the rated life of the lamp. Compact Fluorescent lamp lumens are measured at 25°C (77°F) and 35°C (95°F) |

#### How to Read Ordering Abbreviations

| **CF26DD/830** | | **CF32DT/E/IN/835/ECO** | | **FT40DL/841/RS/ECO** | | **CF20EL/830/MED/ECO** | |
|---|---|---|---|---|---|---|---|
| CF | Compact Fluorescent | CF | Compact Fluorescent | FT | Fluorescent Twin | CF | Compact Fluorescent |
| 26 | Nominal lamp wattage | 32 | Nominal lamp wattage | 40 | Nominal lamp wattage | 20 | Nominal lamp wattage |
| DD | DULUX® Double | DT | DULUX Triple | DL | DULUX Long | EL | Electronic Lamp |
| 8 | 82 CRI | E | Electronic or dimming operation | 8 | 82 CRI | 8 | 82 CRI |
| 30 | 3000K CCT | IN | Amalgam | 41 | 4100K CCT | 30 | 3000K CCT |
| | | 8 | 82 CRI | RS | Rapid Start | MED | Medium screw base |
| | | 35 | 3500K CCT | ECO | ECOLOGIC | ECO | ECOLOGIC |
| | | ECO | ECOLOGIC | | | | |

**Figure 42–13.** Compact fluorescent lamp designations.

- Protection in amperes
- Who will furnish each equipment item (furnished by others or by contractor)
- Installation requirements
- Remarks column for any detailed specific information required

A *receptacle schedule* is valuable when a number of special or specific receptacle types are found on the electrical drawings. If one is not provided, you should make one to expedite the installation time required and reduce the chance of installation error. A receptacle schedule should include:

- The symbol designation used by the designer
- Amperage rating

- Number of wires and poles
- Voltage rating
- NEMA type
- The configuration of the blades or slots
- A manufacturer catalog number reference
- Special information (duplex, single, three phase, etc.)

**Note:** Receptacle information may be found on an *equipment schedule.*

The local plans review and the utility company typically require a *connected load schedule.* This type of schedule includes:

- Type of load
- Building or area designation

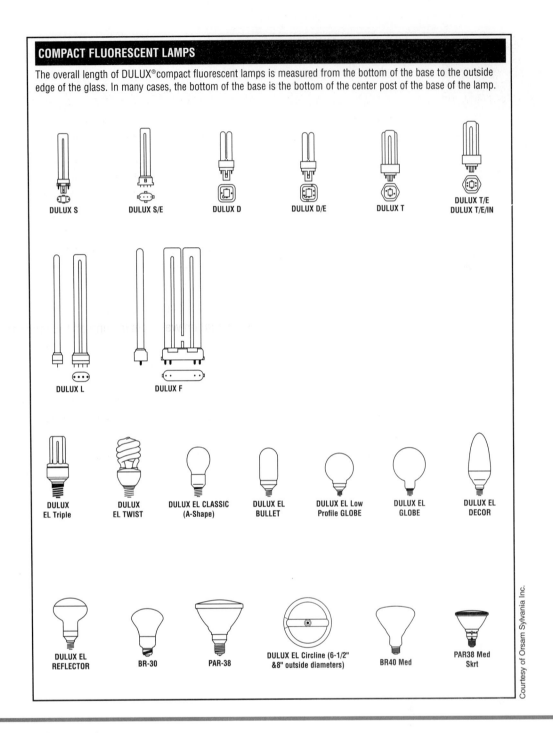

**Figure 42–13.** (continued)

- Size of load (kilowatts or horsepower)
- Total electrical load by type or area
- Notes explaining any special methods used in the load calculations

## Specifications

The drawings and the **specifications** are the items that establish the intended design and the construction requirements required by the owner, architect, and engineers. The contractors and their installers must review the **contract** documents for conflicts and/or discrepancies between the contract, the specifications, and the drawings. The drawings should be reviewed for conflicts between sections (architectural, mechanical, plumbing, structural, civil, HVAC, electrical, etc.). The drawings in your specific section should be reviewed for conflicts sheet by sheet.

The electrical specifications give the quality of materials intended to be used and the installation and testing requirements. Sometimes a specific manufacturer or catalog number is specified with no substitution or "equal" allowed. This usually inflates the cost of the project. There is no standard specification. The contractors and their installers should read through all project specifications and become knowledgeable of their content prior to starting any installation.

### USING WHAT YOU LEARNED

A basic skill for reading electrical drawings is knowing which circuits feed each electrical device and the corresponding panel for that circuit. There are two duplex receptacles shown in the elevator area on the first floor plan. Are those receptacles in the elevator car or elsewhere? From which panel are they fed? What circuit are they on? The receptacles are labeled as being "in pit," so they are not in the elevator car. The numbers 15 and L10 indicate that they are on circuit 15 which is fed from panel L10. This can be verified on the panel schedule E-11 of Sheet E7.

## Assignment

Refer to the School Addition drawings in your textbook packet to complete this assignment.

1. What are the four basic methods of showing electrical circuits?
2. How is a home run circuit shown?
3. What is a reflected ceiling plan?
4. In a low-voltage lighting system, what does the light switch activate?
5. Where do you find the industry standardized electrical symbols?
6. What are the four basic types of electrical drawings?
7. What is the name of the method used to show the path of wiring or raceway from one level of a building to another?
8. What is a schematic wiring diagram?
9. What is used to systematically list equipment, loads, devices, and information?
10. Where do you find the quality of material intended to be used on a project?
11. How many F17T8/SP30 lamps are required on the first floor of the addition?
12. Explain why one of the switches in room 109 is listed as $S_{3M}$ and the other is simply $S_3$.
13. How are the lights turned on and off in the first-floor corridor?
14. What circuit carries the lights for stair #2?
15. What is the approximate total wattage of the lamps in the boys' toilet room on the first floor?
16. What is the total load for the circuit that serves the lights in the boys' toilet room on the first floor?
17. What is indicated by the D in a square near the doors from the existing building into the addition?
18. What is on circuit L10, 24?
19. Where are the devices on circuit L10, 15?
20. Explain what each of the colored terminals on a classroom lighting control is to be connected to:

Green \_\_\_\_
Orange \_\_\_\_
Black \_\_\_\_
White \_\_\_\_
Blue (inner terminal) \_\_\_\_
Blue (outer terminal) \_\_\_\_

# PART IV Test

A. Which of the symbols shown in Column II is used on the School Addition drawings to represent the objects or materials listed in Column I?

| I | II |
|---|---|
| 1. Room number | a.  |
| 2. Building section | b.  |
| 3. Rigid insulation | c.  |
| 4. Window type | d.  |
| 5. Door number | e.  |
| 6. Interior elevation | f.  |
| 7. Wood blocking | g. 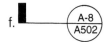 |
| 8. Detail key | h.  |

B. Refer to the door schedule on School Addition Sheet A001 to write the door numbers from Column II that are associated with the material in Column I.

| I | II |
|---|---|
| 1. Wood | 216 |
| 2. Steel hollow metal | 104 |
| 3. Plastic | 101 |

**C. Name the material for each of the numbered items in the drawing of a bond beam.**

1. ___

2. ___

3. ___

4. ___

5. ___

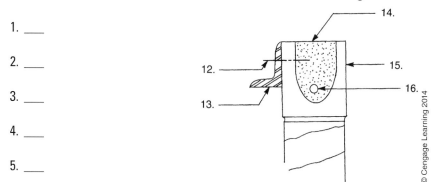

**D. Describe the material at each of the following locations in the School Addition.**

1. Door frame 103
2. South wall of corridor 106
3. Structural portion of west wall of classroom 104
4. Exterior surface of west wall at corners
5. Exterior surface of west wall below windows
6. Stair treads
7. Vertical reinforcement in elevator pit walls
8. Beam at the top of the west wall of classroom 204
9. Lintel over the windows in the west wall
10. Door frame at entrance to stair #1

**E. Refer to the School Addition drawings and write the elevation at the following locations.**

1. Second finish floor
2. Top of roof
3. Top of bricks beneath window in classroom 103
4. Top of concrete blocks in first floor corridor walls
5. Top of tackboard on north wall of classroom 103
6. Finish floor in elevator pit
7. Top of foundation wall at typical locations
8. Top of footing for column F3
9. Top of beams in second floor framing between columns in row C

**F. Refer to the School Addition drawings to answer the following questions.**

1. What is the height of the rough opening for the grille that allows outside air to enter classroom 103?
2. What supports the outer few inches of the roofing membrane in the area where the roof-edge coping meets it?
3. What size are the concrete masonry units in the north wall?
4. What are the three materials that make up the roof decking on the canopy over door 101?
5. What are the drawing and sheet numbers for information about typical column base plates?
6. Give as much information as possible about the welds that fasten the roof framing bar joists to the beams at the tops of the north and south walls.

7. On what branch circuit are the light fixtures in stair #1?
8. How are the light fixtures in the second floor corridor turned on and off?
9. How many lamps are required for all the fixtures in the second floor corridor?
10. What branch circuit supplies the convenience outlets in the second floor corridor?
11. Which terminal on the corridor lighting control unit is connected to the neutral leg of the supply?
12. What is the diameter of the sanitary drain where it leaves the building?
13. What are the pipe sizes that storm water passes through as it flows from the roof drain on the north canopy to the point where it exits the building?
14. Where is the nearest cleanout to the roof drain at the west end of the addition?
15. What type of unit provides heat in storage room 109?

# Appendix A

# School Addition Master Keynotes

## MASTER KEYNOTES

| | | |
|---|---|---|
| 2.11 | Demolition, Removals & Relocation |
| 2.20 | Site Preparation & Earthwork |
| 2.20A | Select Fill - Bank Run Gravel |
| 2.20B | Select Granular Material |
| 2.20D | Topsoil |
| 2.20F | Crushed Gravel |
| 2.20G | 4" perforated P.v.c. |
| 2.20I | Compacted Subgrade |
| 2.22 | Structural Excavation, Backfill and Compaction (Building Area) |
| 2.60 | Pavement and Walks |
| 2.60A | Vehicular Area Sub-Base Course Granular |
| 2.60B | Asphaltic Concrete - Binder Course |
| 2.60C | Asphaltic Concrete Top - Wearing Course |
| 2.60D | Concrete Walk/Paving |
| 260E | Precast Concrete Curb |
| 2.60G | Reinforcing Mesh |
| 2.60H | Expansion Joint Filler |
| 2.60I | Control Joint - 4'-0" O.C. Maximum Saw Cut or Tooled |
| 2.60J | Stabilization Fabric |
| 2.60P | Asphaltic Concrete-Base Course |
| 2.60Q | Detectable Warning Pavers |
| 2.61 | Pavement Markings |
| 2.61B | Painted ANSI Handicap Symbol |
| 2.61C | Painted Traffic Control Lines |
| 2.80 | Landscaping |

| | |
|---|---|
| 2.80A | Seeding |
| 2.83 | Chain Link Fence |
| 2.83D | 2.875" O.C. Corner Post |
| 2.83E | 2.875" O.C. Line Post |
| 2.83F | 1–1/2" Top Rail |
| 2.83G | 2" Mesh Fabric |
| 2.83H | 7 Gauge Tension Wire Continuous |
| 2.83I | Post Cap |
| 2.83J | 11 Gauge Rail Clamps |
| 2.83K | Sleeve |
| 2.83L | 11 Gauge Bands |
| 2.83M | 1/4" × 3/4" Turnbuckle |
| 2.83N | Stretcher Rod |
| 2.83O | 3/8" Diameter Truss Rod |
| 2.83P | Pass Thru Cap |
| 2.83Q | Gate Frame |
| 2.83R | Top Hinge |
| 2.83S | Hinge |
| 2.83T | Fork Latch |
| 3.30 | Concrete |
| See Structural Drawings* | |
| 3.30A | Cast-In-Place-Concrete |
| 3.30B | Reinforcing Bar |
| 3.30D | Welded Wire Mesh, Size as Noted |
| 3.32 | Concrete Slab on Grade |
| 3.325 | Concrete Slab on Metal Deck |
| 3.40 | Precast Concrete |
| 3.40D | Precast Dry Pit |
| 3.40E | Precast Concrete Wall Unit |
| 3.40F | Precast Concrete Cap Unit |
| 3.40G | Geogrid |
| 4.10 | Mortar and Masonry Grout |

| | |
|---|---|
| 4.10A | Grout Solid |
| 4.10B | Mortar Fill |
| 4.20 | Unit Masonry |
| 4.20A | Face Brick - Standard Modular |
| 4.20C | Concrete Masonry Units |
| 4.20D | Horizontal Joint Reinforcing |
| 4.20E | Flexible Masonry Anchors |
| 4.20G | Premolded Control Joint Strips |
| 4.20I | Reinforcing Bars |
| 4.20J | Flexible Masonry Column Tie Anchors |
| 4.20K | Weep Holes |
| 4.20M | Control Joint |
| 4.20P | Slate Sill - 1/4" Chamfer at Exposed Edges |
| 4.20S | Precast Concrete Sill |
| 4.20Y | Precast Concrete Band |
| 4.20A | Expansion Joint Closure |
| 5.10 | Structural Steel |

See Structural Drawings*

| | |
|---|---|
| 5.10A | Steel Beam |
| 5.10B | Steel Angle |
| 5.10D | Anchor Bolt |
| 5.10E | Steel Tube Column |
| 5.10F | Miscellaneous Steel - Shape and Size as Noted |
| 5.10G | Steel Tee |
| 5.20 | Steel Joists |

See Structural Drawings*

| | |
|---|---|
| 5.30 | Metal Decking |

See Structural Drawings*

| | |
|---|---|
| 5.30A | Metal Roof Decking |
| 5.50 | Miscellaneous Metals |
| 5.50A | Miscellaneous Steel - Shape and Size as Noted |
| 5.50B | Steel Angle |

Appendix A

| | |
|---|---|
| 5.50C | 1–1/4" Nominal Steel pipe (1.66" O.D.) Post 4'-0" O.C. Maximum |
| 5.50D | 10 Gauge Metal Riser, Tread and Landing Pan |
| 5.50E | MC 12 × 10.6 Stringer |
| 5.50F | 3–1/2" × 2–1/2" × 1/4" - LLV - Pan Support at 3"-0" O.C. Maximum |
| 5.50G | 1–1/2" × 1–1/2" × 1/4" Steel Angle Carriers |
| 5.50M | Galvanized Steel Bent Plate |
| 5.50N | Galvanized Steel Flat Bar Stock |
| 5.50O | 3/4" Diameter Galvanized Steel Rungs |
| 5.50S | Steel Bollards - Fill With Concrete - Bevel Concrete at Top |
| 5.50U | 1–1/4" Nominal Steel Pipe (1.66" O.D.) |
| 5.50Y | Countersunk Machine Head Screw/Expansion |
| 5.50A | FSteel Plate |
| 5.50A | M1" Nominal Steel Pipe Handrail (1.315" O.D.) |
| 5.50A | N1" Nominal Steel Pipe Bracket |
| 5.50A | PSteel Wall Bracket, Julius Blum #622, or Equal |
| 5.50A | XAluminum Sleeve (.050") |
| 5.50A | YSteel Plate, Size as Noted |
| 5.80 | Expansion Joints |
| 5.80A | Expansion Joint Cover |
| 6.10 | Rough Carpentry |
| 6.10A | Wood Blocking |
| 6.10B | Plywood |
| 6.10C | Wood Furring |
| 6.10I | Wood Framing 16" O.C. |
| 6.10S | Joist Hanger |
| 6.10T | Pressure Treated Bail; Size as Noted |
| 6.10U | Pressure Treated Post; Size as Noted |
| 6.20 | Finish Carpentry |
| 6.2 | OF Softwood Trim |
| 7.15 | Dampproofing |
| 7.20 | Insulation |
| 7.20B | Batt Insulation |

| | |
|---|---|
| 7.20C | Cavity Wall Insulation |
| 7.20H | Metal Z Furring |
| 7.41 | Preformed Wall Panels |
| 7.41C | Vertical Siding |
| 7.41F | Trim to Match Siding |
| 7.50 | Roofing System (EPDM) |
| 7.50A | Roofing Membrane |
| 7.50B | Membrane Flashing |
| 7.50C | Insulation |
| 7.50D | Thermal Barrier |
| 7.50E | Tapered Insulation |
| 7.50F | Vapor Barrier/Air Seal |
| 7.50G | Ballast and Protective Mat |
| 7.50H | Fascia System |
| 7.50K | Fascia Extender |
| 7.50M | Aluminum Counterflashing |
| 7.50N | Sealant |
| 7.50P | Termination Bar |
| 7.50Q | Expansion Joint Support |
| 7.50R | Closed Cell Backer Rod |
| 7.50T | Insert Drain |
| 7.50V | Recovery Board |
| 7.50W | Aluminum Trim -.032" Thickness |
| 7.51 | Roofing System (Silicone) |
| 7.51A | Silicone Coating |
| 7.51B | Urethane Insulation |
| 7.51D | Sealant |
| 7.51E | Aluminum Foam Stop - 0.050" Thickness |
| 7.51J | Aluminum Trim - 0.32" Thickness |
| 7.51K | Insert Drain |
| 7.53 | Roofing System (EPDM-Foam Adhesive) |
| 7.53A | Roofing Membrane |

| | |
|---|---|
| 7.53B | Foam Adhesive |
| 7.53C | Membrane Flashing |
| 7.53D | Insulation |
| 7.53H | Fascia System |
| 7.53K | Fascia Extender |
| 7.53O | Sealant |
| 7.53Q | Termination Bar |
| 7.53R | Closed Cell Backer Rod |
| 7.53T | Insert Drain |
| 7.60 | Flashing and Sheet metal |
| 7.60 | A Thru-wall Flashing - Turn-Up at Ends at Wall Openings to Form Dam |
| 7.72 | Roof Accessories |
| 7.90 | Caulking and Sealants |
| 7.90A | Sealant and Backing Material (Size as Required to Fill Void) |
| 7.90B | Foam Joint Filler - Width Shown × Size Required to Fill Void) |
| 7.90C | Sealant Bead |
| 7.90G | Expansion Joint Filler |
| 8.10 | Metal Doors and Frames |
| 8.10A | Steel Hollow Metal Door Frame |
| 8.10B | Steel Hollow Metal Door |
| 8.10E | Metal Access Door |
| 8.10F | 16 Gauge Hollow Metal Pipe Enclosure |
| 8.15 | Plastic Doors |

See Door Schedule*

| | |
|---|---|
| 8.15A | Plastic Door |
| 8.1 | 5B Aluminum Door Frame System |
| 8.15C | .090" Break Metal to Match Door Frame System |
| 8.15G | Aluminum Angle (Size as Noted) |
| 8.20 | Wood Doors |

See Door Schedule*

| | |
|---|---|
| 8.20A | Wood Door |
| 8.33 | Rolling Counter Fire Door |

| | |
|---|---|
| 8.33B | Vertical Sliding Pass Window |
| 8.33C | Stainless Steel Sill |
| 8.50 | Metal Windows |
| 8.50A | Aluminum Windows |
| 8.50B | 1/8" Break Metal to Match Windows |
| 8.50C | Extruded Aluminum Sill and Clip |
| 8.50D | Extruded Aluminum Head/Jamb Panning |
| 8.50E | Extruded Aluminum Trim |
| 8.50K | Fixed Aluminum Window |
| 8.50L | Horizontal Rolling Aluminum Window |
| 8.50R | Foam Tape |
| 8.50S | Foam Insulation |
| 8.50T | Double Hung Window |
| 8.70 | Hardware and Specialties |
| See Door Schedule* | |
| 8.70D | Aluminum Threshold Set in Mastic |
| 8.80 | Glazing |
| 8.80A | Tempered Insulating Glass |
| 8.80B | Insulating Glass |
| 8.80C | Safety Wire Glass |
| 8.80D | Tempered Glass |
| 8.80I | Muntins Inside Insulated Glass |
| 8.80K | Obscure Insulating Glass |
| 09805 | Encapsulation of Asbestos Containing Material |
| 9.25 | Gypsum Wallboard |
| 9.25A | 5/8" Type 'X' Gypsum Wallboard |
| 9.25B | "J" Casing Bead |
| 9.25C | Metal Stud System |
| 9.25D | Metal Furring |
| 9.25G | Metal Stud Runner - 2" Deep to Provide 1" Expansion for Studs |
| 9.25J | Gypsum Sheathing |
| 9.25k | Carrying Channel |

| | |
|---|---|
| 9.25L | 5/8" Exterior Gypsum Ceiling Board |
| 9.25N | Metal Angle Runner |
| 9.30 | Tile |

See Room Finish Schedule*

| | |
|---|---|
| 9.30A | Glazed Wall Tile |
| 9.30B | Unglazed Ceramic Mosaic Tile |
| 9.30D | Glazed Ceramic Tile Wall Base |
| 9.30E | Marble Thresholds - See Door Schedule* |
| 9.30F | Accent Tile - Continuous Around Room |
| 9.30Q | Unglazed Ceramic Mosaic Tile Base |
| 9.40 | Terrazzo |
| 9.40A | Thin Set Terrazzo |
| 9.40C | Terrazzo Cove Base |
| 9.50 | Acoustical Treatment |

See Room Finish Schedule*

| | |
|---|---|
| 9.50A | Suspended Ceiling System |
| 9.65 | Resilient Flooring |

See Room Finish Schedule*

| | |
|---|---|
| 9.65A | Vinyl Composition Tile |
| 9.65B | Vinyl Cove Base |
| 9.65D | Rubber Stair Treads and Risers |
| 9.65E | Molded Rubber Tile |
| 9.65J | Rubber Base |
| 9.80 | Special Coating System |
| 9.80A | Special Coating System |
| 9.80B | Finish Coat |
| 9.80C | Base Coat |
| 9.80D | Reinforcing Mesh |
| 9.80E | Insulation |
| 9.80F | Gypsum Sheathing |
| 9.80G | Metal Studs |
| 9.80H | Sealant and Backer Rod |

| | |
|---|---|
| 9.80I | Waterproof Base Coat |
| 9.80J | Routed Joint |
| 9.80K | Insulation Board Below Grade |
| 9.80L | Below Grade Waterproofing |
| 9.80M | Slide Clip |
| 9.80N | Bent Galvanized Metal (Size and Gauge as Noted) |
| 9.80O | Expansion Joint |
| 9.90 | Painting |

See Room Finish Schedule*

| | |
|---|---|
| 9.90E-1 | Paint E-1 |
| 9.90E-2 | Paint E-2 |
| 9.90P-1 | Paint P-1 |
| 9.90P-2 | Paint P-2 |
| 9.90P-3 | Paint P-3 |
| 9.90P-4 | Paint P-4 |
| 9.90P-6 | Paint P-6 |
| 9.90P-8 | Paint P-8 |
| 10.10 | Chalkboards and Tackboards |
| 10.10B | Liquid Marker Board |
| 10.10C | Tackboard |
| 10.10F | Projection Screen |
| 10.25 | Firefighting Devices |
| 10.25A | Fire Extinguisher Cabinet. Paint P-2. Model * 2409-R 2 By Larson Manufacturing Co., or Approved Equal |
| 10.42 | Signage and Graphics |
| 10.42A | 12″ × 18″ Aluminum Handicap Sign |
| 10.50 | Lockers |
| 10.50A | Single Tier Lockers |
| 10.50G | Sloping Top |
| 10.50H | Metal Base |
| 10.50I | Recessed Trim |
| 10.62 | Folding Partitions |

10.62A  Folding Partitions

10.62B  Tracks, Support Brackets, Hangar Rods and Finish Trim

10.62D  Liquid Marker Board

10.80   Toilet Accessories

10.80D  Grab Bars

11.46   Unit Kitchens

11.46A  Compact Kitchen Unit

12.17   Entrance Mats

12.17A  Recessed Entrance Mat

12.30   Casework

See Drawings for Casework

12.30A  Plastic Laminate Counter with 4" Backsplash

12.30D  Filler Strip - Size as Required

12.30G  Base Cabinet

12.30I  Lab Countertop

12.30K  Plastic Laminate Top

12.30L  Wall Cabinet

14.20   Elevators

14.20A  Elevator

*See reference noted for information elaborating on or work in addition to materials noted here. Provide all materials as required for a complete and proper installation.

# Appendix B

## MATH REVIEWS

1. Fractions and Mixed numbers—Meanings and Definitions — 352
2. Adding Fractions — 353
3. Adding Combinations of Fractions, Mixed Numbers, and Whole Numbers — 353
4. Subtracting Fractions from Fractions — 354
5. Subtracting Fractions and Mixed Numbers from Whole Numbers — 354
6. Subtracting Fractions and Mixed Numbers from Mixed Numbers — 355
7. Multiplying Fractions — 355
8. Multiplying Any Combination of Fractions, Mixed Numbers, and Whole Numbers — 356
9. Dividing Fractions — 356
10. Dividing any Combination of Fractions, Mixed Numbers, and Whole Numbers — 357
11. Rounding Decimal Fractions — 357
12. Adding Decimal Fractions — 358
13. Subtracting Decimal Fractions — 358
14. Multiplying Decimal Fractions — 358
15. Dividing Decimal Fractions — 359
16. Expressing Common Fractions as Decimal Fractions — 360
17. Expressing Decimal Fractions as Common Fractions — 360
18. Expressing Inches as Feet and Inches — 361
19. Expressing Feet and Inches as Inches — 361
20. Expressing Inches as Decimal Fractions of a Foot — 361
21. Expressing Decimal Fractions of a Foot as Inches — 362
22. Area Measure — 362
23. Volume Measure — 363
24. Finding an Unknown Side of a Right Triangle, Given Two Sides — 366

Appendix B © Cengage Learning 2014

# MATH REVIEW 1  Fractions and Mixed Numbers—Meanings and Definitions

○ A *fraction* is a value that shows the number of equal parts taken from a whole quantity. A fraction consists of a numerator and a denominator.

$\dfrac{7}{16}$ ←Numerator
←Denominator

○ *Equivalent fractions* are fractions that have the same value. The value of a fraction is not changed by multiplying the numerator and denominator by the same number.

**Example** Express $\dfrac{5}{8}$ as thirty-seconds.
Determine what number the denominator is multiplied by to get the desired denominator. (32 ÷ 8 = 4)
Multiply the numerator and denominator by 4.

$\dfrac{5}{8} = \dfrac{?}{32}$

$\dfrac{5}{8} \times \dfrac{4}{4} = \dfrac{20}{32}$

○ The *lowest common denominator of two* or more fractions is the smallest denominator that is evenly divisible by each of the denominators of the fractions.

**Example 1** The lowest common denominator of, $\dfrac{3}{4}, \dfrac{5}{8}, \dfrac{13}{32}$ is 32, because 32 is the smallest number evenly divisible by 4, 8, and 32.

$32 \div 4 = 8$
$32 \div 8 = 4$
$32 \div 32 = 1$

**Example 2** The lowest common denominator of, $\dfrac{2}{3}, \dfrac{1}{5}, \dfrac{7}{10}$ is 30, because 30 is the smallest number evenly divisible by 3, 5, and 10.

$30 \div 3 = 10$
$30 \div 5 = 6$
$30 \div 10 = 3$

○ *Factors* are numbers used in multiplying. For example, 3 and 5 are factors of 15.

$3 \times 5 = 15$

○ A fraction is in its *lowest terms when* the numerator and the denominator **do not** contain a common factor.

**Example** Express $\dfrac{12}{16}$ in lowest terms.
Determine the largest common factor in the numerator and denominator. The numerator and the denominator can be evenly divided by 4.

$\dfrac{12 \div 4}{16 \div 4} = \dfrac{3}{4}$

○ A *mixed number* is a whole number plus a fraction.

$6 \quad \dfrac{15}{16}$
Whole Number ↑      ↑ Fraction

$6 + \dfrac{15}{16} = 6\dfrac{15}{16}$

○ *Expressing fractions as mixed numbers.* In certain fractions, the numerator is larger than the denominator. To express the fraction as a mixed number, divide the numerator by the denominator. Express the fractional part in lowest terms.

**Example** Express $\dfrac{38}{16}$ as a mixed number.16
Divide the numerator 38 by the denominator 16.
Express the fractional part $\dfrac{6}{16}$ in lowest terms.1
Combine the whole number and fraction.

$\dfrac{38}{16} = 2\dfrac{6}{16}$

$\dfrac{6 \div 2}{16 \div 2} = \dfrac{3}{8}$

$\dfrac{38}{16} = 2\dfrac{3}{8}$

- *Expressing mixed numbers as fractions.* To express a mixed number as a fraction, multiply the whole number by the denominator of the fractional part. Add the numerator of the fractional part. The sum is the numerator of the fraction. The denominator is the same as the denominator of the original fractional part.

$$\frac{7 \times 4 + 3}{4} = \frac{31}{4}$$

or

$$\frac{7}{1} \times \frac{4}{4} = \frac{28}{4}$$

$$\frac{28}{4} + \frac{3}{4} = \frac{31}{4}$$

**Example** Express $7\frac{3}{4}$ as a fraction.

Multiply the whole number 7 by the denominator 4 of the fractional part ($7 \times 4 = 28$). Add the numerator 3 of the fractional part to 28. The sum 31 is the numerator of the fraction. The denominator 4 is the same as the denominator of the original fractional part.

## MATH REVIEW 2 Adding Fractions

- Fractions must have a common denominator in order to be added.
- To add fractions, express the fractions as equivalent fractions having the lowest common denominator. Add the numerators and write their sum over the lowest common denominator. Express the fraction in lowest terms.

**Example** Add: $\frac{3}{8} + \frac{3}{4} + \frac{3}{16} + \frac{1}{32}$

Express the fractions as equivalent fractions with 32 as the denominator. Add the numerators.

$$\frac{3}{8} = \frac{3}{8} \times \frac{4}{4} = \frac{12}{32}$$

$$\frac{1}{4} = \frac{1}{4} \times \frac{8}{8} = \frac{8}{32}$$

$$\frac{3}{16} = \frac{3}{16} \times \frac{2}{2} = \frac{6}{32}$$

$$+\frac{1}{32} = \frac{1}{32}$$

$$\frac{27}{32}$$

- After fractions are added, if the numerator is greater than the denominator, the fraction should be expressed as a mixed number.

**Example** Add: $\frac{1}{2} + \frac{3}{4} + \frac{15}{16} + \frac{11}{16}$

Express the fractions as equivalent fractions with 16 as the denominator. Add the numerators.

Express $\frac{46}{16}$ as a mixed number in lowest terms. 16

$$\frac{1}{2} = \frac{1}{2} \times \frac{8}{8} = \frac{8}{16}$$

$$\frac{3}{4} = \frac{3}{4} \times \frac{4}{4} = \frac{12}{16}$$

$$\frac{15}{16} = \frac{15}{16}$$

$$+\frac{11}{16} = \frac{11}{16}$$

$$\frac{46}{16}$$

$$\frac{46}{16} = 2\frac{14}{16} = 2\frac{7}{8}$$

## MATH REVIEW 3 Adding Combinations of Fractions, Mixed Numbers, and Whole Numbers

- To add mixed numbers or combinations of fractions, mixed numbers, and whole numbers, express the fractional parts of the numbers as equivalent fractions having the lowest common denominator. Add the whole numbers. Add the fractions. Combine the whole number and the fraction and express in lowest terms.

**Example 1** Add: $3\frac{7}{8} + 5\frac{1}{2} + 9\frac{3}{16}$

Express the fractional parts as equivalent fractions with 16 as the common denominator. Add the whole numbers. Add the fractions. Combine the whole number and the fraction. Express the answer in lowest terms.

$$3\frac{7}{8} = 3\frac{14}{16}$$
$$5\frac{1}{2} = 5\frac{8}{16}$$
$$+9\frac{3}{16} = 9\frac{3}{16}$$
$$\overline{17\frac{25}{16}} = 17 + 1\frac{9}{16} = 18\frac{9}{16}$$

**Example 2** Add $6\frac{3}{4} + \frac{9}{16} + 7\frac{21}{32} + 15$

Express the fractional parts as equivalent fractions with 32 as the common denominator. Add the whole numbers. Add the fractions. Combine the whole number and the fraction. Express the answer in lowest terms.

$$6\frac{3}{4} = 6\frac{24}{32}$$
$$\frac{9}{16} = \frac{18}{32}$$
$$7\frac{21}{32} = 7\frac{21}{32}$$
$$+15 = 15$$
$$\overline{28\frac{63}{32}} = 28 + 1\frac{31}{32} = 29\frac{31}{32}$$

## MATH REVIEW 4  Subtracting Fractions from Fractions

○ Fractions must have a common denominator in order to be subtracted.

○ To subtract a fraction from a fraction, express the fractions as equivalent fractions having the lowest common denominator. Subtract the numerators. Write their difference over the common denominator.

**Example** Subtract $\frac{3}{4}$ from $\frac{15}{16}$.

Express the fractions as equivalent fractions with 16 as the common denominator. Subtract the numerator 12 from the numerator 15. Write the difference 3 over the common denominator 16.

$$\frac{15}{16} = \frac{15}{16}$$
$$-\frac{3}{4} = -\frac{12}{16}$$
$$\overline{\phantom{xx}\frac{3}{16}}$$

## MATH REVIEW 5  Subtracting Fractions and Mixed Numbers from Whole Numbers

○ To subtract a fraction or a mixed number from a whole number, express the whole number as an equivalent mixed number. The fraction of the mixed number has the same denominator as the denominator of the fraction that is subtracted. Subtract the numerators of the fractions and write their difference over the common denominator. Subtract the whole numbers. Combine the whole number and fraction. Express the answer in lowest terms.

**Example 1** Subtract $\frac{3}{8}$ from 7

Express the whole number as an equivalent mixed number with the same denominator as the denominator of the fraction that is subtracted $\left(7 = 6\frac{8}{8}\right)$.

Subtract $\frac{3}{8}$ from $\frac{8}{8}$

Combine whole number and fraction.

$$
\begin{aligned}
7 &= 6\frac{8}{8} \\
-\frac{3}{8} &= -\frac{3}{8} \\
\hline
&\phantom{=}6\frac{5}{8}
\end{aligned}
$$

**Example 2** Subtract $5\frac{15}{32}$ from 12

Express the whole number as an equivalent mixed number with the same denominator as the denominator of fraction that is subtracted $\left(12 = 11\frac{32}{32}\right)$.
Subtract fractions.
Subtract whole numbers.
Combine whole number and fraction.

$$
\begin{aligned}
12 &= 11\frac{32}{32} \\
-5\frac{15}{32} &= -5\frac{15}{32} \\
\hline
&\phantom{=}6\frac{17}{32}
\end{aligned}
$$

## MATH REVIEW 6  Subtracting Fractions and Mixed Numbers from Mixed Numbers

○ To subtract a fraction or a mixed number from a mixed number, the fractional part of each number must have the same denominator. Express fractions as equivalent fractions having a common denominator. When the fraction subtracted is larger than the fraction from which it is subtracted, one unit of a whole number is expressed as a fraction with the common denominator. Combine the whole number and fractions. Subtract fractions and subtract whole numbers.

**Example 1** Subtract $\frac{7}{8}$ from $4\frac{3}{16}$

Express the fractions as equivalent fractions with the common denominator 16.

Since 14 is larger than 3, express one unit of $4\frac{3}{16}$ as a fraction and combine whole number and fractions.

$\left(4\frac{3}{16} = 3 + \frac{16}{16} + \frac{3}{16} = 3\frac{19}{16}\right)$.

Subtract.

$$
\begin{aligned}
4\frac{3}{16} &= 4\frac{3}{16} = 3\frac{19}{16} \\
-\frac{7}{8} &= \frac{14}{16} = -\frac{14}{36} \\
\hline
&\phantom{=}3\frac{5}{16}
\end{aligned}
$$

**Example 2** Subtract 13 from 20
Express the fractions as equivalent fractions with the common denominator 32.
Subtract fractions.
Subtract whole numbers.

$$
\begin{aligned}
20\frac{15}{32} &= 20\frac{15}{32} \\
-13\frac{1}{4} &= -13\frac{8}{32} \\
\hline
&\phantom{=}7\frac{7}{32}
\end{aligned}
$$

## MATH REVIEW 7  Multiplying Fractions

○ To multiply two or more fractions, multiply the numerators. Multiply the denominators. Write as a fraction with the product of the numerators over the product of the denominators. Express the answer in lowest terms.

**Example 1** Multiply $\frac{3}{4} \times \frac{5}{8}$
Multiply the numerators.
Multiply the denominators.
Write as a fraction.

$$\frac{3}{4} \times \frac{5}{8} = \frac{15}{32}$$

**Example 2** Multiply
Multiply the numerators.
Multiply the denominators.
Write as a fraction and express answer in lowest terms.

$$\frac{1}{2} \times \frac{2}{3} \times \frac{4}{5} = \frac{8}{30} = \frac{4}{15}$$

## MATH REVIEW 8 Multiplying any Combination of Fractions, Mixed Numbers, and Whole Numbers

○ To multiply any combination of fractions, mixed numbers, and whole numbers, write the mixed numbers as fractions. Write whole numbers over the denominator 1. Multiply numerators. Multiply denominators. Express the answer in lowest terms.

**Example 1** Multiply $3\frac{1}{4} \times \frac{3}{8}$

Write the mixed number $3\frac{1}{4}$ as the fraction $\frac{13}{4}$.
Multiply the numerators.
Multiply the denominators.
Express as a mixed number.

$$3\frac{1}{4} \times \frac{3}{8} = \frac{13}{4} \times \frac{3}{8} = \frac{39}{32} = 1\frac{7}{32}$$

**Example 2** Multiply $2\frac{1}{3} \times 4 \times \frac{4}{5}$

Write the mixed number $2\frac{1}{3}$ as the fraction $\frac{7}{3}$.
Write the whole number 4 over 1.
Multiply the numerators.
Multiply the denominators.
Express as a mixed number.

$$2\frac{1}{3} \times 4 \times \frac{4}{5} = \frac{7}{3} \times \frac{4}{1} \times \frac{4}{5} = \frac{112}{15}$$

$$\frac{112}{15} = 7\frac{7}{15}$$

## MATH REVIEW 9 Dividing Fractions

○ Division is the inverse of multiplication. Dividing by 4 is the same as $\frac{1}{4}$. multiplying by j. 4 is the inverse of $\frac{1}{4}$.and is the inverse of 4.

The inverse of is $-\frac{5}{16}$ is $\frac{16}{5}$.

○ To divide fractions, invert the divisor, change to the inverse operation and multiply. Express the answer in lowest terms.

362 Appendix B

**Example** Divide: $\dfrac{7}{8} \div \dfrac{2}{3}$

Invert the divisor $\dfrac{2}{3}$

$\dfrac{2}{3}$ inverted is $\dfrac{3}{2}$.

Change to the inverse operation and multiply.
Express as a mixed number.

$$\dfrac{7}{8} \div \dfrac{2}{3} = \dfrac{7}{8} \times \dfrac{3}{2} = \dfrac{21}{16} = 1\dfrac{5}{16}$$

## MATH REVIEW 10 Dividing any Combination of Fraction, Mixed Numbers, and Whole Numbers

○ To divide any combination of fractions, mixed numbers, and whole numbers, write the mixed number as fractions. Write whole numbers over the denominator 1. Invert the divisor. Change to the inverse operation and multiply. Express the answer in lowest terms.

**Example 1** Divide: $6 \div \dfrac{7}{10}$.

Write the whole number 6 over the denominator 1.

Invert the divisor $\dfrac{7}{10}$; $\dfrac{7}{10}$ inverted is $\dfrac{10}{7}$.

Change to the inverse operation and multiply.
Express as a mixed number.

$$\dfrac{6}{1} \div \dfrac{7}{10} =$$

$$\dfrac{6}{1} \times \dfrac{10}{7} = \dfrac{60}{7} = 8\dfrac{4}{7}$$

**Example 2** Divide: $\dfrac{3}{4} \div 2\dfrac{1}{5}$

Write the mixed number divisor $2\dfrac{1}{5}$ as the fraction $\dfrac{11}{5}$.

Invert the divisor $\dfrac{11}{5}$; $\dfrac{11}{5}$ inverted is $-\dfrac{5}{11}$

Change to the inverse operation and multiply.

$$\dfrac{3}{4} \div \dfrac{11}{5} =$$

$$\dfrac{3}{4} \times \dfrac{5}{11} = \dfrac{15}{44}$$

**Example 3** Divide: $4\dfrac{5}{8} \div 7$

Write the mixed number $4\dfrac{5}{8}$ as the fraction $\dfrac{37}{8}$

Write the whole number divisor over the denominator 1.

Invert the divisor $\dfrac{7}{1}$; $\dfrac{7}{1}$ inverted is $\dfrac{1}{7}$.

Change to the inverse operation and multiply.

$$\dfrac{37}{8} \div \dfrac{7}{1} =$$

$$\dfrac{37}{8} \times \dfrac{1}{7} = \dfrac{37}{56}$$

## MATH REVIEW 11 Rounding Decimal Fractions

○ To round a decimal fraction, locate the digit in the number that gives the desired number of decimal places. Increase that digit by 1 if the digit that directly follows is 5 or more. Do not change the value of the digit if the digit that follows is less than 5. Drop all digits that follow.

**Example 1** Round 0.63861 to 3 decimal places.  
Locate the digit in the third place (8). The fourth decimal-place digit, 6, is greater than 5 and increases the third decimal-place digit 8, to 9. Drop all digits that follow.

$0.63\underline{8}61 \approx 0.639$

**Example 2** Round 3.0746 to 2 decimal places.  
Locate the digit in the second decimal place (7). The third decimal-place digit 4 is less than 5 and does not change the value of the second decimal-place digit 7. Drop all digits that follow.

$3.07\underline{4}6 \approx 3.07$

## MATH REVIEW 12 Adding Decimal Fractions

○ To add decimal fractions, arrange the numbers so that the decimal points are directly under each other. The decimal point of a whole number is directly to the right of the last digit. Add each column as with whole numbers. Place the decimal point in the sum directly under the other decimal points.

**Example** Add: 7.65 + 208.062 + 0.009 + 36 + 5.1037  
Arrange the numbers so that the decimal points are directly under each other.  
Add zeros so that all numbers have the same number of places to the right of the decimal point.  
Add each column of numbers.  
Place the decimal point in the sum directly under the other decimal points.

```
   7.6500
 208.0620
   0.0090
  36.0000
 + 5.1037
 ─────────
 256.8247
```

## MATH REVIEW 13 Subtracting Decimal Fractions

○ To subtract decimal fractions, arrange the numbers so that the decimal points are directly under each other. Subtract each column as with whole numbers. Place the decimal point in the difference directly under the other decimal points.

**Example** Subtract: 87.4 − 42.125  
Arrange the numbers so that the decimal points are directly under each other. Add zeros so that the numbers have the same number of places to the right of the decimal point.  
Subtract each column of numbers.  
Place the decimal point in the difference directly under the other decimal points.

```
  87.400
 -42.125
 ───────
  45.275
```

## MATH REVIEW 14 Multiplying Decimal Fractions

○ To multiply decimal fractions, multiply using the same procedure as with whole numbers. Count the number of decimal places in both the multiplier and the multiplicand. Begin counting from the last digit on the right of the product and place the decimal point the same number of places as there are in both the multiplicand and the multiplier.

**Example** Multiply: 50.216 × 1.73
Multiply as with whole numbers.
Count the number of decimal places in the multiplier (2 places) and the multiplicand (3 places).
Beginning at the right of the product, place the decimal point the same number of places as there are in both the multiplicand and the multiplier (5 places).

```
                  Multiplicand
    50.216 ←      (3 places)
   × 1.73 ←       Multiplier
   150648         (2 places)
   351512
    50216
   86.87368       (5 places)
```

○ When multiplying certain decimal fractions, the product has a smaller number of digits than the number of decimal places required. For these products, add as many zeros to the left of the product as are necessary to give the required number of decimal places.

**Example** Multiply: 0.27 × 0.18
Multiply as with whole numbers.
The product must have 4 decimal places.
Add one zero to the left of the product.

```
     0.27 (2 places)
    ×0.18 (2 places)
      216
       27
   0.0486 (4 places)
```

## MATH REVIEW 15  Dividing Decimal Fractions

○ To divide decimal fractions, use the same procedure as with whole numbers. Move the decimal point of the divisor as many places to the right as necessary to make the divisor a whole number. Move the decimal point of the dividend the same number of places to the right. Add zeros to the dividend if necessary. Place the decimal point in the answer directly above the decimal point in the dividend. Divide as with whole numbers. Zeros may be added to the dividend to find the number of decimal places required in the answer.

**Example 1**  Divide: 0.6150 ÷ 0.75
Move the decimal point 2 places to the right in the divisor.
Move the decimal point 2 places in the dividend.
Place the decimal point in the answer directly above the decimal point in the dividend.
Divide as with whole numbers.

```
                        0.82
   Divisor → 0 75. )0 61.50  ← Dividend
                    60 0
                     1 50
                     1 50
```

**Example 2**  Divide: 10.7 ÷ 4.375. Round the answer to 3 decimal places.
Move the decimal point 3 places to the right in the divisor.
Move the decimal point 3 places in the dividend, adding 2 zeros.
Place the decimal point in the answer directly above the decimal point in the dividend.
Add 4 zeros to the dividend. One more zero is added than the number of decimal places required in the answer.
Divide as with whole numbers.

```
                  2.4457 ≈ 2.446
   4 375. )10 700.0000
            8 750
            1 950 0
            1 750 0
              200 00
              175 00
               25 00
               21 875
                3 1250
                3 0625
                  625
```

## MATH REVIEW 16 Expressing Common Fractions as Decimal Fractions

○ A common fraction is an indicated division. A common fraction is expressed as a decimal fraction by dividing the numerator by the denominator.

**Example** Express 5/8 as a decimal fraction.
Write 5/8 as an indicated division.
Place a decimal point after the 5 and add zeros to the right of the decimal point.
Place the decimal point for the answer directly above the decimal point in the dividend.
Divide.

$$8\overline{)5}$$
$$8\overline{)5.000}$$
$$8\overline{)5.000}$$
$$8\overline{)5.000}^{\,0.625}$$

○ A common fraction which will not divide evenly is expressed as a repeating decimal.

**Example** Express $\frac{1}{3}$ as a decimal fraction.
Write $\frac{1}{3}$ as an indicated division.
Place a decimal point after the 1 and add zeros to the right of the decimal point. Place the decimal point for the answer directly above the decimal point in the dividend.
Divide.

$$3\overline{)1}$$
$$3\overline{)1.0000}$$
$$3\overline{)1.0000}$$
$$3\overline{)1.0000}^{\,0.3333}$$

## MATH REVIEW 17 Expressing Decimal Fractions as Common Fractions

○ To express a decimal fraction as a common fraction, write the number after the decimal point as the numerator of a common fraction. Write the denominator as 1 followed by as many zeros as there are digits to the right of the decimal point. Express the common fraction in lowest terms.

**Example 1** Express 0.9 as a common fraction.
Write 9 as the numerator.
Write the denominator as 1 followed by 1 zero. The denominator is 10.

$$\frac{9}{10}$$

**Example 2** Express 0.125 as a common fraction.
Write 125 as the numerator.
Write the denominator as 1 followed by 3 zeros. The denominator is 1,000.
Express the fraction in lowest terms.

$$\frac{125}{1000}$$
$$\frac{125}{1000} = \frac{1}{8}$$

## MATH REVIEW 18 Expressing Inches as Feet and Inches

- There are 12 inches in 1 foot.
- To express inches as feet and inches, divide the given length in inches by 12 to obtain the number of whole feet. The remainder is the number of inches in addition to the number of whole feet. The answer is the number of whole feet plus the remainder in inches.

**Example 1** Express $176\frac{7}{16}$ inches as feet and inches.

Divide $176\frac{7}{16}$ inches by 12.

There are 14 feet plus a remainder of $8\frac{7}{16}$ inches.

$$\begin{array}{r} 14 \quad \text{(feet)} \\ 12\overline{)176\frac{7}{16}} \\ \underline{12\phantom{0}} \\ 56 \\ \underline{48} \\ 8\frac{7}{16} \leftarrow \text{Reminder} \\ \text{(feet)} \end{array}$$

$14' - 8\frac{7}{16}''$

**Example 2** Express 54.2 inches as feet and inches.
Divide 54.2 inches by 12.
There are 4 feet plus a remainder of 6.2 inches.

$$\begin{array}{r} 4 \quad \text{(feet)} \\ 12\overline{)54.2} \\ \underline{48\phantom{.0}} \\ 6.2 \leftarrow \text{Reminder} \\ \text{(inches)} \end{array}$$

4 feet 6.2 inches

## MATH REVIEW 19 Expressing Feet and Inches as Inches

- There are 12 inches in 1 foot.
- To express feet and inches as inches, multiply the number of feet in the given length by 12. To this product, add the number of inches in the given length.

**Example** Express 7 feet $9\frac{3}{4}$ inches as inches.

Multiply 7 feet by 12. There are 84 inches in 7 feet.

Add $9\frac{3}{4}$ inches to 84 inches.

$7 \times 12 = 84$
7 feet $= 84$ inches
$84$ inches $+ 9\frac{3}{4}$ inches $=$
$93\frac{3}{4}$ inches

## MATH REVIEW 20 Expressing Inches as Decimal Fractions of a Foot

- An inch is $\frac{1}{12}$ of a foot. To express whole inches as a decimal part of a foot, divide the number of inches by 12.

**Example** Express 7 inches as a decimal fraction of a foot.
Divide 7 by 12.

$7 \div 12 = 0.58$
0.58 feet

- To express a common fraction of an inch as a decimal fraction of a foot, express the common fraction as a decimal, and then divide the decimal by 12.

    **Example 1** Express $\frac{3}{4}$ inch as a decimal fraction of a foot. Expresses $\frac{3}{4}$ a decimal. Divide the decimal by 12.

    $3 \div 4 = 0.75$
    $0.75 \div 12 = 0.06$
    0.06 feet

    **Example 2** Express $4\frac{3}{4}$ inches as a decimal fraction of a foot.

    Express $4\frac{3}{4}$ as a decimal.

    Divide the decimal inches by 12.

    $4 + \frac{3}{4} = 4 + 0.75 = 4.75$
    $4.75 \div 12 = 0.39$
    0.39 feet

## MATH REVIEW 21 Expressing Decimal Fractions of a Foot as Inches

- To express a decimal part of a foot as decimal inches, multiply by 12.

    **Example:** Express 0.62 foot as inches. Multiply 0.62 by 12.

    $0.62 \times 12 = 7.44$
    7.44 inches

- To express a decimal fraction of an inch as a common fraction, see Math Review 17.

## MATH REVIEW 22 Area Measure

- A surface is measured by determining the number of surface units contained in it. A surface is two-dimensional. It has length and width but no thickness. Both length and width must be expressed in the same unit of measure. Area is expressed in square units. For example, 5 feet × 8 feet equals 40 square feet.

- *Equivalent Units of Area Measure:*
    1 square foot (sq ft) = 12 inches × 12 inches = 144 square inches (sq in)
    1 square yard (sq yd) = 3 feet × 3 feet = 9 square feet (sq ft)

- To express a given unit of area as a larger unit of area, divide the given area by the number of square units contained in one of the larger units.

    **Example 1** Express 648 square inches as square feet. Since 144 sq in = 1 sq ft, divide 648 by 144.

    $648 \div 144 = 4.5$
    648 square inches = 4.5 square feet

    **Example 2** Express 28.8 square feet as square yards. Since 9 sq ft = 1 sq yd, divide 28.8 by 9.

    $28.8 \div 9 = 3.2$
    28.8 square feet = 3.2 square yards

- To express a given unit of area as a smaller unit of area, multiply the given area by the number of square units contained in one of the larger units.

**Example 1** Express 7.5 square feet as square inches.
Since 144 sq in = 1 sq ft, divide 7.5 by 144.

7.5 × 144 = 1080
23 square yards = 207 square feet

**Example 2** Express 23 square yards as square feet.
Since 9 sq ft = 1 sq yd, multiply 23 by 9.

23 × 9 = 207
23 square yards = 207 square feet

○ *Computing Areas of Common Geometric Figures:*

1. **Rectangle** A rectangle is a four-sided plane figure with four right (90°) angles.

   The area of a rectangle is equal to the product of its length and its width. Area = length × width (A = l × w)

   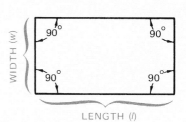

   **Example** Find the area of a rectangle 24 feet long and 13 feet wide.
   A = l × w
   A = 24 ft × 13 ft
   A = 312 square feet

2. **Triangle** A triangle is a plane figure with three sides and three-angles. The area of a triangle is equal to one-half the product of its base and altitude.
   $A = \frac{1}{2}$ base × altitude (A = b × a)

   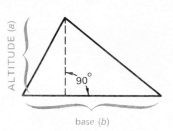

   **Example** Find the area of a triangle with a base of 16 feet and an altitude of 12 feet.
   $A = \frac{1}{2} b \times a$
   $A = \frac{1}{2} \times 16 \text{ ft} \times 12 \text{ ft}$
   A = 96 square feet

3. **Circle** The area of a circle is equal to *p* times the square of its radius. Area = π × radius² (A = π × r²)
   Note: π (pronounced "pi") is approximately equal to 3.14. Radius squared (r²) means r × r.

   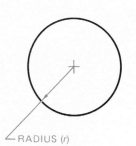

   **Example** Find the area of a circle with a 15-inch radius.
   A = π × r²
   A = 3.14 × (15 in)²
   A = 3.14 × 225 sq in
   A = 706.5 square inches

## MATH REVIEW 23 Volume Measure

○ A solid is measured by determining the number of cubic units contained in it. A solid is three-dimensional; it has length, width, and thickness or height. Length, width, and thickness must be expressed in the same unit of measure. Volume is expressed in cubic units. For example, 3 feet × 5 feet × 10 feet = 150 cubic feet.

- *Equivalent Units of Volume Measure:*
  1 cubic foot (cu ft) =
    12 in × 12 in × 12 in = 1728 cubic inches (cu in)
  1 cubic yard (cu yd) =
    3 ft × 3 ft × 3 ft = 27 cubic feet (cu ft)

- To express a given unit of volume as a larger unit of volume, divide the given volume by the number of cubic units contained in one of the larger units.

  **Example 1**  Express 6,048 cubic inches as cubic feet
  Since 1728 cu in = 1 cu ft, divide 6048 by 1728.

  6048 ÷ 1728 = 3.5
  6048 cubic inches = 3.5 cubic feet

  **Example 2**  Express 167.4 cubic feet as cubic yards.
  Since 17 cu ft = 1 cu yd, divide 167.4 by 27.

  167.4 ÷ 27 = 6.2
  167.4 cubic feet = 6.2 yards

- To express a given unit of volume as a smaller unit of volume, multiply the given volume by the number of cubic units contained in one of the larger units.

  **Example 1**  Express 1.6 cubic feet as cubic inches.
  Since 1728 cu in = 1 cu ft, divide 1.6 by 1728.

  1.6 × 1728 = 2764.8
  1.6 cubic feet = 2764.8 cubic inches

  **Example 2**  Express 8.1 cubic yards as cubic feet.
  Since 27 cu ft = 1 cu yd, divide 8.1 by 27.

  8.1 × 27 = 218.7
  8.1 cubic yards = 218.7 cubic feet

- *Computing Volumes of Common Solids*

- A prism is a solid that has two identical faces called bases and parallel lateral edges. In a right prism, the lateral edges are perpendicular (at 90°) to the bases. The altitude or height ($h$) of a prism is the perpendicular distance between its two bases. Prisms are named according to the shapes of their bases.

- The volume of any prism is equal to the product of the area of its base and altitude or height.
  Volume = area of base 3 altitude ($V = AB \times h$)

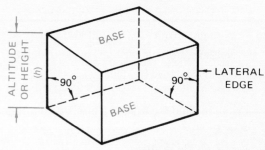

- *Right Rectangular Prism:*
  A right rectangular prism has rectangular bases.
  Volume = area of base 3 altitude
    $V = A_B \times h$

  **Example**  Find the volume of a rectangular prism with a base length of 20 feet, a base width of 14 feet, and a height (altitude) of 8 feet.
  $V = A_B \times h$
  Compute the area of the base ($A_B$):
  Area of base = length × width
  $A_B = 20 \text{ ft} \times 14 \text{ ft}$
  $A_B = 280 \text{ sq ft}$

Compute the volume of the prism:
$V = A_B \times h$
$V = 280 \text{ sq ft} \times 8 \text{ ft}$
$V = 2{,}240 \text{ cu ft}$

○ *Right Triangular Prism:*
A right triangular prism has triangular bases.
Volume = area of base × altitude
$V = A_B \times h$

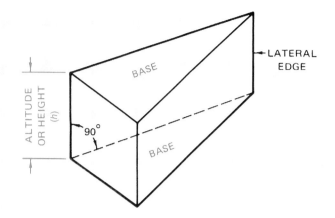

**Example** Find the volume of a triangular prism in which the base of the triangle is 5 feet, the altitude of the triangle is 3 feet, and the altitude (height) of the prism is 4 feet. Refer to the accompanying figure.
Volume = area of base × altitude
$V = A_B \times h$
Compute the area of the base:
Area of base = $\frac{1}{2}$base of triangle × altitude of triangle
$A_B = \frac{1}{2} b \times a$
$A_B = \frac{1}{2} \times 5 \text{ ft} \times 3 \text{ ft}$
$A_B = 7.5 \text{ sq ft}$
Compute the volume of the prism:
$V = A_B \times h$
$V = 7.5 \text{ sq ft} \times 4 \text{ ft}$
$V = 30 \text{ cubic feet}$

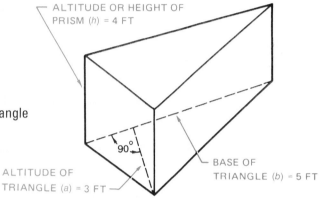

○ *Right Circular Cylinder:*
A right circular cylinder has circular bases.
Volume = area of base × altitude
$V = A_B \times h$

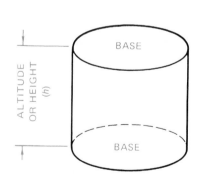

**Example** Find the volume of a circular cylinder 1 foot in diameter and 10 feet high.

Note: Radius = $\frac{1}{2}$ Diameter; Radius = 1 × 1 ft = 0.5 ft.
$V = A_B \times h$
Compute the area of the base:
Area of base = π × radius squared
$A_B = 3.14 \times (0.5)^2$
$A_B = 3.14 \times 0.5 \text{ ft} \times 0.5 \text{ ft}$
$A_B = 3.14 \times 0.25 \text{ sq ft}$
$A_B = 0.785 \text{ sq ft}$
Compute the volume of the cylinder:
$V = A_B \times h$
$V = 0.785 \text{ sq ft} \times 10 \text{ ft}$
$V = 7.85 \text{ cubic feet}$

# MATH REVIEW 24 Finding an Unknown Side of a Right Triangle, Given Two Sides

○ If one of the angles of a triangle is a right (90°) angle, the figure is called a *right triangle*. The side opposite the right angle is called the *hypotenuse*. In the figure shown, c is opposite the right angle; c is the hypotenuse.

○ In a right triangle, the square of the hypotenuse is equal to the sum of the squares of the other two sides:

$$c^2 = a^2 + b^2$$

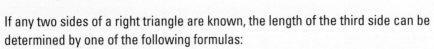

If any two sides of a right triangle are known, the length of the third side can be determined by one of the following formulas:

$$c = \sqrt{a^2 + b^2}$$
$$a = \sqrt{c^2 - b^2}$$
$$b = \sqrt{c^2 - a^2}$$

**Example 1** In the right triangle shown, $a = 6$ ft, $b = 8$ ft, find $c$.

$$c = \sqrt{a^2 + b^2}$$
$$c = \sqrt{6^2 + 8^2}$$
$$c = \sqrt{36 + 64}$$
$$c = \sqrt{100}$$
$$c = 10 \text{ feet}$$

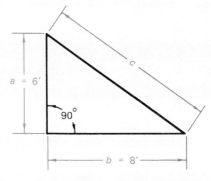

**Example 2** In the right triangle shown, $c = 30$ ft, $b = 20$ ft, find $a$.

$$a = \sqrt{c^2 - b^2}$$
$$a = \sqrt{30^2 - 20^2}$$
$$a = \sqrt{900 - 400}$$
$$a = \sqrt{500}$$
$$a = 22.36 \text{ feet (to two decimal places)}$$

**Example 3** In the right triangle shown, $c = 18$ ft, $a = 6$ ft, find $b$.

$$b = \sqrt{c^2 - a^2}$$
$$b = \sqrt{18^2 - 6^2}$$
$$b = \sqrt{324 - 36}$$
$$b = \sqrt{288}$$
$$b = 16.97 \text{ feet (to two decimal places)}$$

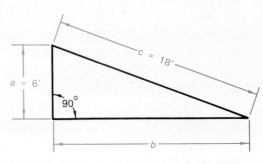

# Appendix C

## MATERIAL SYMBOLS IN SECTIONS

EARTH

ROCK

GRAVEL OR CRUSHED STONE

CONCRETE

CONCRETE BLOCK

FACE BRICK OR COMMON BRICK

FIRE BRICK

ROUGH WOOD-CONTINUOUS

WOOD BLOCKING

FINISH WOOD

STRUCTURAL STEEL

REINFORCING BARS

GENERAL METAL

BATT INSULATION

RIGID INSULATION

PLASTER OR GYPSUM BOARD

THIN SHEET MATERIALS (PLASTIC FILM, SHEET METAL, PAPER, ETC.)

# Appendix D

## PLUMBING SYMBOLS

**PIPING**

- DRAIN OR WASTE ABOVE GROUND
- DRAIN OR WASTE BELOW GROUND
- VENT
- COLD WATER
- HOT WATER
- HOT WATER HEAT SUPPLY —— HW —— HW ——
- HOT WATER HEAT RETURN —— HWR —— HWR ——
- GAS —— G —— G ——
- PIPE TURNING DOWN OR AWAY
- PIPE TURNING UP OR TOWARD
- BREAK—PIPE CONTINUES

| FITTINGS | SOLDERED | SCREWED |
|---|---|---|
| TEE | | |
| WYE | | |
| ELBOW – 90° | | |
| ELBOW – 45° | | |
| CAP | | |
| UNION CLEANOUT | | |
| STOP VALVE | | |

# Appendix E

## ELECTRICAL SYMBOLS

| | | | |
|---|---|---|---|
| CEILING FIXTURE | ⊕ | TELEPHONE | ◁ |
| CEILING FIXTURE WITH PULL SWITCH | ⊕PS | INTERCOM | ◁ |
| WALL MOUNTED FIXTURE | ⊢○ | TELEVISION ANTENNA | ◁TV |
| RECESSED CEILING FIXTURE – OUTLINE SHOWS SHAPE | [○] | SMOKE DETECTOR | ⊙ |
| FLOURESCENT FIXTURE | [○___] | DISTRIBUTION PANEL | ▬ |
| FAN OUTLET | F or ◐ | JUNCTION BOX | J |
| CONVENIENCE DUPLEX OUTLET | ⊖ | SINGLE-POLE SWITCH | S |
| SPLIT WIRED DUPLEX OUTLET | ⊖ | THREE-WAY SWITCH | $S_3$ |
| WEATHERPROOF OUTLET | ⊖WP | SWITCH WITH PILOT LIGHT | $S_p$ |
| OUTLET WITH GROUND FAULT INTERRUPTER | ⊖GFI | WEATHERPROOF SWITCH | $S_{WP}$ |
| SPECIAL-EQUIPMENT OUTLET | ◀ | SWITCH WIRING | ─ ─ or ─ ─ ─ |
| RANGE OUTLET | ⊖R | | |
| PUSH BUTTON | [•] | | |
| CHIME | [CH] | | |
| TRANSFORMER | [T] | | |

# Appendix F

## ABBREVIATIONS

**Note:** These standard abbreviations can appear on construction drawings and like documentation with or without periods (e.g., AC or A.C.).

**AB**—anchor bolt
**AC**—air conditioning
**AL or ALUM**—aluminum
**BA**—bathroom
**BLDG**—building
**BLK**—block
**BLKG**—blocking
**BM**—beam
**BOTT**—bottom
**BPL**—base plate
**BR**—bedroom
**BRM**—broom closet
**BSMT**—basement
**CAB**—cabinet
**CL**—centerline
**CLNG or CLG**—ceiling
**CMU**—concrete masonry unit (concrete block)
**CNTR**—center or counter
**COL**—column
**COMP**—composition
**CONC**—concrete
**CONST**—construction
**CONT**—continuous
**CORRUG**—corrugated

**CRNRS**—corners
**CU**—copper
**d**—penny (nail size)
**DBL**—double
**DET**—detail
**DIA or ⌀**—diameter
**DIM**—dimension
**DN**—down
**DO**—ditto
**DP**—deep or depth
**DR**—door
**DW**—dishwasher
**ELEC**—electric
**ELEV**—elevation
**EQ**—equal
**EXP**—exposed or expansion
**EXT**—exterior
**FG**—fuel gas
**FIN**—finish
**FL or FLR**—floor
**FOUND or FDN**—foundation
**FP**—fireplace
**FT**—foot or feet
**FTG**—footing
**GAR**—garage
**GFCI**—ground fault circuit interrupter
**GI**—galvanized iron
**GL**—glass
**GRD**—grade

**GYPBD**—gypsum board
**HC**—hollow core door
**HCW**—hollow core wood
**HDR**—header
**HM**—hollow metal
**HORIZ**—horizontal
**HT or HGT**—height
**HW**—hot water
**HWM**—high water mark
**IN**—inch or inches
**INSUL**—insulation
**INT**—interior
**JSTS**—joists
**JT**—joint
**LAV**—lavatory
**LH**—left hand
**LIN**—linen closet
**LT**—light
**MANUF**—manufacturer
**MAS**—masonry
**MATL**—material
**MAX**—maximum
**MIN**—minimum
**MTL**—metal
**NAT**—natural
**N/F**—now or formerly
**NIC**—not in contract
**o/**—overhead or over
**OC**—on centers
**OH Door**—overhead door

**OSB**—Oriented strand board
**PERF**—perforated
**PL**—plate
**PLYWD**—plywood
**PT**—pressure-treated lumber
**R**—risers
**REF**—refrigerator
**REINF**—reinforcement
**REQ**—requirement
**RH**—right hand
**RM**—room
**ROB**—run of bank (gravel)
**ROW**—right of way

**SCRND**—screened
**SHT**—sheet
**SHTG**—sheathing
**SHWR**—shower
**SIM**—similar
**SL**—sliding
**S&P**—shelf and pole
**SPEC**—specifications
**SQ** or ⌧—square
**STD**—standard
**STL**—steel
**STY**—story
**T&G**—tongue and groove

**THK**—thick
**T'HOLD**—threshold
**TYP**—typical
**VB**—vapor barrier
**w/**—with
**WARD**—wardrobe
**WC**—water closet
**WD**—wood
**WDW**—window
**WH**—water heater
**WI**—wrought iron

# Glossary

## A

**Addendum**—A change or modification to the bid documents, plans, and specifications that is made prior to the contractor's bid date

**Aggregate**—Hard materials such as sand and crushed stone used to make concrete

**Ampere (amp)**—Unit of measure of electric current

**Anchor Bolt**—A bolt placed in the surface of concrete for attaching wood framing members

**Apron**—Concrete slab at the approach to a garage door—also the wood trim below a window stool

**Architect's Scale**—A flat or triangular scale used to measure scale drawings

**Ash Dump**—A small metal door in the bottom of a fireplace

**Awning Window**—A window that is hinged near the top so the bottom opens outward

## B

**Backfill**—Earth placed against a building wall after the foundation is in place

**Backsplash**—The raised lip on the back edge of a countertop to prevent water from running down the backs of the cabinets

**Balloon Framing**—Type of construction in which the studs are continuous from the sill to the top of the wall—upper floor joists are supported by a let-in ribbon

**Balusters**—Vertical pieces that support a handrail

**Balustrade**—The complete assembly of railings, balusters, and newel posts around a stair or balcony

**Batt Insulation**—Flexible, blanket-like pieces, usually of fiber-glass, used for thermal or sound insulation

**Batten**—Narrow strip of wood used to cover joints between boards of sheet materials

**Batter Boards**—An arrangement of stakes and horizontal pieces used to attach lines for laying out a building

**Beam**—Any major horizontal structural member

**Beam Pocket**—A recessed area to hold the end of a beam in a concrete or masonry wall

**Board Foot**—144 cubic inches of wood or the amount contained in a piece measuring 12″ × 12″ × 1″

**Bottom Chord**—The bottom horizontal member in a truss

**Box Sill**—The header joist nailed across the ends of floor joists at the sill

**Branch Circuit**—The electrical circuit that carries current from the distribution panel to the various parts of the building

**British Thermal Unit (BTU)**—The amount of heat required to raise the temperature of 1 pound of water 1° Fahrenheit

**Building Lines**—The outside edge of the exterior walls of a building

## C

**Casement Window**—A window that is hinged at one side so the opposite side opens outward

**Casing**—The trim around a door or window

**Centerline**—An actual or imaginary line through the exact center of any object

**Change Order**—A change or modification to the contract documents, plans, and specifications that is made after the contract has been awarded to the selected trade contractor

**Cleanout**—A pipe fitting with a removable plug that allows for cleaning the run of piping in which it is installed, or an access door at the bottom of a chimney

**Collar Beam**—Horizontal members that tie opposing rafters together, usually installed about halfway up the rafters

**Column**—A metal post to support an object above

**Common Rafter**—A rafter extending from the top of the wall to the ridge

**Concrete**—Building material consisting of fine and coarse aggregates bonded together by Portland cement

**Conductor**—Electrical wire—a cable may contain several conductors

**Contour Lines**—Lines on a topographic map or site plan to describe the contour of the land

**Contract**—Any agreement in writing for one party to perform certain work and the other party to pay for the work

**Convenience Outlet**—Electrical outlet provided for convenient use of lamps, appliances, and other electrical equipment

**Cornice**—The construction that encloses the ends of the rafters at the top of the wall

**Cornice Return**—The construction where the level cornice meets the sloping rake cornice

**Course**—A single row of building units such as concrete blocks or shingles

**Cove Mold**—Concave molding used to trim an inside corner

## D

**Damper**—A door installed in the throat of a fireplace to regulate the draft

**Dampproofing**—Vapor barrier or coating on foundation walls or under concrete slabs to prevent moisture from entering the house

**Datum**—A reference point from which elevations are measured

***Delivery Sheet (Trusses)***—A summary sheet included with a packet of truss drawings to show how many of each type of truss is required

***Detail***—A drawing showing special information about a particular part of the construction—details are usually drawn to a larger scale than other drawings and are sometimes section views

***Dormer***—A raised section in a roof to provide extra headroom below

***Double-hung Window***—A window consisting of two sashes that slide up and down past one another

***Drip Cap***—A wood ledge over wall openings to prevent water from running back under the frame or trim around the opening

***Drip Edge***—Metal trim installed at the edge of a roof to stop water from running back under the edge of the roof deck

***Drywall***—Interior wall construction using gypsum wallboard

## E

***Elevation***—A drawing that shows vertical dimensions—it may also be the height of a point, usually in feet, above sea level

## F

***Fascia***—The part of a cornice that covers the ends of the rafters

***Firestop***—Blocking or noncombustible material between wall studs to prevent vertical draft and flame spread

***Flashing***—Sheet metal used to cover openings and joints in walls and roofs

***Float***—To level concrete before it begins to cure—floating is done with a tool called a *float*

***Floor Plan***—A drawing showing the arrangement of rooms, the locations of windows and doors, and complete dimensions—a floor plan is actually a horizontal section through the entire building

***Flue***—The opening inside a chimney—the flue is usually formed by a terra cotta flue liner

***Flush Door***—A door having flat surfaces

***Footing***—The concrete base upon which the foundation walls are built

***Footing Drain***—(Also called *perimeter drain*) An underground drain pipe around the footings to carry groundwater away from the building

***Frieze***—A horizontal board beneath the cornice and against the wall above the siding

***Frostline***—The maximum depth to which frost penetrates the earth

***Furring***—Narrow strips of wood attached to a surface for the purpose of creating a plumb or level surface for attaching the wall, ceiling, or floor surface

## G

***Gable***—The triangular area between the roof and the top plate walls at the ends of a gable roof

***Gable Studs***—The studs placed between the end rafters and the top plates of the end walls

***Gauge***—A standard unit of measurement for the diameter of wire or the thickness of sheet metal

***Girder***—A beam that supports floor joists

***Grout***—A thin mixture of high-strength concrete or mortar

***Gypsum Wallboard***—Drywall material made of gypsum encased in paper to form boards

## H

***Header***—A joist fastened across the ends of regular joists in an opening, or the framing member above a window or door opening

***Hearth***—Concrete or masonry apron in front of a fireplace

***Hip***—Outside corner formed by intersecting roofs

***Hip Rafter***—The rafter extending from the corner of a building to the ridge at a hip

***Hose Bibb***—An outside faucet to which a hose can be attached

## I

***I-joist***—An engineered wood joist with a vertical web of plywood or OSB and solid wood upper and lower chords

***Insulated Glazing***—Two or more pieces of glass in a single sash with an air space between them for the purpose of insulation

***Invert Elevation***—The elevation at the lowest point inside a pipe

***Isometric Drawing***—A kind of drawing in which horizontal lines are 30° from true horizontal and vertical lines are vertical

## J

***Jack Rafter***—A rafter between the outside wall and a hip rafter or the ridge and a valley rafter

***Jamb***—Side members of a door or window frame

***Joists***—Horizontal framing members that support a floor or ceiling

## L

***Lintel***—Steel or concrete member that spans a clear opening— usually found over doors, windows, and fireplace openings

## M

***Masonry Cement***—Cement that is specially prepared for making mortar

***Mil***—A unit of measure for the thickness of very thin sheets—1 mil equals .001″

***Miter***—A 45° cut so that two pieces will form a 90° corner

***Mortar***—Cement and aggregate mixture for bonding masonry units together

***Mullion***—The vertical piece between two windows that are installed side by side—window units that include a mullion are called *mullion windows*

***Muntins***—Small vertical and horizontal strips that separate the individual panes of glass in a window sash

## N

**Nailer**—A piece of wood used in any of several places to provide a nailing surface for other framing members

**Nominal Size**—The size by which a material is specified—the actual size is often slightly smaller

**Nosing**—The portion of a stair tread that projects beyond the riser

## O

**Orthographic Projection**—A method of drawing that shows separate views of an object

## P

**Panel Door**—A door made up of panels held in place by rails and stiles

**Penny Size**—The length of nails

**Perimeter Drain**—(See *Footing Drain*)

**Pilaster**—A masonry or concrete pier built as an integral part of a wall

**Pitch**—Refers to the steepness of a roof—the pitch is written as a fraction with the rise over the span

**Plan View**—A drawing that shows the layout of an object as viewed from above

**Plate**—The horizontal framing members at the top and bottom of the wall studs

**Platform Framing**—(also called *Western framing*) A method of framing in which each level is framed separately—the subfloor is laid for each floor before the walls above it are formed

**Plenum**—A chamber within a forced-air heating system that is pressurized with warm air

**Plumb**—Truly vertical or true according to a plumb bob

**Ply (trusses)**—If trusses are plied, they are joined together face-to-face to make a stronger unit

**Potable Water**—Water that is safe for drinking

**Portland Cement**—Finely powdered limestone material used to bond the aggregates together in concrete and mortar

## R

**R value**—The ability of a material to resist the flow of heat

**Rafter**—The framing members in a roof

**Rail**—Also the horizontal members in a balustrade, such as the railing around a stairway

**Rake**—The sloping cornice at the end of a gable roof

**Resilient Flooring**—Vinyl, vinyl-asbestos, and other man-made floor coverings that are flexible yet produce a smooth surface

**Ridge Board**—The framing member between the tops of rafters that runs the length of the ridge of a roof

**Rise**—The vertical dimension of a roof or stair

**Riser**—The vertical dimension of one step in a stair—the board enclosing the space between two treads is called a riser

**Rowlock**—Position of bricks in which the bricks are laid on edge

**Run**—The horizontal distance covered by an inclined surface such as a rafter or stair

## S

**Sash**—The frame holding the glass in a window

**Saturated Felt**—Paper-like felt that has been treated with asphalt to make it water resistant

**Screed**—A straight board used to level concrete immediately after it is placed

**Section View**—A drawing showing what would be seen by cutting through a building or part

**Setback**—The distance from a street or front property line to the front of a building

**Sheathing**—The rough exterior covering over the framing members of a building

**Shim**—Thin material, typically wood shingle, used to adjust a small space

**Sill**—The framing member in contact with a masonry or concrete foundation

**Sill Sealer**—Compressible material used under the sill to seal any gaps

**Site Constructed**—Built on the job

**Site Plan**—The drawing that shows the boundaries of the building, its location, and site utilities

**Sliding Window**—A window with two or more sashes that slide horizontally past one another

**Soffit**—The bottom surface of any part of a building, such as the underside of a cornice or lowered portion of a ceiling over wall cabinets

**Soldier**—Brick position in which the bricks are stood on end

**Span**—The horizontal dimension between vertical supports—the span of a beam is the distance between the posts that support it

**Specifications**—Specific written instructions for materials to be used and methods of construction or application

**Square**—The amount of siding or roofing materials required to cover 100 square feet

**Stack**—The main vertical pipe into which plumbing fixtures drain

**Stair Carriage**—The supporting framework under a stair

**Stile**—The vertical members in a sash, door, or other panel construction

**Stool**—Trim piece that forms the finished window sill

**Stop**—Molding that stops a door from swinging through the opening as it is closed—also used to hold the sash in place in a window frame

**Stud**—Vertical framing member in a wall

**Subfloor**—The first layer of rough flooring applied to the floor joists

**Sweat**—Method of soldering used in plumbing

## T

**Termite Shield**—Sheet metal shield installed at the top of a foundation to prevent termites from entering the wood superstructure

**Thermal-break Window**—Window with a metal frame that has the interior and exterior separated by a material with a higher R value than the metal itself

**Thermostat**—An electrical switch that is activated by changes in temperature

**Top Chord**—The top horizontal or sloped member of a truss

**Trap**—A plumbing fitting that holds enough water to prevent sewer gas from entering the building

**Tread**—The surface of a step in stair construction

**Trimmers**—The double framing members at the sides of an opening

**Truss**—A manufactured assembly used to support a load over a long span

**Truss Detail**—An engineering drawing giving detailed specification for the manufacture of one particular truss; a truss drawing packet will usually include a truss detail for each type of truss to be used

**Truss Drawings**—Any of several types of drawings provided by an engineer to give detailed information about the construction and placement of trusses

**Truss Layout Plan**—A plan drawing giving detailed information about where each truss is to be placed, including spacing between trusses

## U

**Underlayment**—Any material installed over the subfloor to provide a smooth surface over which floor covering will be installed

## V

**Valley**—The inside corner formed by intersecting roofs

**Valley Rafter**—The rafter extending from an inside corner in the walls to the ridge at a valley

**Vapor Barrier**—Sheet material used to prevent water vapor from passing through a building surface

**Veneer**—A thin covering; in masonry, a single wythe of finished masonry over a wall; in woodwork, a thin layer of wood

**Vent Pipe**—A pipe, usually through the roof, that allows atmospheric pressure into the drainage system

**Vertical Contour Interval**—The difference in elevation between adjacent contour lines on a topographic map or site plan

**Volt**—The unit of measurement for electrical force

## W

**Water Closet**—A plumbing fixture commonly called a *toilet*

**Watt**—The unit of measurement of electrical power—1 watt is the amount of power from 1 ampere of current with 1 volt of force

**Weep Hole**—A small hole through a masonry wall to allow water to pass

**Western Framing**—(See *Platform Framing*)

**Wythe**—A single thickness of masonry construction

# Index

Page numbers followed by "f" indicate figure.

## A

Abbreviations, electrical, 326
Air admittance valve (AAV), 266
Air-conditioning. *See* Heating, ventilating, and air- conditioning (HVAC)
Air-handling equipment, 315, 320, 320f
Alphabet of lines, 19–24
    centerlines, 22, 22f
    construction drawings and, 19
    cutting-plane lines, 23, 24f
    dashed lines, 20, 20f, 21f
    dimension lines, 21–22, 21f, 22f
    extension lines, 21–22, 21f, 22f
    hidden lines, 20
    leaders, 22–23, 23f
    object lines, 19, 19f
    phantom lines, 20, 21f
American Welding Society, 240, 286
Amperes (amps), 274, 274f
Anchor bolts, 86, 86f
Applied stop molding, 182
Apron, pitch of, 93
Architect, 3–4, 142, 258, 270, 342
Architect's scale, reading, 15, 17, 17f
Architectural drawing sheets, 29
Area-separation wall, 258
Arrow side, of reference line, 286, 290f
Ash dump, 208, 208f
Ash pit, 208
Asphalt roofing materials, 154
Attic ventilation, 173, 174f
Authority Having Jurisdiction (AHJ), 5

## B

Backfilled excavation, 92
Balloon diagrams, 4
Balloon framing, 96, 98f
Balusters, 216
Balustrade, 216, 216f
Band joist, 100
Bar joists, 306, 307f
Base moldings, 225f–230f
Battens, 185, 186f
Batter boards, 66f, 67
Beam pockets, 88, 88f, 103f
Beams, 99
    collar, 154
    grade, 291
    joists, 108
    lintels, 304
    LVL (laminated veneer lumber), 110f, 118
    plywood box, 99
    steel, 87, 109
    wooden, 87
Bearing, for floor joists, 118
Bearing angle, 60, 60f
Bearing plate, 108, 109f
Bird's mouth, 140, 154
Black iron plumbing materials, 263
Blanket insulation, 220, 222f
Bolts, 286, 289f
Bond beam, 307, 310f
Bond patterns, 190, 191f
Bottom chord, 138
Bottom (sole) plate, 100
Boundary lines, 60–61
Box cornice, 168, 169f, 170f
    fascia, 168
    narrow, 168, 170f
    parts of, 169f
    sloping, 168, 169f
    soffit, 168
    wide, 168, 171f
Box sill, 115, 116f. *See also* Sill construction
Branch circuits, 275, 277f
Brick positions, 191f
Brown coat, 190
Builder's level, 69
Building codes
    footings, 79, 80
    model codes and, 5
    party walls, 258
    purpose of, 5–6
    seismic codes, 82f
    state and local, 5
Building layout, 37–38
Building lines, 65–67, 65f
Building permit, 5
Building sewer, 71, 72–73
Bus ducts and wireway symbols, 328f–332f

## C

Cabinets
    on drawings, 232–234, 232f, 233f, 234f
    elevations, 50, 51f, 232, 232f
    in floor plans, 37, 39
    letter/number designations for, 234f
    manufacturers' literature on, 234
    shelves/valances/filler pieces/trim, 234
    typical designations for, 234f
Cabinet unit heaters, 313, 314f
Cantilevered framing, 118, 124f
Casing, 178
Cast iron plumbing materials, 73, 263
Centerlines, 22, 22f
Ceramic tile, 223
Certificate of occupancy, 5
Chain dimensions, 21–22
Chase (for metal chimney), 210, 210f, 212f
Check rails, 180
Chimney, 209–210, 209f, 210f
Civil construction, 2, 3f
Cleanout door (fireplace), 208, 208f
Cleanouts, 264
Collar beams, 154. *See also* Collar ties
Collar ties, 154. *See also* Collar beams
Column footings, 80f, 291, 304
Column pads, 80, 80f
Columns, 108–111, 109f, 111f
Commercial construction, 2, 2f, 3f, 283–294, 286f
    architectural style, 285, 286f
    coordination of drawings, 296–303
    foundations, 288–291, 292f–294f
    indexes, 296, 297f–299f
    masonry reinforcement, 307
    material keying, 296, 299–300, 300f
    mental walk-through, 301–303
    metal floor joists, 116
    metal framing, 100–101, 104f, 105f
    reference symbols on drawings, 300–301, 301f
    shop fabrication, 287–288, 291f
    structural drawings, 304, 306
    structural grid coordinate, 301, 302f
    structural steel, 285–286, 287f, 288f, 289f
    welding symbols, 286, 290f–291f
Commercial electrical symbols, 328f–332f
Common rafters, 152–158
    finding length of, 156, 156f
    International Residential Code and, 152
    roof construction and, 152, 152f, 154, 154f
    roof covering and, 154, 156f
    roof openings, 157
    shed roof framing, 156–157, 157f
Composite products, 195
Concrete foundations, 84–86, 283
    anchor bolts and, 86, 196
    reinforced, 283
    structural concrete and, 85

Concrete slabs, 93–94
Conductor, of electric current, 275, 276f
Conduit, 275
Connected load schedule, 341
Construction classifications, 2
Construction industry, 1, 2, 19
Contour lines, 62, 63f
   interpolation and, 62
   topographic, 62
   vertical contour interval, 62
Contract documents, 223, 342
Contractor, 5–6, 80
Copper plumbing, 73, 263
Corner boards, 188, 188f
Cornice returns, 171, 172f, 173–174, 173f
   framing for, 173f
   level box cornice and, 171
   style of, 171, 172f
   types of, 173f
Cornices, 168–175
   box, 168, 169f, 170f
   close, 171, 172f
   cornice returns, 171, 172f, 173–174, 173f
   end (rake) rafters and, 168
   open, 168, 171f
   types of, 168, 171
   ventilation, 173–174
Couplings, 264
Cove molding, 216, 224
Cripple studs, 135, 136f, 137f
Cross section, 47, 213, 213f
Current, 274–276, 274f
Cut (earth removal), 62
Cutting-plane lines, 23, 23f

## D

Damper (fireplace), 209, 209f
Dashed lines, 20, 20f, 21f, 44f
Datum symbols, 81
Decks, 194–197
   anchoring, 196–197, 196f
   composite materials, 195
   on floor plan, 195, 195f
   railings, 197, 197f
   support, 194–196, 194f, 195f
   typical construction of, 194f
Delta Energy Center, 3f
Depth, of rafters and joists, 152
Design and construction team, 6, 6f
Design process, 3–6, 4f
Details
   large scale, 50, 51f
   orienting, 52, 53f
Dimension lines, 21–22, 21f, 22f
Dimensions, 38, 38f, 39f
   on an elevation, 45f
   cabinet, 233
   chain, 21–22

column footing, 291
finding, 129–130, 130f
footing, 79
nominal, 27
of rough openings, 133, 133f, 135
Do (ditto), 306
Doors
   construction, 180–181, 181f
   details of, 182, 182f
   dimensions of, 38
   floor plans and, 37–38
   flush, 181f
   hollow core, 181
   jamb detail (exterior), 182f
   measurement of, 27f
   molded, 181, 181f
   panel, 180, 181f
   pocket, 136, 136f
   reading catalogs of, 182
   schedules, 27, 133, 133f
   symbols, 25, 25f, 27
Dormer, 147
Double-hung window, 180
Double top plate, 95, 96f
Dowels, 291
Drainage, 90, 92
   purpose of, 266–267
   septic system and, 74
Drainage waste and vent system, 266–268, 269f
Drain field, 73, 73f
Drains
   floor drains, 90, 93
   footing drains, 90, 90f, 91f
   pitch of, 71
   plastic drain pipe, 90, 91f
   sanitary drain, 321
   storm drain, 308–309
Drawing to scale, 15
Drip cap molding, 187

## E

Earthquake protection, 5, 86, 86f
Elbows, 264, 265f
Electrical drawings, 324–343
   distribution symbols, 328f–332f
   lighting circuits, 325
   numbering of, 29–30, 237
   plan views, 324–326, 326f
   power circuits, 324–325, 325f
   riser diagrams, 333–334, 334f, 335f, 336f
   schedules, 337–339, 339f, 340f, 341–342, 341f–342f
   schematic diagrams, 334–336, 337f
   single-line diagrams, 326–327, 333, 333f
   specifications, 342–343
   symbols, 328f–332f
Electrical outlets, cabinets and, 232
Electrical raceway installations, 94

Electrical service, 74–75, 75f, 334f
   overhead, 75, 75f
   underground, 75, 75f
Electrical symbols, 328f–332f
Electricity, 74, 272, 274, 275
   circuits, 324–327, 333–334, 336
   current/voltage/resistance/watts, 274–278
   electrical symbols on plans, 277, 277f, 278f
Elevation drawings (showing height), 41–45
   information given on, 42, 44–45, 44f, 45f
   orienting, 42, 42f, 43f
Elevation reference symbol, 300–301, 301f
Elevations (denoting vertical position), 62
   defined, 62
   interpolating, 62–63, 62f, 63f
Energy saving techniques, 101–102, 105f
Engineers, 5, 35, 116, 276, 288, 291, 342
Engineer's scale, and architect's scale compared, 15, 17
Equipment schedule, 341
Excavation, 67, 69, 69f, 76, 90
Exhaust fan (EF), 315, 320
Expansion joints, 93, 260, 304
Extension jambs, 180
Extension lines, 21, 21f
Exterior wall coverings, 185–192
   corner and edge treatment, 188–189
   corner boards, 188, 188f
   drip caps and flashing, 187–188, 187f, 188f
   exposure, 185, 186f
   fiber cement siding, 186–187, 186f
   horizontal siding, 185, 185f, 186f, 188
   masonry veneer, 190, 190f
   metal and plastic siding, 189, 189f
   stucco, 189–190, 189f, 190f
   wood siding, 185, 185f, 186f

## F

Fabrication drawings, 287. *See also* Shop drawings
Fan coil units (FCU), 313
Fascia, 168
Fiber cement siding, 186–187, 186f
Field weld symbol, 288, 291f
Fillet weld, 286
Financing, 6
Finished grade, 62, 63f, 90, 200
Fire alarm riser diagram, 333–334, 335f
Fire damper, 320, 320f
Fireplace drawings, 213, 213f, 214f
Fireplaces, gas-burning, 210, 211, 211f, 212f
Fireplaces, prefabricated metal, 210, 210f, 211f
Fireplaces, wood-burning, 207–210
   ash dump, 208, 208f
   basic construction of, 207–210
   chimney and, 209–210, 209f, 210f

clay flue, 209
cleanout door, 208, 208f
damper, 209, 209f
drawings, 213, 213f, 214f
efficiency of, 210
firebox, 208–209, 208f
foundation, 208
hearth, 209
masonry, 208–209, 208f, 210
theory of operation of, 207–210
throat, 209, 209f
zones, 207, 207f
Fire-rated construction, 258–259, 260f, 261f
Fire-resistant materials, 258–259
Firestops, 259, 261f
Fixture schedule, 337, 338, 338f
　information shown on, 337–338
　School Addition, 338, 338f
Flashing, 164, 185, 187
　battens and, 185
　drip caps and, 187–188, 187f, 188f
　valley, 164
　Z-flashing, 185
Floor drain, 90, 91f, 93
Floor framing, 87f, 96, 108, 108f, 115–125
　cantilevered, 118, 124f
　double-line plan, 121f
　floor joists, 116–117, 117f
　for irregular-shaped house, 120f
　joist-and-girder, 108f
　Lake House, 117–118, 121f, 122f
　at openings, 119, 125f
　sill construction, 115, 115f
Floor plans
　building layout, 37–38
　decks, 195, 195f
　defined, 37
　dimensions, 128
　doors and windows in, 37, 39
　features of, 39
　Lake House, 117, 123f, 127
　scale, 15
　as section view, 12
Floors, finished, 223–224
Foamed-in-place insulation, 221f
Footing-and-wall-type foundation, 36
Footing drains, 90, 91f. *See also* Footings
Footings, 76–82
　building codes for, 79, 80
　column pads, 80, 80f
　depth of, 80–81, 81f, 82f
　dimensions of, 79, 81
　drains, 90, 91f
　elevations indicated, 81
　footing schedule, 294f
　reinforcement, 80, 80f
　slab-on-grade, 76, 76f, 77f
　spread footings, 77, 78f, 79, 79f
　stepped footings, 82, 82f

Forced-air system, 272f
Foundation
　for commercial buildings, 304
　for commercial construction, 288, 291
　detail, 293f
　dimensions, 37, 79, 84f
　fireplace, 208
　footing-and-wall-type, 36
　laying out, 84–86, 84f, 85f, 86f
　materials, 85
　permanent wood, 88, 88f
　pile, 291
　preventing seepage in, 92
　section through, 85f
　sill construction and, 115
　slab-on-grade, 36, 36f, 76, 76f
　special features, 87–88, 87f, 88f
　walls, 84–89
Foundation plans, 36–37, 37f, 68f
　detail, 47
　dimensioning on, 84f
　foundation layout, 304
Framing systems, 95–105
　balloon, 96, 98f
　energy-saving techniques, 101–102, 105f
　metal framing, 100–101, 104f, 105f
　platform framing, 95, 95f, 96f
　post-and-beam, 98–100, 100f, 101f
　timber fastenings, 102f
Front elevation, 253. *See also* Elevation drawings (showing height)
Frostline, 81
Furring, 223, 223f

## G

Gable ends, 149, 149f
Gable roof, 138
Gambrel roof, 138
Garages, 258, 275
Gas pipes, 74
General contractor, 5, 6
GFCI. *See* Ground-fault circuit interrupter (GFCI or GFI)
GFI. *See* Ground-fault circuit interrupter (GFCI or GFI)
Girders, 108, 109, 110
Grade beams, 291
Grade line, 44, 90
Grading, 61–62, 203
Grass, 202
Grilles (ventilation), 315
Grinder pump, 71, 72f
Ground conductor, 275, 276f
Ground-fault circuit interrupter (GFCI or GFI), 275
Grout (foundation), 86
Grout (masonry reinforcement), 307
Gusset, 138, 139f
Gypsum wallboard, 222

## H

Hammering, of pipes, 265–266
Handrail, for stairs, 216
Hardiplank, 186
Hash mark, 28
Haunch, 76, 77f
Haunched slab, 288
Header, 45, 100
Header, door and wall openings, 135
Head jamb
　door, 180
　window, 178
Heating, ventilating, and air-conditioning (HVAC), 272–273, 312–322
　air-handling equipment, 315, 320, 320f
　forced-air system, 272f
　heating piping, 315, 319f
　plans, 312
　symbols, 316f–319f
　types of heating systems, 273
　unit ventilators, 313, 313f, 314f
Heat sink, 94
Hidden line, 20
Hidden Valley, 236, 250–251
　construction drawings table of contents, 254f
　first phase of, 250
　kitchen, 255f
　living room, 256f
　photos, 255–256
　reversed plan, 250, 253f
　site plan for, 251f
　stairs, 255f
High tide areas, 100
Hip jack rafters, 161, 161f, 164–165, 165f
Hip rafters, 159–161, 161f
Hip roof, 138
Hold-down straps, 86, 86f
Hollow core doors, 181
Hoover Dam, 3f
Housed stringers, 216, 216f
Housing starts, and economy, 2
Hurricane areas, 100
HVAC. *See* Heating, ventilating, and air-conditioning (HVAC)

## I

ICC. *See* International Code Council (ICC)
Ice dam, 173, 174f
I-joists, 116, 174
Incandescent lamp designations, 339f
Industrial construction, 2, 3f, 116
Industrial electrical symbols, 328f–332f
Institutional electrical symbols, 328f–332f
Insulated doors, 181, 182
Insulating glass (for windows), 180
Insulation, 92–93, 92f, 220–222, 221f, 222f
　batt, 220
　blanket, 220, 222f

condensation and, 173
fiberglass, 23
fire-resistant fiberglass, 259, 260
foamed-in-place, 221f
of foundation/concrete slab, 220
pouring insulation, 220
raised heel rafters and, 154
rigid boards, 220
rigid plastic foam, 92f
R-value of, 220
sound, 259–260, 261f
thermal, 85, 99, 216, 221f
vapor barrier, 221
Integral stop molding, 182
Interior elevation symbols, 301, 301f
Interior molding, 224
International Building Code®, on stair risers, 217
International Code Council (ICC), 5
International Residential Code®, 5, 5f, 79, 79f, 81, 82f, 109, 152, 209
 footings, 79, 79f, 81
 girders and, 109
 roof construction, 152
International Residential Code® for One- and Two-Family Dwellings, 5, 5f
Interpolation, 62
Inverted-T foundation, 288
Invert elevation, 74, 74f, 321
Isometric drawings, 8, 9f, 10f. *See also* Pictorial drawing

## J

Jack rafters, 161, 164–166, 164f, 165f
Joist-and-girder floor framing, 108f
Joist headers, 117, 122f
Joists
 attaching to girders, 111f
 bar, 306, 307f
 ceiling, 154
 defined, 108
 floor, 116–117, 117f, 118f
 I-joists, 116, 174
 metal floor, 116–117, 117f
 open-web steel, 306, 307f
 types of, 116

## K

Kitchen elevations, 232f
Kitchen equipment schedule, 339

## L

Lake House
 bedroom, 118
 Construction Specifications Institute's MasterFormat (CSI format) for, 236, 237f
 floor framing, 117–118, 121f, 122f
 foundation plan, 85

foundation plan for, 85
 kitchen and dining room, 128f
 living room, 128f
 loft, 129f
 playroom, 128
 post-and-beam framing, 99
 wall dimensions, 130
Laminated veneer lumber (LVL), 109, 110, 110f
Lamp characteristic designations, 338, 339f, 340f
Landing (stair), 215, 218f
Landscaping, 6, 67, 201, 203
Laser level, 69
Lath, for stucco, 189
Lawns, 199, 201
Leach lines, 73
Left overhang (LOH), 138, 142
Let-in ribbon boards, 96
Leveling instrument, 81
Light fixtures, 277, 325, 338
Lighting circuits, 325
Lighting symbols, 328f–332f
Lintels
 defined, 304
 lintel schedule, 307, 308f
 loose steel categorization, 306–307
 masonry wall openings, 306
Lite (glass), 177
Longitudinal sections, 47, 49f
Lookouts, 168, 170f
Louvered grille, 315
L shaped stairs, 128, 217
LVL. *See* Laminated veneer lumber (LVL)

## M

Mansard roof, 138
*Manual of Steel Construction* (AISC), 286
Masonry construction, 38, 39f, 306
Masonry foundations, 69, 81, 84, 86
Masonry opening, 27, 38, 88
Masonry veneer, 38, 188, 190, 190f
Mast head, 75. *See also* Service head
Material keying, 296, 299–300, 300f
Material legend, 286
Material symbols, 27, 28f
Measuring line (roof), 140
Mechanical drawings, 312
 air-handling equipment, 315, 320, 320f
 heating piping, 315
 HVAC, 312
 plumbing, 320–321, 322f
 symbols, 316f–319f
 unit ventilators, 313, 313f, 314f
Mechanical symbols, 316f–319f
Meetings rails, 180
Mental walk-through, 301–303
Metal fireplaces, prefabricated, 210, 210f

Metal framing, 100–101, 104f, 105f, 136
Metal siding, 189, 189f
Metal tile roofs, 154
Metal windows, 178
Mils (measurement), 92
Model codes, 5
Molded doors, 181, 181f
Mortar, in masonry joints, 86, 307, 309f
Mortgage, 6
Motor control schematic, 336, 337f
Mullion, 180
Multiconductor feeders, 75
Multifamily buildings, 250. *See also* Town house construction
Multifamily construction, 249
Multifamily dwellings, 250
Municipal sewers, 71, 71f, 73–74
Muntins, 177

## N

Narrow box cornice, 168, 170f
National Electrical Manufacturers Association (NEMA), 327, 341f
National Electric Code®, 275
National Fire Protection Association, 276
Natural features, 61
Natural grade, 62. *See also* Finished grade
NEMA. *See* National Electrical Manufacturers Association (NEMA)
NEMA specifications, 327
Newel post, 216
Nominal size, 88
Nonmetallic-sheathed cable, 101–102, 105f, 116
Nosing (stair), 215–216

## O

Oblique drawings, 9, 10f
Open cornice, 168, 171f
Open stringers, 216–217, 216f
Open-web steel joists, 306, 307f
Oriented-strand board (OSB), 154
Orthographic projection, 9–12, 10f
OSB. *See* Oriented-strand board (OSB)
Other side, of reference line, 286
Overhang, 138, 142
 left. *See* Left overhang (LOH)
 right. *See* Right overhang (ROH)
Overhead construction, shown in floor plans, 20, 39
Overhead electrical service, 74–75, 75f

## P

Panelboard symbols, 328f–332f
Panel doors, 180, 181f
Panel schedule, 339
Panel strips mullion casings, 225f–230f
Partitions. *See* Walls and partitions

Party wall, 258, 259f
  area-separation wall and, 258
  defined, 258, 259f
  fire-rated, 258, 260f
  fire-rated construction and, 258–259
  sound insulation and, 259–260
  Town House, 258
Paved areas, 200, 201f
Percolation, 74
Percolation test, 73
Permanent wood foundations, 88, 88f
Phantom lines, 20, 21f
Pictorial drawing, 8, 9. *See also* Isometric drawings
Piers, 108–109
Pilasters, 79
Piles, 291
Pipe columns, 108
Pitch
  of concrete slab, 92, 93, 93f
  of roof, 138, 142, 143f
  of sewers, 71, 91f
Planters, 200, 202f, 203f
Plantings, 201–202, 203f, 204f
Plan views, 11–12, 11f, 35–39
  building layout, 37–38
  in construction drawings, 11–12, 11f
  dimensions of, 38, 38f, 39f
  electrical drawings and, 324–326, 326f
  floor plans, 37
  foundation plans, 36–37, 36f, 37f
  lighting circuits and, 325
  site plans, 35–36, 35f
  of a two-flue chimney, 209–210, 209f
Plastic plumbing materials, 263
Plastic siding, 189, 189f
Platform framing, 95–96, 95f, 96f.
  *See also* Western framing
  balloon framing and, 98–99
  characteristics of, 95–96
Plumbing
  drainage waste and vent system, 266–268, 268f, 269f
  fittings, 263–264
  gas piping sleeve, 271f
  isometric drawings of, 319f
  materials, 263
  plans, 270–272, 270f, 271f
  riser diagram, 321, 322f
  rough-in sheet, 270f
  single-line isometric, 9f, 271f
  supply piping design, 264–266, 267f
  symbols, 30f, 316f–319f
  traps, 267–268, 268f
  vents, 268, 268f, 269f
Plumbing wall, 127
Pocket doors, 136
Pole structures, 100
Portland cement, 80, 86

Posts, metal or wooden, 108
Power equipment symbols, 328f–332f
Power riser diagram, 333, 334f. *See also* Riser diagrams
Prefabricated metal fireplaces, 210, 210f
Property line, 35, 60
  direction of a, 60
  site plans and, 35

## R

Rabbet joints, 185
Raceways, 94, 259
  electrical, 94
  sound transmission and, 259
Radius of an arc, 22, 22f
Rafters, 140, 152–157
  hip, 159–161, 160f, 161f
  jack, 164–166, 164f, 165f
  length of common, 156, 156f
  raised heel, 154
  rake, 168, 169f
  size and spacing of, 152, 153f
  as structural members in common frame roof, 152
  valley, 161–164, 162f, 163f
Rafter table, 156, 156f, 160f, 164–165
Rails
  door, 180
  meetings or check, 180
  window, 177
Rake rafters, 168, 169f
Rebars, 80, 80f
Receptacle outlet symbols, 328f–332f
Receptacle schedule, 341
Reference symbols, 300–301, 301f, 328f–332f
Remote control station symbols, 328f–332f
Residential Code, 5, 5f, 79, 79f, 81, 82f, 109, 209
Residential construction, 2
  floor joists in, 116
  footings used in, 76
  girders used in, 109
  plumbing plans and, 270
  types of roofs used in, 138, 139f
Resistance (electrical), 274
Retaining walls, 199–200, 199f, 200f
Reversed plans, 250–251, 253f
Ridge board, 140, 152, 154f
Right overhang (ROH), 138, 142
Rigid boards, 220
Rise, of stairs, 215
Riser (drainage), 92
Riser diagrams, 333–334
  fire alarm, 333–334, 335f
  power, 333, 334f
  special, 333
  telephone, 334, 336f
Risers (stair), 215, 217

Roofs
  common rafters, 152–157
  construction of, 152, 154
  construction terms, 138–141, 141f
  coverings, 154
  elements of, 152f
  fire-resistant materials, 258
  framing openings, 157
  framing plan, 154, 155f
  hip and valley framing, 159–166
  ridge boards and rafters, 140, 152–157, 159–165
  spacing of rafters, 148, 153f
  trusses, 140, 142–149
  types of, 138, 139f
  ventilation, 173–174, 174f
  weather protection, 154
Room finishing
  floors, 223–224
  interior molding, 224, 225f–230f
  schedule, 223, 224f
  wall and ceiling covering, 222–223, 223f, 224f
Rough opening, 133–135
Run (roof), 140
Run, of stairs, 215

## S

Sanitary drain, 321
Sash (window), 177
Scale drawings, 15
Schematic wiring diagram, 334, 336, 337f
School Addition
  fire alarm riser diagram for, 333, 335f
  fixture schedule, 338, 338f
  heating plans, 313
  panel schedule for, 339
  plumbing, 320–321
  structural steel, 286
  telephone riser diagram for, 336f
Scratch coat, of stucco, 189–190, 190f
Section views, 12, 23, 49, 50f
  cutting-plane lines, 23, 23f, 24f
  in elevation, 213, 214f
  orienting, 52, 53f
  symbols, 300–301, 301f
  transverse (cross section), 47, 49f
  typical wall, 30–31, 47, 49, 77f, 190f
Septic systems, 71f, 73–74, 73f
Service head, 75, 75f. *See also* Mast head
Settling, preventing, 76
Sewers
  building sewer, 71, 71f, 72–73
  drains, 71
  municipal, 73–74, 73f
  uphill flow of, 71, 72f
Sheathing, 52, 96
  board, 52
  gypsum-based structural, 52

horizontal or diagonal boards as, 52
magnesium oxide structural, 52
Oriented-strand board (OSB) and, 52
plywood, 52, 96
wall, 189
Shed roof, 11f, 138, 157f
framing, 156–157, 157f
rake and, 168
Shed roof framing, 156–157, 157f
Sheetrock™, 222–223
Shingle panel molding, 169f, 170f
Shingles, 185, 188, 188f
Shop drawings, 287, 306. *See also*
Fabrication drawings
Shrubs, 199, 201–202, 203f
Side jambs
door, 180
window, 178
Siding
exposure of, 185, 186f
fiber cement, 186–187, 186f
horizontal, 185, 185f, 188
metal, 189, 189f
plastic, 189, 189f
vertical, 185, 186f
wood, 185, 185f, 186f
Sill (door), 180
Sill (window), 178
Sill construction, 115, 115f, 116f.
*See also* Box sill
Sill plate, 115, 115f
Sill sealer, 115
Site
clearing, 61, 61f
grading, 61–62, 63f, 203
Site elevation, 44, 44f
Site plans, 35–36, 35f, 74f, 199f, 201f, 251f
building position, 36, 65
Hidden Valley, 250–251, 251f, 252f
Lake House, 201f
paved areas on housing sites in, 200, 201f
retaining walls on, 199f, 200f
reversed, 250–251, 253f
sewer drains on, 71, 72f
utilities, 74f
visualizing, 251–253, 256
Site utilities, 71–75
building sewer, 72–73
electrical service, 74–75, 75f
municipal sewers, 73–74
septic systems, 73–74, 73f, 74f
sewer drains, 71, 71f, 72f
utility piping other than sewers, 74, 74f
6-8-10 method, 66, 66f
Slab-on-grade construction, 77f
Slab-on-grade footing, 76
Slab-on-grade foundation, 36, 36f, 76, 76f

Slater's felt, 154
Slope
contour lines and, 62
hip roof, 138
of roof, 138–139, 140f, 154
stepped footings and, 82
Sloped sites, stepped footings and, 82
Sloping box cornice, 168, 169f
Smoke shelf, 209, 209f
Soffit, 168, 169f, 171f
Sole plate, 95
Solid wood header, 135f
Sound insulation, 259–260, 261f
Sound transmission classification (STC), 259–260, 261f
Span, of roof, 138, 142
Special riser diagram, 333
Spiral stairs, 217
Spot elevations, 62, 62f, 69f
Spread footings, 77–79, 78f, 79f, 288
Stair cove, 216
Stairs, 215–219
basic parts of, 215–216, 215f, 216f
calculating risers and treads, 217, 218f, 219f
cove, 216
trim and balustrade, 216, 216f
types of, 216–217, 216f, 219f
STC. *See* Sound transmission classification (STC)
Stepped footings, 82, 82f
Stiles, 177, 177f, 180
Stock plans, 6, 6f
Stop molding, 178
Storm drain, 321
Straight stairs, 217
Stringers, 215, 216–217
housed, 216, 216f
open, 216, 216f
Structural drawings, commercial buildings, 304
Structural grid coordinate, 301, 302f
Structural steel, 285–286
framing, 304, 306–307, 306f
Structural steel framing, 304, 306–307, 306f
Studs, 38, 95
in balloon framing, 96
cripple, 135, 136f, 137f
door and window framing and, 135–136, 135f, 136f
firestops and, 259, 261
in platform framing, 95, 95f
roof framing and, 149
sound insulation and, 260
wracking, 96
Stud wall construction, 135
Subcontractors, 6, 312
Subfloor, 127, 224
Superstructure, 86, 86f, 288

Switchboard symbols, 328f–332f
Switches, on circuits, 275–276, 276f
Switch wiring, on plans, 277, 277f
Symbols, 25–33
abbreviations, 31
door and window, 25, 25f, 26f, 27, 27f
electrical and mechanical, 27–29, 29f
elevation reference, 301f
material, 27, 28f
plumbing, 30f, 316f–319f
reference marks, 29–31, 31f, 32f
section view, 300–301, 301f
welding, 286, 289f, 290f

## T

Tail (roof), 138
Target rod, for leveling instrument, 69, 69f
Tees and wyes, 264, 265f
Telephone riser diagram, 334, 336f.
*See also* Riser diagrams
Termite areas, and metal framing, 115, 116f
Termite shield, 115, 116f
Terra-cotta roof tiles, 154
Thermal-break windows, 178
Thermal insulation, 220–222, 221f
Timber fastenings, 102f
TJIs. *See* I-joists
Top chord, 138, 139f
Top plate, 95–96, 96f
Town house construction, 258–261
fire-rated, 258–259, 260f
Hidden Valley site plan, 250–251, 251f, 255f, 256f
party wall, 258–260, 259f, 260f
sound insulation, 259–260, 261f
Transom bar, 180
Transverse section, 47, 49f
Traps, 267–268, 268f
Treads (stair), 217, 218f, 219f
Trees, 61, 61f, 199, 201–202, 201f, 203f
Tripod, for leveling instrument, 69
Trusses, 140, 142–149
delivery sheet, 142, 145f
engineering drawings, 142, 145f, 146f, 147, 147f, 148f, 149
gable ends, 149, 149f
information on drawings, 142, 143f, 144f
truss detail, 148f, 149
truss layout plan, 146f, 147
Typical wall sections, 30–31, 32f, 47, 77f

## U

Underlayment, 223–224
Union, 264, 264f
Unit rise, 138–139, 140f
Unit ventilators, 313, 313f, 314f

Index 387

U shaped stairs, 217, 217f
Utility installations, 71
Utility systems coordination, 94

## V

Valley jack rafters, 160f, 164
Valley rafters, 161–164, 162f, 163f
Valves, 264
Vapor barriers, 92
Veneer, 109
  brick, 85
  masonry, 190, 190f, 191f
Ventilation, cornices and, 173–174. *See also* Heating, ventilating, and air-conditioning (HVAC)
Vent pipe, 321
Vertical contour interval, 62
Voltage, 274, 274f
Volts, 274, 274f, 338

## W

Wallboard
  fire-rated, 258
  gypsum, 222–223, 258, 260f

Walls and partitions, 127–130
  finding dimensions, 129–130, 129f, 130f
  framing openings in, 135–136, 135f, 136f
  rough opening dimensions, 133, 133f, 134f, 135
  visualizing layout of, 127–129, 128f, 129f
Warm-air heating systems, 312. *See also* Heating, ventilating, and air-conditioning (HVAC)
Water distribution system, 263, 265, 267f
Water hammer arrester, 266, 268f
Water supply pipes, 74, 74f
Weather stripping, on windows, 180
Web (roof), 138, 139f
Weep holes, 200
Welded wire fabric (WWF), 93, 93f
Welded wire fabric reinforcement, 93
Welded wire mesh (WWM), 93, 93f
Welding symbol, 286, 289f
Weld symbol, 286, 290f
Western framing, 95, 95f, 96f. *See also* Platform framing
Wide box cornice, 168, 170f
Windows
  construction, 177–180

  details of, 178–180, 178f, 179f, 180f
  dimensions of, 38, 38f, 39f
  in foundation, 1
  framing, 135
  hinged, 27
  metal, 178
  parts of, 177–178, 177f, 179f
  reading catalogs of, 182, 183f
  sash, 12f, 177
  symbols, 25–27, 26f
  thermal-break, 178
  wood, 177–178
Wiring symbols, 28, 328f–332f. *See also* Electrical drawings
Wood drip cap molding, 187
Wood shingles, 154, 185
Wood siding, 185, 185f, 186f
Wracking, 96, 98f, 99f
WWF. *See* Welded wire fabric (WWF)
WWM. *See* Welded wire mesh (WWM)

## Z

Z flashing, 185
Zoning laws, 5